Tropical Agriculture in Transition – Opportunities for Mitigating
Greenhouse Gas Emissions?

Tropical Agriculture in Transition – Opportunities for Mitigating Greenhouse Gas Emissions?

Edited by:

REINER WASSMANN

Institute for Meteorology and Climate Research (IMK-IFU),
Forschungszentrum Karlsruhe, Garmisch-Partenkirchen,
Germany

and

PAUL L.G. VLEK

Center for Development Research (ZEF),
University of Bonn,
Germany

Reprinted from *Environment, Development and Sustainability*, Volume 6, Nos. 1–2, 2004

A C.I.P. Catalogue record for this book is available from the library of Congress.

ISBN 978-90-481-6341-0

Published by Kluwer Academic Publishers,
P.O. Box 17, 3300 AA Dordrecht, The Netherlands

Sold and distributed in North, Central and South America
by Kluwer Academic Publishers,
101 Philip Drive, Norwell, MA 02061, U.S.A.

In all other countries, sold and distributed
by Kluwer Academic Publishers
P.O. Box 322, 3300 AH Dordrecht, The Netherlands

Cover photos courtesy of Georgia Carvalho

Printed on acid-free paper

TABLE OF CONTENTS

MITIGATING GREENHOUSE GAS EMISSIONS FROM TROPICAL AGRICULTURE: SCOPE AND RESEARCH PRIORITIES

REINER WASSMANN[1]* and PAUL L.G. VLEK[2]

[1] *Institute for Meteorology and Climate Research (IMK-IFU), Forschungszentrum Karlsruhe, Garmisch-Partenkirchen, Germany;* [2] *Center for Development Research (ZEF), University of Bonn, Germany*
(* *author for correspondence, e-mail: reiner.wassmann@imk.fzk.de; fax: -49-8821-183296; tel.: -49-8821-183139*)

Abstract. The scope of mitigation options in tropical agriculture is discussed for three different activities (a) agroforestry, (b) rice-based production systems and (c) pasture/animal husbandry. The prevention of deforestation – and the re-forestation of degraded land – could become a key elements to national climate protection programs of some developing countries. Agroforestry may offer economically viable windows of opportunity for sustainable use of tropical forests whereas additional funds (e.g. through "Clean Development Mechanism") will be required to make re-forestation programs profitable. Alternative management practices in rice-based systems may offer win–win options to reduce emissions and – at the same time – to obtain another improvement in the production system, namely through optimized timing of nitrogen fertilizer, temporary drainage in irrigated rice fields and integrated residue management. Introducing pasture in degraded land can sequester substantial amounts of carbon (similar to re-forestation).

Future research has to include participation of stakeholders from all conceivable levels, i.e. farmers' cooperatives, non-governmental organizations, national agricultural research centers and extension services, to devise simple and financially interesting incentives for reducing emissions. The feasibility of environmentally friendly production techniques has to be disseminated to the public through 'success stories' (documented in public media) and demonstration farms.

Key words: agroforestry, biomass burning, efficiency, participatory research, pasture, policy, residue, rice.

1. Introduction

The workshop entitled "Tropical Agriculture in Transition – Opportunities for Mitigating Greenhouse Gas Emissions?" was jointly organized by Center for Development Research, Bonn, and the Fraunhofer Institute for Atmospheric Environmental Research, Garmisch-Partenkirchen (recently integrated as IMK-IFU within Research Center Karlsruhe). The workshop was held in Bonn from November 9 to 11 2001 and was attended by fifty scientists from national and international research institutes.

The topic of the workshop has gained special attention through the Kyoto Protocol that introduces new mechanisms to finance the mitigation of greenhouse gas (GHG) emissions in developing countries. Production from tropical agricultural systems will need to increase drastically, because of rising demands in quantity and quality of food. Meeting this demand with current technologies of intensification (i.e. increasing inputs on given land) and/or extensification (i.e. increasing land area used for agriculture) may lead to enhanced GHG production. The workshop aimed at assessing the pros and cons of strategies to meet future demands for food, fiber and renewable energy.

The workshop was concluded by a wrap-up session providing a forum for an intensive information exchange in which the workshop participants reached broad consensus on the scope of mitigation options and research priorities for tropical agriculture. The purpose of this paper is to summarize the basic findings reached during these discussions at the workshop.

2. Rationale for mitigation options

Researcher are often confronted with the fundamental question – posed by politicians and farmers alike – to what extent emissions from agriculture are tolerable and in what quantity they have to be regarded as harmful. Obviously, the nature of the greenhouse effect defies the definition of a clear threshold between acceptable and unacceptable emission levels for agriculture. Being a gradual process caused by composite sources, the overarching question about harmful emissions has to be split into two sub-questions: (a) the acceptable level of global warming and (b) the individual share allotted to agriculture within the entire suite of sources. Whatever answers are given, they will in both cases reflect more social considerations than scientific arguments.

The point of departure for justifying mitigation options in tropical/subtropical agriculture should be the statements in the United Nations Framework Convention on Climate Change (UNFCCC, 1992). All signatories acknowledge "that the global nature of climate change calls for the widest possible cooperation by all countries and their participation in an effective and appropriate international response, in accordance with their common but differentiated responsibilities and respective capabilities and their social and economic conditions . . .". The cooperation by all countries is called for irrespective of the recognition that "per capita emissions in developing countries are still relatively low" (UNFCCC, 1992). Since almost all developing countries have signed and ratified the UNFCCC, mitigation options should be implemented within the "the sovereign right of states to own environmental and developmental policies" (UNFCCC, 1992). Thus, mitigation of GHG emissions from tropical agriculture may not need additional justification as long as production systems offer win–win options for increased productivity and reduced GHG emissions.

3. The scientific base

3.1. INDIVIDUAL SOURCE STRENGTHS OF AGRICULTURAL ACTIVITIES

Tropical agriculture comprises an enormous variety of activities that directly or indirectly affect GHG emissions. Globally, the most significant activities were identified as (a) deforestation for reclaiming new agricultural land as a source of carbon dioxide, (b) rice-based production systems (incl. rice–wheat rotation) as sources of methane and nitrous oxide (as well as carbon dioxide when repeated straw burning leads to depletion of the soil carbon pool) and (c) animal husbandry as a source of methane. Among these activities, deforestation has the largest emission potential for GHGs. The prevention of deforestation – and the re-forestation of degraded land – could become one of the key elements to national climate protection programs of some developing countries. Agroforestry may offer economically viable windows of opportunity for sustainable use of tropical forests whereas additional funds (e.g. through the "Clean Development Mechanism" (CDM) introduced in the Kyoto Protocol) will be required to make re-forestation programs profitable.

The scope for mitigating emissions from rice-based systems and animal husbandry is rather small as compared to emissions deriving from fossil fuel consumption. However, alternative management practices offer win–win options to reduce emissions and – at the same time – to obtain another improvement in the production system (see below).

3.2. NATIONAL INVENTORIES

One of the stipulations of UNFCCC (1992) is that each signatory compiles a "national inventory of anthropogenic emissions by sources and removals by sinks of all greenhouse gases". Although national estimates have been submitted by most countries, these estimates of sources and sinks were largely based on IPCC methodology; i.e. multiplying country-specific "activity data" with default emission factors and algorithms. Such a generic approach can only achieve a very limited accuracy, because region-specific features of agricultural production are ignored. In fact, measurement programs aimed at upscaling of measured emissions revealed pronounced heterogeneity of basically all production systems investigated so far. In the case of rice paddies, for example, it was shown that permanent water logging (as practiced in some areas of Central China) can trigger excessively high methane emissions and should not be pooled with other rice systems (see Wassmann et al., this issue). Likewise, spatial and temporal variability of carbon stocks yields pronounced differences in net C emissions through deforestation as documented in several field studies (see Palm et al., this issue). Default-based inventories will inevitably omit such hot-spots of GHG emissions.

Another source of errors attached to national inventories is the incomplete statistics on agricultural activities. Only those activities associated with pecuniary transactions, e.g. fertilizer consumption, are recorded reasonably well in statistics. However, GHG emissions are largely determined by residue management, an activity not shown in agricultural statistics. Moreover, assessment of net Global Warming Potential (GWP) should include off-site emissions (e.g. CO_2 emissions through fertilizer production and transport) and life cycle assessment (e.g. the fate of timber products).

3.3. MECHANISMS OF EMISSION/SEQUESTRATION

The level of understanding CH_4, N_2O and CO_2 fluxes from tropical agriculture has substantially increased due to recent field, greenhouse and laboratory studies (see Mosier et al., this issue). However, some fundamental questions still remain unanswered:

- As for methane emissions, the major problem of synthesizing emission records into a 'bottom line' is the enormous spatial variability. At this point, the priority should be on integrating GIS data bases and models to capture the heterogeneity (in natural factors and crop management) rather than providing more field data.
- As for nitrous oxide emissions, the field data base is not as exhaustive as for methane. Apart from spatial variation, the nature of N_2O emissions also causes methodological problems in capturing temporal variations. The temporal patterns of N_2O emission typically show distinct intervals with very high emissions that may be missed out by discontinuous measurements. Continuous emission records by automated systems showed that such individual episodes, e.g. after heavy rainfall, can account for the bulk annual N_2O emissions.
- As for carbon sequestration, the mechanisms and limits of this process in tropical systems are only poorly understood. While the sequestration potential of an ecosystem is often perceived as carbon fixation in the vegetation, carbon sequestration in the soils may in fact become the more effective component in the long-term CO_2 balance (see Batjes, this issue). Temperate forests have been identified as a net-sink for carbon, but the role of tropical forests and agroforestry in the global carbon budget is still unclear (see Palm et al., this issue). Likewise, carbon sequestration in crop land has been studied in many temperate systems whereas the data base on tropical land use affecting carbon fluxes is still very scarce (see Mosier et al., this issue; Wassmann et al., this issue).

For future studies, more emphasis should be given to the interaction of CH_4, N_2O and CO_2 emissions. Controlling factors for GHG source and sink functions have to be identified based on the GWP balance (see Robertson and Grace, this issue), and not only on one GHG.

4. Systems design and optimization

4.1. IMPROVING EFFICIENCIES

In a broader context of sustainability, mitigation options should aim at improving input/output use efficiency of energy, carbon, nutrients and water (see Vlek et al., this issue). Optimizing the use of these valuable resources offers various synergies with mitigation GHG emissions, e.g.:

- Timing of nitrogen fertilizer according to actual needs reduces nitrous oxide emissions.
- Temporary drainage in irrigated rice fields can reduce both, water consumption and GHG emissions (see Wassmann et al., this issue).
- The common practice of biomass burning causes in many cases a depletion of soil organic carbon.
- Introducing pasture in degraded land can sequester substantial amounts of carbon (see Fisher and Thomas, this issue).

Efficiency increases can be targeted in both, intensive (high input) and extensive (low input) agricultural systems. Coordination with the energy sector will integrate mitigation efforts in "high input" agriculture into a broader scheme of energy policies. Rapid technological development in these highly productive systems demands an "early warning" framework for detrimental impact of new technologies, including excessive emissions of GHG. For low-input agriculture, the major thrusts should be (a) investigating agroforestry systems and (b) reversing the trend of declining soil fertility as observed in many marginal areas.

4.2. RESIDUE MANAGEMENT/BURNING

The fate of residues is a crucial component to determine the sustainability of agroforestry and cereal production systems. In large parts of the tropics, plant residues are burnt to clear the field for the next cropping cycle. This concerns wood and litter in slash-and-burn systems as well as straw in cereal production. The common practice of residue burning not only is a waste of a valuable carbon source, but also depletes the native carbon pool and has adverse impacts on local air quality. Moreover, substantial amounts of organic material are often wasted during processing of agricultural produce, e.g. milling.

The total biomass that is annually burnt corresponds to app. 4000 Tg C including app. 900 Tg C deriving from the burning of agricultural residues (Levine, 1994). Asia is the continent with the highest intensity of residue burning and accounts for app. 68% of residue-borne emissions (Koopmans and Koppejan, 1998). Composition and availability of agricultural wastes show pronounced variation within different production systems. Table I compiles "residue to product ratios" (RPR) and estimated annual biomass for some important crops in Asia. Farming

TABLE I. "Residue to product ratios" (RPR) and estimated annual amounts for some important crops in Asia (Koopmans and Koppejan, 1998).

Crop residue	RPR	Σ_{Asia} (Tg)	Crop residue	RPR	Σ_{Asia} (Tg)
Rice straw	1.76	885	Cassava stalks[1]	0.062	3
Rice husks	0.27	135	Sugar cane tops	0.30	159
Wheat straw	1.75	371	Sugar cane bagasse	0.29	154
Maize stalks	2.00	274	Oil palm fiber[2]	0.15	10
Maize cob	0.27	37	Oil palm shells[2]	0.065	5
Maize husks	0.2	27	Oil palm bunches[2]	0.23	17

[1]Excluding fuel consumption of wood and leaves.
[2]Excluding biomass at clearing of mature trees.

residues are usually burnt in smoldering fires and emit relatively less CO_2 than other fires, e.g. 40% less CO_2 emissions as compared to the burning of charcoal (Andreae and Merlet, 2001). In turn, however, smoldering fires emit more CO (app. 500%) and CH_4 (app. 270%) than flames. CO has adverse effects on local air chemistry and CH_4 has a strong GWP. Fire-borne organic compounds and nitrogen oxides lead to tropospheric ozone formation; high ozone concentrations coincide with the peak of the residue burning season in Asia.

The immediate alternative to residue burning is plowing of organic material into the soil. Soil incorporation of fresh residues with high C/N ratios results in temporary nitrogen immobilization. Therefore, these materials should either be blended with other fertilizers or should be composted before incorporation. While the principles and technical aspects for composting agricultural residues are well known, this technique has only scarcely been adopted in Asia. A further alternative is to leave the straw as mulch, an option that is largely restricted to no-till systems, which have spread to around 300.000 ha in the Indo-Gangetic plains.

For oil palm plantations, mechanical clearing followed by mulching of the organic material may allow substantial recycling of plant material. This technique developed for an environmentally sound practice of shifting cultivation (Kato et al., 1999) could be exploited to reduce the rampant haze development in regions dominated by oil palm plantations.

Apart from recycling, the use of plant residues for energy consumption has a great potential that is insufficiently tapped at present. For example, bagasse is commonly used as biofuel in USA, whereas its use in Asia is rather inefficient. Innovative biofuel technologies may yield new prospects for tropical agriculture in the future. Initial demonstration projects show that straw could successfully be used for ethanol production or directly be fed into electricity generators.

The practice of residue management is closely intertwined with other management practices such as tillage, fertilization etc. Thus, alternative approaches of plant residue management have to be incorporated in a system approach providing economic benefits to the farmers. The objective of residue management (in all systems) should be to strike a sound balance between effective use of energy and sufficient retention of organic matter and nutrients in the soil.

5. Comprehensive research approaches

5.1. FARMERS' PARTICIPATORY RESEARCH

There was broad consensus among all workshop participants that participatory research has been largely neglected so far in the context of mitigation efforts. Farmer's participation appears indispensable for technology transfer of any kind, including management changes aimed at sustainable production systems (see Carvalho et al., this issue). Scientist should start the dialogue with farmer groups about GHG concerns through various pathways:

- Improving the understanding of farmers' perceptions and decision making to classify different target groups for specific mitigation strategies.
- Conducting research on farmer's fields or community forest (instead of research stations) as a "reality check" for suggested improvements.
- Developing alternative management options (e.g. to residue burning) in close collaboration with farmers, preferably derived from indigenous knowledge on sustainable management practices.
- Focussing on farm households rather than individual production systems and evaluating the economic benefit to the farmer, e.g. affordability *versus* profitability.
- Packaging scientific knowledge in practical and user-friendly forms (e.g. easy decision-support tools).
- Establishing continuous feed-back on mitigation strategies over longer time spans, e.g. farmers perception on water pricing may vary according to recent weather events.
- Educating farmers and rural communities by through schools, farmer cooperatives and local newspaper/radio stations.

5.2. POLICY RESEARCH

In most developing countries, the public awareness of GHG emissions as an environmental problem is still rudimentary. Research findings have to be communicated not only to farmers, but also to decision-makers at all conceivable levels. Researchers have to identify information needs of the policy making process and stress the mitigation potential of tropical agriculture. The costs of mitigation options in agriculture have to be set against mitigation costs in other sectors. Technical education on mitigation options should target national agricultural research agencies and extension services.

Mitigation efforts have to become an integral component of environmental policy and capitalize on synergies with other environmental targets. Improving air quality is a fundamental argument against biomass burning. Water saving will become top priority in many rice growing areas and will thus work indirectly in favor of low methane emissions. In turn, national policies should aim at providing (simple and financially interesting) incentives to farmers to apply 'best management practices'.

Ecological research has to team up with agronomic, economic and social research to define commonly accepted methodologies for comprehensive cost/benefit analyses of mitigation options. Exploring profitable uses of straw/ agricultural debris and timber/non-timber forest products will rely on profound knowledge of local markets and industries. Only interdisciplinary research can advise policy makers on

- institutional reforms,
- clear legal regulations and
- economic incentives e.g. prize stabilization and taxation.

National and international institutes have to be linked to research networks on mitigation to avoid redundancies and to ensure best possible outputs.

Public awareness programs and multimedia campaigns on recycling should promote community action to eliminate unnecessary emissions. Many Non-Governmental Organization (NGOs) in rural areas aim at improving the livelihood of farmers through sustainable agriculture. These NGOs could become genuine partners for establishing 'success stories' on sustainable (and low-emitting) production systems. In the next step, these show-cases can be used to demonstrate how the CDM and/or private sponsorship may fund mitigation efforts.

References

Andreae, M.O. and Merlet, P.: 2001, 'Emission of trace gases and aerosols from biomass burning', *Global Biogeochemical Cycles* **15**, 955–966.

Batjes, N.H.: (this issue), 'Estimation of soil carbon gains upon improved management within croplands and grasslands of Africa', *Environment, Development and Sustainability*.

Carvalho, G., Moutinho, P., Nepstad, D. Mattos, L. and Santilli, M.: (this issue), 'An Amazon perspective on the forest–climate connection: opportunity for climate mitigation, conservation and development?', *Environment, Development and Sustainability*.

Fisher, M.J. and Thomas, R.J.: (this issue), 'Implications of land use changes in the central lowlands of tropical South America for carbon stocks', *Environment, Development and Sustainability*.

Kato, M.S.A., Kato, O.R., Denich, M. and Vlek, P.L.G.: 1999, 'Fire-free alternatives to slash-and burn for shifting cultivation in the Eastern Amazon region: the role of fertilizers', *Field Crops Research* **62**, 225–237.

Koopmans, A. and Koppejan, J.: 1998, 'Agricultural and forest residues – generation, utilization and availability', in *Proceedings of the Regional Expert Consultation on Modern Applications of Biomass Energy*, FAO Regional Wood Energy Development Programme in Asia, Report No. 36, Bangkok. pp. 23.

Levine, J.S.: 1994, 'Biomass burning and the production of greenhouse gases', in R.G. Zepp (ed.), *Climate Biosphere Interaction: Biogenic Emissions and Environmental Effects of Climate Change*, New York, John Wiley and Sons, pp. 139–159.

Mosier A, Wassmann R, Verchot L. and Palm C.: (this issue) 'Greenhouse gas emissions from tropical agriculture: sources, sinks and mechanisms', *Environment, Development and Sustainability*.

Palm, C. Tomich, T., van Noordwijk, M., Vosti, S., Gockowski, J., Alegre, J. and Verchot, L.: (this issue), 'Mitigating GHG emissions in the humid tropics: case studies from the alternatives to slash and burn program (ASB)', *Environment, Development and Sustainability*.

Robertson, G.P. and Grace, P.R.: (this issue) 'Greenhouse gas fluxes in tropical agriculture: the need for a full-cost accounting of Global Warming Potentials', *Environment, Development and Sustainability*.

Vlek, P.L.G., Rodríguez-Kuhl, G. and Sommer, R.: (this issue), 'Energy use and CO_2 production in tropical agriculture and means and strategies for reduction or mitigation', *Environment, Development and
Sustainability.*

UNFCCC (United Nations Framework Convention on Climate Change): 1992, *Full text of the Convention,*
Climate Change Secretariat, Bonn, http://unfccc.int/resource/docs/convkp/conveng.pdf.

Wassmann, R., Neue, H.U., Ladha, J.K. and Aulakh, M.S.: (this issue), 'Mitigating greenhouse gas emissions
from rice–wheat cropping systems in Asia', *Environment, Development and Sustainability.*

METHANE AND NITROGEN OXIDE FLUXES IN TROPICAL AGRICULTURAL SOILS: SOURCES, SINKS AND MECHANISMS

ARVIN MOSIER[1]*, REINER WASSMANN[2], LOUIS VERCHOT[3], JENNIFER KING[4] and CHERYL PALM[5]

[1]*United States Department of Agriculture/Agricultural Research Service, Fort Collins, CO, USA;*
[2]*Institute for Meteorology and Climate Research (IMK-IFU), Forschungszentrum Karlsruhe, Garmisch-Partenkirchen, Germany;* [3]*World Agroforestry Centre (ICRAF), Nairobi, Kenya;*
[4]*Department of Soil, Water and Climate, University of Minnesota, St. Paul, MN, USA;*
[5]*Tropical Soil Biology and Fertility Programme (TSBF), Nairobi, Kenya*
(*author for correspondence, USDA/ARS, P.O. Box E. 301 South Howes, Federal Building Room 420, Fort Collins, CO 80522, USA; e-mail: arvin.mosier@ars.usda.gov; fax: 970-490-8213; tel.: 970-490-8250)*

(Accepted in Revised form 15 January 2003)

Abstract. Tropical soils are important sources and sinks of atmospheric methane (CH_4) and major sources of oxides of nitrogen gases, nitrous oxide (N_2O) and NO_x ($NO+NO_2$). These gases are present in the atmosphere in trace amounts and are important to atmospheric chemistry and earth's radiative balance. Although nitric oxide (NO) does not directly contribute to the greenhouse effect by absorbing infrared radiation, it contributes to climate forcing through its role in photochemistry of hydroxyl radicals and ozone (O_3) and plays a key role in air quality issues. Agricultural soils are a primary source of anthropogenic trace gas emissions, and the tropics and subtropics contribute greatly, particularly since 51% of world soils are in these climate zones.

The soil microbial processes responsible for the production and consumption of CH_4 and production of N-oxides are the same in all parts of the globe, regardless of climate. Because of the ubiquitous nature of the basic enzymatic processes in the soil, the biological processes responsible for the production of NO, N_2O and CH_4, nitrification/denitrification and methanogenesis/methanotropy are discussed in general terms. Soil water content and nutrient availability are key controls for production, consumption and emission of these gases.

Intensive studies of CH_4 exchange in rice production systems made during the past decade reveal new insight. At the same time, there have been relatively few measurements of CH_4, N_2O or NO_x fluxes in upland tropical crop production systems. There are even fewer studies in which simultaneous measurements of these gases are reported. Such measurements are necessary for determining total greenhouse gas emission budgets. While intensive agricultural systems are important global sources of N_2O and CH_4 recent studies are revealing that the impact of tropical land use change on trace gas emissions is not as great as first reports suggested. It is becoming apparent that although conversion of forests to grazing lands initially induces higher N-oxide emissions than observed from the primary forest, within a few years emissions of NO and N_2O generally fall below those from the primary forest. On the other hand, CH_4 oxidation is typically greatly reduced and grazing lands may even become net sources in situations where soil compaction from cattle traffic limits gas diffusion.

Establishment of tree-based systems following slash-and-burn agriculture enhances N_2O and NO emissions during and immediately following burning. These emissions soon decline to rates similar to those observed in secondary forest while CH_4 consumption rates are slightly reduced. Conversion to intensive cropping systems, on the other hand, results in significant increases in N_2O emissions, a loss of the CH_4 sink, and a substantial increase in the global warming potential compared to the forest and tree-based systems. The increasing intensification of crop production in the tropics, in which N fertilization must increase for many crops to sustain production, will most certainly increase N-oxide emissions. The increase, however, may be on the same order as that expected in temperate crop production, thus smaller than some have predicted. In addition, increased attention to management of fertilizer and water may reduce trace gas emissions and simultaneously increase fertilizer use efficiency.

Key words: CH_4, greenhouse gases, land use change, NO, N_2O, rice fields.

Environment, Development and Sustainability **6:** 11–49, 2004.
© 2004 Kluwer Academic Publishers.

1. Introduction

The purpose of this paper is to provide a general background for discussions of human induced trace gas emissions from tropical agricultural soils. We use the geographical area as the equatorial belt of the earth between the Tropic of Cancer at 23.5°N and the Tropic of Capricorn at 23.5°S. Tropics are generally considered to lie at low elevations within the equatorial zone, and the subtropics extend to both sides of the equator to 30°N and 30°S latitudes, at low elevations. Fifty one percent of world soils are in the tropics and subtropics. Lowland tropics have mean annual temperature >24°C and the mean monthly temperature in winter is >18°C. In the low elevation subtropics, the mean annual temperature is between 17°C and 24°C and the coldest month averages between 10°C and 18°C. Temperatures in the tropics decrease with the increase in elevation at the rate of about 2°C for each 300 m in elevation. Because of lower solar radiation, a tropical climate is restricted to elevations of no greater than 300 m on the Tropic of Cancer and the Tropic of Capricorn (Donahue et al., 1983).

Tropical soils are important sources and sinks of atmospheric methane (CH_4) and major sources of oxides of nitrogen gases, nitrous oxide (N_2O) and NO_x (NO + NO_2), which are present in the atmosphere in trace amounts (Bouwman, 1990, 1998; Breuer et al., 2000; Matson and Vitousek, 1990; Potter et al., 1996;). These gases are important to atmospheric chemistry and earth's radiative balance because of the long atmospheric life times of CH_4 and N_2O (~10 yr for CH_4 and 120 yr for N_2O) and infrared absorption properties in the troposphere. Both species absorb terrestrial thermal radiation and thus contribute to greenhouse warming of the atmosphere. On a per-molecule basis and 100-year time frame, CH_4 is about 21 and N_2O is about 310 times more effective at trapping infrared radiation than CO_2 (IPCC, 1996). Although nitric oxide (NO) does not directly contribute to the greenhouse effect by absorbing infrared radiation, it contributes to climate forcing through its role in photochemistry of hydroxyl radicals (OH^-) and ozone (O_3) and plays a key role in air quality issues.

1.1. CH_4

The current global average atmospheric concentration of CH_4 is 1780 ppbv, more than double its pre-industrial value of 800 ppbv (Dlugokencky, 2001). The concentration of CH_4 in the Northern Hemisphere is about 100 ppbv more than in the Southern Hemisphere, indicating either greater sources or lower sink strength in the Northern Hemisphere (Houghton et al., 1992). The rate of increase in atmospheric CH_4 slowed from about 15 ppbv yr^{-1} in the 1980s to near zero in 1999 (Dlugokencky, 2001). Since 1990 the annual rate of CH_4 increase in the atmosphere has varied between less than zero to 15 ppbv. The reason or reasons for this change is not known (Rudolph, 1994).

About 70% of CH_4 production arises from anthropogenic sources and about 30% from natural sources (Table I). Biological generation in anaerobic environments

TABLE I. Estimated sources and sinks of CH_4 (adapted from Mosier et al., 1998b).

	Tg CH_4 yr^{-1}
Sources	
Natural	
Wetlands	100–200
Termites	10–50
Oceans	5–20
Freshwater	1–25
CH_4 hydrate	0–5
Anthropogenic	
Coal mining, natural gas and pet industry	70–120
Rice paddies	20–150
Enteric fermentation	65–100
Animal wastes	10–30
Domestic sewage treatment	25
Land fills	20–70
Biomass burning	20–80
Sinks	
Atmospheric (tropospheric plus stratospheric) removal	420–520
Removal by soils	15–45
Atmospheric increase	28–37

(natural and human made wetlands, enteric fermentation, and anaerobic waste processing) is the major source of CH_4, although losses associated with coal and natural gas industries are also significant. The primary sink for CH_4 is reaction with (OH^-) radicals in the troposphere (Crutzen and Andreae, 1990; Fung et al., 1991). Aerobic soils oxidize 10–20% of net CH_4 production, while a stratospheric sink of about 10% of production is suggested (IPCC, 1994).

CH_4 emissions from tropical agricultural soils constitute a significant portion of annual global CH_4 emissions. Rice paddies, termites, biomass burning and enteric fermentation are the main contributors. In our discussions, here we focus only on the soil sources and sinks associated with agricultural production of food, forage and fiber.

1.2. N_2O

The current atmospheric N_2O concentration, ~317 ppbv, has increased from ~275 ppbv in 1900. Most of this increase has occurred during the past 50 years with a near linear increase of ~0.7 ppbv yr^{-1} (CMDL, 2001). An increase of 0.2–0.3% in atmospheric concentrations would contribute about 5% to greenhouse warming (Cicerone and Oremland, 1988). N_2O is also involved in the depletion of the O_3 layer in the stratosphere, which protects the biosphere from the harmful effects of solar ultraviolet radiation (Crutzen, 1981). Doubling the concentration of N_2O in the atmosphere would result in an estimated 10% decrease in the O_3 layer and this

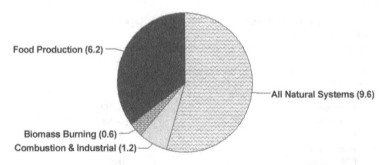

Figure 1. Estimate of global annual N_2O (17.6 Tg N_2O–N) emissions in 1994 (adapted from Kroeze et al., 1999).

would increase the ultraviolet radiation reaching the earth by 20% (Crutzen and Ehhalt, 1977). This could result in increased skin cancer and other health problems (Lijinsky, 1977). Given the relatively long atmospheric lifetime for N_2O (Hao et al., 1987; Ko et al., 1991; Volk et al., 1997), there are justifiable reasons for concern.

The concentration of N_2O in the atmosphere is increasing, as a result of biotic and anthropogenic activities, at the rate of about 4 Tg N_2O–N yr^{-1} (1 Tg = 10^{12} g) above the rate of decomposition in the stratosphere (IPCC, 1996). Kroeze et al. (1999) estimated that total global N_2O emissions in 1994 were approximately 17.6 Tg N, with ~55% (9.6 Tg) of the total arising from relatively natural terrestrial and aquatic systems and ~8 Tg derived from anthropogenic sources (Figure 1). Of the human initiated N_2O emissions, ~70% is thought to result from emissions from agriculture, both crop and livestock production.

Data drawn from temperate agricultural crop production systems demonstrates that N_2O is emitted in response to N fertilization. The emissions from specific crops vary greatly and the range of observations within each crop is large. Mosier and Kroeze (1999) grouped N_2O emissions by crop in which the measurements were made from the 35 temperate studies (Table II). No distinct trends are apparent between crops fertilized with 100–500 kg N ha^{-1}. The data provide an idea of the range of N_2O flux measurements observed in temperate agricultural systems. N_2O emissions resulting from N fertilizer use and biological N-fixation needed to meet food demands for the projected world population are projected to continue to increase during the next 100 years (Hammond, 1990; World Development Report, 1992). Worldwide synthetic fertilizer consumption was ~85 Tg N in 1999 with ~56 Tg N consumed in developing countries (FAO, 2001). Global N fertilizer consumption is projected to reach 105 Tg N by 2030 (IFA, 2000). Thus, we anticipate that the current linear trend of increasing atmospheric N_2O concentrations is likely to continue over the next several decades.

1.3. NO_x

The release of nitrogen oxides (NO_x = NO + NO_2) has accelerated during the last few decades primarily through the increase in fossil fuel combustion

TABLE II. Relationship of type of crop to measured N_2O emissions in temperate agricultural systems from 35 studies published since 1994 (Mosier and Kroeze, 1999). N_2O emitted is that in excess of unfertilized control.

Crop	N_2O emitted as % of N applied (range)	Number of studies
Maize	0.5–7.3	9
Spring barley or wheat	0.3–6.8	7
Other crops	0.2–2.5	4
Grasslands	0.2–12	11
Fallow fields	0.1–1.0	4

(Galloway et al., 1995; Holland and Lamarque, 1997) and to a lesser extent through tropical forest clearing (Verchot et al., 1999). NO is a short-lived species in the troposphere (hours to days) and thus has mostly local or regional effects on photochemistry. NO is a strong oxidant and contributes indirectly to climate forcing through a series of concentration-dependent cycles with other atmospheric oxidants, O_3 and OH^- radicals during the oxidation of carbon monoxide (CO), CH_4, and non-methane hydrocarbons. Increasing NO emissions are contributing substantially to observed increases in O_3 concentrations in the Northern Hemisphere (Chameides et al., 1994; Holland et al., 1997). Detailed descriptions of the complex interactive cycles of NO in the atmosphere can be found in many references (e.g. Ehhalt et al., 1992; Holland and Lamarque, 1997; Liu et al., 1992; Williams et al. 1992).

Several authors have used an inventory approach to estimate that about 20–40% of global NO emitted to the atmosphere is derived from soils (Davidson, 1991; Davidson and Kingerlee, 1997). Estimates of annual soil-derived NO emissions range from 5.5 (Delmas et al., 1997; Holland et al., 1997), to 21 Tg N (Davidson, 1991; Davidson and Kingerlee, 1997). Yienger and Levy (1995) used a combination of inventory and modeling to estimate global NO emissions from soils to be $10.2 \, \text{Tg yr}^{-1}$. Potter et al. (1996) also used a modeling approach and estimated the global flux as $10 \, \text{Tg yr}^{-1}$. As Davidson and Kingerlee (1997) pointed out, the estimate of Yienger and Levy is prc.,ably a bit low for some biomes, particularly tropical savannas. Therefore, it is probably best to think of this estimate as a reasonable lower bound.

Estimates of soil fluxes do not represent biosphere fluxes to the atmosphere. NO rapidly oxidizes to NO_2, which adsorbs readily onto surfaces in the plant canopy. Taking this adsorption into account the estimate of Yienger and Levy decreased to $5.45 \, \text{Tg yr}^{-1}$ of NO_x ($NO + NO_2$) emitted to the atmosphere from ecosystems. Applying this same adsorption coefficient to the Davidson and Kingerlee (1997) estimate would lower their estimate to $13 \, \text{Tg yr}^{-1}$. Thus the best estimate available right now would suggest that soil emissions are between 10 and $20 \, \text{Tg yr}^{-1}$ and that ecosystem emissions to the free atmosphere are at most 40% lower that the soil emissions.

A. MOSIER ET AL.

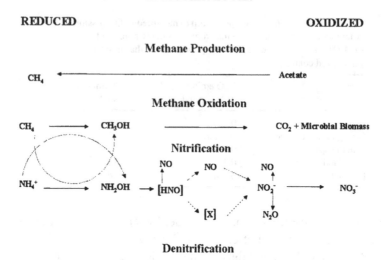

Figure 2. Diagram of CH_4 production and oxidation and nitrification and denitrification processes (adapted from Knowles, 1993; Firestone, 1982; Firestone and Davidson, 1989).

While the magnitude of the soil flux may be difficult to pin down with any precision at the moment, the likely future trend of emissions is less difficult to predict. The microbial processes responsible for NO emissions in soils are the same processes responsible for N_2O emissions, namely nitrification and denitrification (Figure 2). NO emissions are related to N turnover rates and increases in N cycling in soils have contributed to the increases in NO emissions during the past century. We fully expect this trend to continue. Other factors have increased N cycling rates such as conversions of forests and grasslands to croplands. Changes in global climate may affect temperature and moisture, which will directly influence N cycling. Increased in N input into the soil system via atmospheric deposition will also increase N cycling (Vitousek et al., 1997). All of these changes are likely to increase soil NO emissions in the coming decades.

2. Production and consumption of CH_4, N_2O and NO

The soil microbial processes responsible for the production and consumption of CH_4 and production of N-oxides are the same in all parts of the globe, regardless of climate. Different species may predominate in specific situations and organisms adapted to a specific climate may respond differently when exposed to changing environmental conditions. Because of the ubiquitous nature of the basic enzymatic processes, we can discuss nitrification/denitrification and methanogenesis/methanotropy in general and not be confined to discussing the process for specific climatic conditions (Figure 2). Soils organisms responsible for trace gas consumption and production that are adapted to tropical climates typically respond near linearly to temperature between 20°C and 40°C (Holland et al., 2000). Discussion of flux rates or trace gas exchange in this paper will follow the convention that fluxes to the atmosphere are positive while fluxes from the atmosphere to the land surface are negative.

2.1. CH_4

Atmospheric CH_4 is produced by a wide variety of natural and anthropogenic processes in which agriculture plays a major part. CH_4 emissions from tropical agriculture, namely from rice paddies, termites, biomass burning and enteric fermentation, constitute a significant portion of annual global CH_4 emissions. Upland agricultural soils are a CH_4 sink. The net flux (emission or consumption) of CH_4 will vary depending on the nature of the agricultural system and the management practice adopted in the system. Measurements made at various locations of the world show that there are large temporal variations of CH_4 flux which differ with soil type, application of organic matter and mineral fertilizer, and soil water regime.

2.1.1. CH_4 production in soils

Methane production occurs only under highly anaerobic conditions such as typically occur in natural wetlands and lowland rice fields. There are many distinguishing factors which differentiate wetlands from upland, aerobic soil systems. The main processes that occur in flooded soils can be regarded as a series of successive reduction and oxidation (redox) reactions mediated by different types of microorganisms. Flooding alters the character of the microbial flora in soils by decreasing O_2 concentration. Fermentation is one of the major biochemical processes responsible for organic matter degradation in flooded soils. The main products of the fermentation process in flooded soil are ethanol, acetate, lactate, propionate, butyrate, H_2, N_2, CH_4 and CO_2. The latter three gases usually constitute the largest portion of the gas phase of flooded soils.

The major pathways of CH_4 production in flooded soil are the reduction of CO_2 with H_2, with fatty acids or alcohols as the hydrogen donor, and the transmethylation of acetic acid or methyl alcohol by CH_4 producing bacteria (Takai, 1970; Conrad, 1989). In paddy fields, the kinetics of reduction processes are strongly affected by the composition and texture of soil and it's content of inorganic electron acceptors. The period between flooding soil and the onset of methanogenesis can vary with different soils. However, it is unclear if soil type also affects the rates of methanogenesis and CH_4 emission when steady state conditions have been reached (Conrad, 1989). After flooding the redox potential (E_h) declines to reach a point where the E_h is low enough for CH_4 production (Takai et al., 1956). For a given soil, there is typically a correlation between the soil E_h and CH_4 emission (Patrick et al., 1981; Cicerone et al., 1983; Yagi and Minami, 1990). Addition of decomposable organic matter to paddy fields generally accelerates the decline in E_h to increases CH_4 emission compared with an application of compost prepared from rice straw or chemical fertilizer (Yagi and Minami, 1990).

2.1.2. CH₄ transport

There are three mechanisms for the transfer of CH_4 from rice paddy soil to the atmosphere. CH_4 may be lost through ebullition (in bubbles from paddy soils); it may diffuse across the water surface; or it may be transported through the rice plant aerenchyma. Release to the atmosphere through the shoot nodes, which are not subject to stomatal control, is generally the most important emission mechanism accounting for more than 90% of the total CH_4 emission from rice paddies (Seiler et al., 1984; Cicerone et al., 1983; Minami, 1993; Nouchi et al., 1990). During the course of the rice-growing season a 20–90% of the CH_4 produced in the flooded soil is oxidized before being released to the atmosphere (Schütz et al., 1989; Sass et al., 1992). Although CH_4 flux rates are a function of the total amount of CH_4 in the soil, there is the possibility that the gas may be consumed in the thin oxidized layer close to the soil surface and in deep floodwater. Methanotrophic bacteria can grow with CH_4 as their sole energy source, and other soil bacteria, e.g., *Nitrosomonas* species consume CH_4 (Seiler and Conrad, 1987). As a small amount of CH_4 is dissolved in water, and is leached to ground water.

2.1.3. CH₄ oxidation

Knowles (1993) describes in detail microbial pathways of CH_4 oxidation in soil methanotrophic organisms. He notes that all methanotrophs isolated thus far are obligate aerobes. This seems reasonable since the enzyme responsible for the initial step in CH_4 oxidation is a monooxygenase enzyme (MMO) that requires molecular O_2. Knowles also describes the biochemical pathway for CH_4 oxidation and ammonium oxidation. The affinity of a methanotroph for CH_4 governs its ability to compete for CH_4 at low concentrations. Bender and Conrad (1993) suggest the existence of two different types of methanotrophic bacteria in soil. One functions at high CH_4 mixing ratios and has a relatively low affinity for CH_4 and the other population which exhibit low thresholds and high affinity for CH_4 oxidizes atmospheric CH_4 at mixing ratios at or below atmospheric concentrations. These organisms can actively oxidize CH_4 at concentrations as low as 0.1 ppmv.

Most methanotrophs are mesophiles, but some such as *M. capsulatus* are tolerant of high temperatures up to 50°C (Knowles, 1993). The temperature quotients (Q_{10}) for consumption were in the range of 1.4–2.1 in contrast to much higher values for CH_4 production. The low activation energy required by methanotrophs buffers temperature effects and explains the relatively high rates of CH_4 oxidation observed in cold soils (Mosier et al., 1991; Dorr et al., 1993).

Most organisms that have MMO appear to have the ability to co-oxidize ammonia and to contribute to nitrification. The MMO of methanotrophs and the ammonia monooxygenase (AMO) of nitrifiers have similar substrate specificities, and apparently, CH_4 and NH_3 are competitive substrates for both enzymes (Knowles, 1993). The autotrophic nitrifiers and methanotrophs occur in similar soil habitats and may compete for O_2, CH_4 and NH_3. Knowles (1993) notes that methanotrophs can also participate in gaseous N oxide metabolism. Schnell and King (1994) found a strong non-competitive inhibition of nitrite on CH_4 consumption in forest soils, which indicates a different type of linkage between CH_4 uptake and N dynamics. Likewise, Mosier et al. (1991) found a differential inhibitory effect of NH_4^+ at different positions on the landscape. N application in mid-slope and upland pasture inhibited CH_4 uptake, but no inhibition was found in soils at the bottom of the slope (Swale). These observations can likely be explained as an effect of cation exchange capacity (CEC) on CH_4 oxidation. Since cation exchange characteristics affect the behavior of NH_4^+ ions, and NH_4^+ in turn affects CH_4 oxidation, Visscher et al. (1998) hypothesized a relationship between CEC and CH_4 oxidation capacity of a soil. Using a modeling approach to simulate the fate of added NH_4^+, these authors showed that factoring in a CEC effect was superior to predicting experimental observations than a simple model of competitive inhibition. Thus, to come back to the results of Mosier et al. (1991), the Swale sites most likely had a higher CEC because of higher clay content and perhaps higher organic matter content. This higher CEC resulted in greater NH_4^+ adsorption and suppressed the inhibitory effect of NH_4^+.

Methane uptake is controlled by the interplay of biotic and abiotic factors. The major factors controlling total consumption rate are potential biological demand and diffusion. Thus, prediction and explanation of CH_4 consumption rate hinges on understanding these two processes. Diffusion rate is regulated by physical factors, while biological demand is regulated by both the physical and chemical environments. Under any particular situation, either biotic or abiotic factors may provide the proximate limitation on CH_4 uptake rate.

Using a database that included measurement of the soil-atmosphere exchange of CH_4 from a variety of studies in temperate and tropical native and managed ecosystems, Del Grosso et al. (2000a) observed that oxidation of CH_4 in the soil is clearly dependent upon soil water filled pore space (WFPS). The optimum soil WFPS was dependent on soil texture with the optima of about 7.5% for coarse textured soils and 13% for fine textured soils. It is clear that CH_4 oxidation is controlled by gas diffusivity and biological activity. When soil water content is greater than the optimum, gas diffusivity limits CH_4 oxidation, but when soil water content is below the optimum, water stress limits biological activity (DelGrosso et al., 2000a).

2.2. N_2O AND NO

Research during the past several decades has improved our understanding of how NO and N_2O are produced, factors that control production, source/sink relationships, and gas movement processes. However, despite extensive knowledge of processes involved we are only beginning to be able to predict the fate of a unit of N that is applied or deposited on a specific agricultural field (e.g., Davidson et al., 2000; Del Grosso et al., 2000b; Frolking et al., 1998; Li et al., 1992; Plant and Bouwman, 1999; Potter et al., 1998). Studies of emissions of NO_x and N_2O from presumably similar agricultural and natural systems show highly variable results in both time and space. The complex interaction of the physical and biological processes involved must be understood before reliable predictive capability can be developed (Mosier and Bouwman 1993; Mosier et al., 1998a).

As we stated above, NO and N_2O are produced primarily from the microbial processes of nitrification and denitrification in soil (Figure 2) (see Schmidt 1982; Firestone, 1982; Hutchinson and Davidson 1993 for more detailed reviews). Nitrification, the oxidation of ammonia to nitrite and then nitrate (Figure 2) occurs in essentially all terrestrial, aquatic and sedimentary ecosystems. Although ecologically ubiquitous, a surprisingly few different chemoautotrophic bacteria have been identified and considered to accomplish most of the ammonium oxidation that occurs (Schmidt, 1982). *Nitrosomonas* and *Nitrosospira* are the main soil and water bacteria which oxidize ammonia to nitrite while *Nitrobacter* is the principal genus of bacteria identified to oxidize nitrite to nitrate (Figure 2; Schmidt, 1982). Denitrification, the microbial reduction of nitrate or nitrite to gaseous nitrogen with NO and N_2O being produced as intermediate reduction compounds (Firestone, 1982) (Figure 2), is performed by a diverse and also widely distributed group of aerobic, heterotrophic bacteria that are facultative anaerobes. The general requirements for denitrification are: the presence of bacteria possessing the metabolic capacity, suitable electron donors such as organic C compounds, reduced S compounds or molecular hydrogen, anaerobic conditions or restricted oxygen availability, and N oxides as terminal electron acceptors (Figure 2; Firestone, 1982).

When soils are near field capacity (often ~60% WFPS), N_2O and NO_x emissions from nitrification of ammonium-based fertilizers can be substantial (Bremner and Blackmer, 1978; Duxbury and McConnaughey, 1986; Hutchinson and Davidson, 1993). Other work suggests that N_2O release is a byproduct of nitrification and may occur by denitrification of nitrite by nitrifying organisms under oxygen stress (Poth and Focht, 1985). NO emissions are considered to arise from the soil generally from nitrification (Hutchinson and Davidson, 1993) but can arise from abiotic chemical reactions under specific conditions. In wet soils, where aeration is restricted, denitrification is generally the source of N_2O (Smith, 1990). Typically, low NO emissions are observed under such conditions (Hutchinson and Davidson, 1993). Soil structure and water content, which affect the balance between diffusive escape of N_2O and its further reduction to N_2, are important in determining the proportions of the two gases as well as NO (Smith, 1990; Firestone and Davidson, 1989). Soil WFPS is used to express the interplay of soil water content and microbial activity through most of these discussions-and is the one factor that is generally reflected in the response of trace gas emissions from the soil to changes in soil conditions. This will be repeatedly mentioned in the specific studies noted in following sections of this paper. Linn and Doran (1984) showed that WFPS is closely related to soil microbial activity. As a result, the activity of soil microbial processes responsible for CH_4 production and consumption (DelGrosso et al., 2000a) and production and consumption of nitrogen oxides can be roughly predicted from soil WFPS (Davidson, 1991).

Linn and Doran (1984) demonstrated that aerobic microbial reactions, nitrification and respiration, peaked near field capacity while responses that are inhibited by oxygen, i.e., denitrification, increased greatly above 80% WFPS. Firestone and Davidson (1989) developed the relationships of WFPS and soil microbial activity to describe NO and N_2O emissions. Their conceptual model, the hole-in-the-pipe (HIP) model, is used to describe two levels of regulation of N-oxide emissions from soils: (i) the amount of fluid flowing through the pipe is analogous to the rate of N cycling in general, or specifically to rates of NH_4^+ oxidation by nitrifying bacteria and NO_3^- reduction by denitrifying bacteria; and (ii) the amount of N that 'leaks' out of the pipe as gaseous N-oxides, through one 'hole' for NO and another 'hole' for N_2O, is determined by several soil properties, but most commonly and most strongly by soil water content. This effect of soil water content, and in some cases acidity or other soil factors, determines the relative rates of nitrification and denitrification and, hence, the relative proportions of gaseous end products of these processes. The first level of regulation determines the total amount of N-oxides produced (NO + N_2O) while the second level of regulation determines the relative importance of NO and N_2O as the gaseous end products of these processes. In dry soils, NO is the dominant gas that leaks out of the pipe while in wetter soils N_2O is the more important end product (Davidson and Verchot, 2000; Davidson et al., 2000). Davidson (1991) observed that the largest NO emissions could be expected at WFPS values of 30–60% and the highest N_2O emissions at 50–80% WFPS. Under wetter soil conditions N_2 should be the dominant gaseous N product emitted from the soil.

3. The impact of tropical agriculture on trace gas exchange

3.1. FLOODED RICE

When we think of crop production in the tropics the scenes that may first come to mind are from Asian rice paddies. Rice production certainly is a large part of tropical food production, especially in Asia. In 1995–1997 global rice production area was approximately 154 million ha to which 15.4 Tg of N was applied as fertilizer (IFA, 2000). Although generally considered a tropical crop, rice is grown as far north as 50°N and as far south as 40°S. Because of the unique nature of rice production, typically flooded soils and relatively high N input, the potential for significant emission of CH_4 during flooded periods and N_2O emissions during non-flooded periods exists.

During the past 15 years, a large number of field studies have quantified CH_4 emissions from rice fields during the rice-growing season (e.g. Wassmann et al., 2000a,b,c). These and a host of other studies have shown that emissions are affected by a multitude of different factors related to both natural conditions as well as crop management. In reviewing the numerous field studies, however, it seems justified to distinguish between

(1) primary factors that determine the level of emissions and
(2) secondary factors that modulate emissions within a smaller range i.e. selection of rice cultivars, sulfate fertilizer, etc.

Given the broad scope of this presentation, the following discussion will focus on the primary factors, namely water regime, organic amendments, soil characteristics and climate (as well the natural disposition by soil and climate).

Permanent flooding favors the formation of large amounts of CH_4 whereas even short periods of soil aeration significantly reduce emission rates. Unstable water supply is generic to rainfed rice, so that this rice ecosystem is, in most cases, characterized by lower emission potentials than irrigated rice. In vast parts of equatorial Asia, rainfed rice suffers from dry periods either at the beginning or at the end of the growing season which reduces the overall emissions by ~50% (Setyanto et al., 2000). However, ample and evenly distributed rainfall may create soil conditions comparable to irrigated rice in some rainfed systems; e.g., the rainfed season in Eastern India yielded equally high emission levels as irrigated rice (Adhya et al., 2000).

Conceptually, irrigated rice is characterized by control of the water management which implies the ability to flood and drain the field whenever it is deemed appropriate from agronomic considerations. In practice, however, irrigation schemes have only a limited buffer capacity against variability in rainfall. Irrespective of the rice ecosystem, it is literally impossible to drain fields at the peak of the rainy seasons whereas water shortages may occur in some dry seasons in irrigated rice as well. Superimposed on this natural variation of the water regime in irrigated rice are the diverging flooding patterns applied in different regions and seasons. Permanent flooding of rice fields over the entire annual cycle is found in some remote parts of Central China and leads to extremely high emissions of approximately $900\,kg\,CH_4\,ha^{-1}\,yr^{-1}$. Consistent flooding throughout the growing season, which is relatively common for the wet season crop in wide parts of Southeast Asia, also entails a relatively high emission potential. Numerous field studies under this type of flooding indicate an emission potential ranging from less than 100 upto 500 kg $CH_4\,ha^{-1}$ (Wassmann et al., 2000a,b,c).

In many rice-growing regions of China, the flooding of the fields is interrupted by short drainage periods in the middle of the growing season. Although the reduction effect varied considerably at different locations and in different seasons, this local practice reduced emission rates in most cases between 20% and 40% as compared to permanent flooding. In Northern India, irrigation has to compensate for high percolation losses, so that the floodwater is replenished by frequent flooding pulses, e.g., once a day (Jain et al., 2000). CH_4 emissions are generally below 30 kg $CH_4\,ha^{-1}$ from this type of irrigated rice field (Jain et al., 2000). The second management factor determining the level of emission rates is the dose of organic inputs (Yagi and Minami, 1990; Denier van der Gon and Neue, 1995; Wassmann et al., 2000a).

Traditional agriculture in China encompasses relatively high amounts of manure leading to high emission rates (e.g. Wang et al., 2000). The decline of this practice over the last decades has subsequently led to a major reduction of the CH_4 source strength of Chinese rice fields (Denier van der Gon, 1999). In addition to exogenous organic material such as animal manure, CH_4 emissions are influenced by the management of crop residues, i.e., straw as well as stubble and roots. Soil incorporation of rice straw generally stimulates emissions, but the incremental effect depends on timing of the straw application. The practice of straw addition is rather unpopular among farmers that have access to other fertilizers, so that the remaining plant parts in the field represent the only input of organic material into the soil. Under these conditions of low organic inputs, even the height of the stubbles can have a major impact on the level of emission rates (Wassmann et al., 2000a).

The intercomparison of emissions under identical crop management and measurement protocols clearly demonstrates the strong influence of natural environmental factors on CH_4 emission (Wassmann et al., 2000a). In incubation studies, rice soils showed a wide range of CH_4 production potentials that may be responsible for the pronounced variation of *in situ* emission rates found even within small areas (Wassmann et al., 1998). Climate can also act as a natural determinant for CH_4 emissions. In Northern China, rice is grown in one crop that encompasses very low temperature at the late phase of the growing season that brings emissions down to very low levels. This temporary impediment of emissions occurs during a period of the growing season that represents the bulk of CH_4 release in other rice growing regions. Thus, the temperature regime of this temperate climate exerts a significant reduction of the cumulative emission over the entire season as compared to a tropical climate.

3.1.1. Earlier estimates on rice production

Of the wide variety of sources for atmospheric CH_4, flooded rice fields are considered important because of the about 75% increase in rice harvest area in the world since 1935 (Shearer and Khalil, 1993). The harvested paddy rice area has increased from 86×10^6 ha in 1935 to 148×10^6 ha in 1985. However, in the last few years, the rate of expansion of rice growing area has decreased. About 90% of the world's harvested area of rice paddies is located in Asia. Of the total harvested area in Asia, about 60% is located in India and China.

A review of CH_4 studies in China, India, Japan, Thailand, the Philippines and the USA (Sass, 1994) tightened the range of projected CH_4 emissions. Sass combined the data on total area of rice paddies with the flux estimates published in various chapters of Minami et al. (1994). In the case of China and India, the annual CH_4 flux estimates were specified in Sass (1994). In the other cases, the figures are based on the minimum and maximum reported emission averages. The rice areas in the countries shown in Sass's (1994) estimate represent 63% of the total world rice paddy area and result in a total annual CH_4 emission of 16–34 Tg. Extrapolating these data to the world, Sass (1994) estimates total CH_4 emissions from rice fields to range between 25.4 and 54 Tg yr^{-1}. Participants of a recent symposium on CH$_4$ emissions from rice-based agriculture concluded that current global annual CH$_4$ emissions from rice fields are likely near the lower end of the Sass (1994) estimate (R. Sass personal communication).

3.1.2. Recent advances in upscaling of emissions from flooded rice ecosystems

New emission models and GIS databases may in the future narrow down the enormous uncertainties implied in recent estimates of CH$_4$ source strengths. Initial approaches for upscaling emissions through coupling of emission models and GIS are shown in Matthews et al. (2000b) and Li (2001). The Methane Emissions from Rice Ecosystem (MERES) model that was developed by Matthews et al. (2000a,b) is largely based on a crop yield model (CERES) that was extended by a module describing the steady-state concentrations of CH$_4$ and oxygen in the soil. The combination with an newly compiled GIS database on rice ecosystems, soils and weather allowed computation of national source strengths under different crop management scenarios (Matthews et al. 2000b). The baseline scenario assuming no addition of organic amendments and permanent flooding of the fields during the growing season yielded CH$_4$ emissions of 3.73, 2.14, 1.65, 0.14 and 0.18 Tg CH$_4$ yr^{-1} for China, India, Indonesia, Philippines and Thailand, respectively.

Another modeling approach to derive emission rates of trace gases is the DeNitrification and DeComposition (DNDC) model describing carbon and nitrogen biogeochemistry in agricultural ecosystems and forests (Li et al., 2000). The DNDC model has been applied to compute N$_2$O emissions in USA and China (Li et al., 1996; 2001), but the new versions of this model potentially cover CH$_4$ emissions as well. The DNDC results for N$_2$O emissions from Chinese agriculture were compared to emissions computed by the IPCC methodology; total emissions were similar, but geographical patterns deviated substantially (Li et al., 2001).

3.2. ANNUAL EMISSIONS OF CH$_4$ AND N$_2$O FROM RICE-BASED CROPPING SYSTEMS (RICE-FALLOW-RICE AND RICE-FALLOW-WHEAT)

Relatively, few studies have quantified annual fluxes of CH$_4$ and N$_2$O in rice-based cropping systems. In many rice-based agricultural areas one or two rice crops and an upland crop are grown. Between cropping periods are fallow times when no crops are grown. Since the climate in much of the rice growing area of the world is semi tropical or tropical, temperatures are warm during non-rice production periods and when precipitation occurs potentially important N$_2$O emission events occur during non-flooded times of the year.

Robertson et al. (2000) show that to evaluate greenhouse gas emissions from agricultural systems, total global warming potential (GWP) needs to be considered. For rice-based systems a consideration of both CH$_4$ and N$_2$O is relevant. In our discussions, here we use GWP values of 21 and 310 for a 100-year time frame for CH_4 and N_2O, respectively, compared to CO_2 on a per molecule basis (IPCC, 1996). Of the few studies that have quantified CH_4 and N_2O emission through whole annual cropping sequences, most notable are those conducted at the International Rice Research Institute (IRRI) in the Philippines, in rice-fallow–rice-fallow and rice-fallow–wheat-fallow cropping sequences. In these studies, automated chamber systems were employed which permitted several flux measurements per day on each location throughout several years of measurement (Bronson et al. 1997a, b and Abao et al. 2000; and see Wassmann this issue for detailed discussion).

Data in Table III show the cumulative amount of CH_4 or N_2O (expressed as g CO_2 equivalent m^{-2} per season) emitted during each rice production (Bronson et al., 1997a) or fallow season (Bronson et al., 1997b). The GWP from CH_4 emissions generally exceeded GWP from N_2O from continuously flooded rice. Incorporation of rice straw increased CH_4 emissions during rice cropping but had little effect on N_2O emissions. When rice fields were drained, at midtillering particularly, CH_4 emissions substantially decreased while N_2O emissions increased (Table III). A side-by-side comparison of the effect of midseason drainage on trace gas emissions showed that total GWP was changed little compared to continuously flooded rice. Total GWP for $CH_4 + N_2O$ was 1040 g CO_2 equivalents m^{-2} for the drained plots compared to 800 for the continuously flooded plots. In terms of GWP, the increase in N_2O emissions more than offset the decrease in CH_4 emissions with field drainage.

During fallow periods GWP from N_2O emissions far exceeded that from CH_4 (2.7–500 fold). Tsuruta et al. (1997) found similar trends where CH_4 fluxes were relatively high during intermittently flooded cropping compared to N_2O and vice versa during fallow. Over the ~480 day observation period it is clear that N_2O

TABLE III. Seasonal CH_4 and N_2O emissions (expressed as GWP in terms of CO_2 equivalents where one molecule of CH_4 or N_2O equals 21 or 310 molecules of CO_2, respectively) from rice field fertilized with urea or urea plus rice straw (Bronson et al., 1997a,b).

Season (days)	Urea fertilized*		Urea + rice straw**	
	N_2O	CH_4	N_2O	CH_4
	(g CO_2 Equivalent m^{-2} season^{-1})			
Fallow (46)	832	4.2	1420	7.7
Rice (#DS-PI) (111)	122	296	78	5520
Fallow (36)	2340	4.6	2350	5.5
Rice (##WS-CF) (98)	179	907	81	9840
Fallow (89)	263	98	492	93
Rice (#DS-MT) (97)	870	168	225	5340
Rice (#DS-CF) (97)	353	441	38	9340

*Plots were fertilized with 200 kg urea–N ha^{-1} for dry season and 120 kg N ha^{-1} for wet season in four equal split applications at plot harrowing, midtillering, panicle initiation and flowering.
**Rice straw (5.5 Mg dry straw ha^{-1} (32 kg N ha^{-1}) and 160 and 80 kg urea–N ha^{-1} during dry and wet seasons, respectively.
#DS – dry season crop; PI + floodwater drainage at panicle initiation stage of rice growth; MT = floodwater drainage at midtillering stage of rice growth; CF = continuous flooding of rice field.
##WS = wet season crop values in parentheses are length of specified season in days.

is an important part of rice-based agriculture GWP. Where two rice crops were continuously flooded total GWP totaled $\sim$5800 g CO_2 equivalents m^{-2} with about 70% coming from N_2O. During the three rice growing periods total GWP was $\sim$2300 g CO_2 equivalent m^{-2}, with $\sim$72% CH_4 emissions (Table III).

3.3. TRACE GAS FLUXES FROM TROPICAL UPLAND CROPPING SYSTEMS

There are relatively few measurements of CH_4, N_2O or NO fluxes in upland tropical crop production systems published (as noted in Erickson and Keller, 1997; Crill et al., 2000) and even fewer studies where simultaneous measurements of these gases are reported. As a result, it is difficult to assess the relationship of the main regulating factors, e.g., crop, weather pattern, soils and fertilization on total GWP from trace gases in tropical upland agricultural systems. Because of the paucity of data, we discuss individual gases and describe measurements that have been published during production of various crops.

3.3.1. Sugar cane
Sugar cane is a crop that is grown mainly in the tropics and subtropics. Globally, approximately 16 million ha of sugar cane was grown in 1995 (IFA, 2000), to which about 2.02 Tg of synthetic fertilizer N was applied. The relatively few measurements of CH_4, NO_x and N_2O flux in sugar cane fields indicate that fertilizer placement and crop residue management impact trace gas emissions. Weier et al. (1998) in northeastern Australia found that the presence of cane trash layer ($\sim$10 Mg ha^{-1}) tended to increase N losses. N lost from total denitrification ($N_2 + N_2O$) during an 8-day period following application of 160 kg ha^{-1} of urea-N was 9.2 and 7.6 kg N ha^{-1} with and without the trash layer, respectively. Of this total denitrification 2.8 and 1.9 kg N ha^{-1} (15 and 10.2 μg N m^{-2} h^{-1}) was emitted as N_2O, respectively. In another study, Weier (1996) found high rates of CH_4 oxidation, 0.7–2 kg CH_4–C ha^{-1} d^{-1} (-2.9 to -8.3 μg C m^{-2} hr^{-1}) (Table IV). The highest oxidation rates were from plots that had been fertilized with KNO_3 (160 kg N ha^{-1}). CH_4 uptake rates with a trash layer were generally higher than without trash, which Weier (1996) attributed to retention of soil moisture by the trash layer.

TABLE IV. Effect of a cane trash layer (10 Mg ha^{-1}) and fertilizer N addition (160 kg N ha^{-1}) on oxidation of atmospheric CH_4 and N_2O emission in a sugar cane field in northeastern Australia (Weier, 1996), measured every 8 h for 9 d following N fertilization.

Treatment	No N		Urea		KNO_3	
	N_2O	CH_4	N_2O	CH_4	N_2O	CH_4
			(g C or N ha^{-1} d^{-1})			
+Trash	730 (157)*	−33 (3.8)	750 (242)	−42.5 (2.8)	2000 (190)	−45.2 (3.3)
−Trash	420 (126)	−28.2 (0.7)	720 (147)	−29.4 (6.5)	1360 (450)	−56.7 (22.1)

*Numbers in parentheses are standard error of the mean, $n = 3$.

In other experiments, in Australia the effect of trash layer on fertilizer N losses were variable from year to year and within different soil types (Vallis et al., 1996) and ranged from 4% to 65% of fertilizer N applied.

Matson et al. (1996) observed the influence of soil type and N-fertilizer application method on sugar cane fields on NO and N_2O emissions in two locations in Hawaii, USA. Subsurface drip irrigation/fertilization in Mollisol and Inceptisol soils in Maui was compared to surface broadcast application of fertilizer N in Andosols in Hawaii. Subsurface fertilization decreased NO emissions compared to surface application but increased N_2O emissions. N_2O fluxes were lower from the Mollisol soils compared to Andosol soils at the same soil WFPS. Seasonal NO + N_2O emissions totaled 0.03–0.5% of fertilizer N applied in drip irrigation systems and 1.1–2.5% when surface application of urea was practiced.

3.3.2. Row crops: Cotton, maize, sorghum, wheat, cassava, peanut

Total emissions from tropical and subtropical land areas for upland crops such as cotton, maize, sorghum and wheat are not readily discernable from common databases (FAO, 2001; IFA, 2000). Crop production information for countries such as China are not broken down into geographical (climatic) area, and much of the maize and wheat in China, for example, is grown in temperate areas of the country. Based on the few published studies of trace gas fluxes from tropical upland production of cotton, maize, sorghum, and wheat, fertilization and soil moisture are the dominant controls on N_2O and CH_4 fluxes. Because these upland systems are better drained, CH_4 fluxes are low compared to flooded crop systems.

Mahmood et al. (1998; 2000) quantified fertilizer N loss (based on ^{15}N recovery) and total denitrification (using acetylene block techniques) in irrigated cotton, maize and wheat in a semiarid subtropical region of Pakistan. They found that total fertilizer N losses were typically 30–45% of the N applied and that denitrification accounted for 3–40% of the loss (Table V). Neither N_2O nor NO_x fluxes were measured individually. The authors considered NH_3 volatilization to be a main loss mechanism when denitrification was limited. The high denitrification rates in cotton were ascribed to high August temperatures and concomitant large rainfall events. Temperatures were lower and rainfall less abundant during the maize–wheat cropping periods (Mahmood et al., 1998).

TABLE   V. Total fertilizer N and denitrification N losses from irrigated cotton, maize and wheat crops in subtropical Pakistan (Mahmood et al., 1998; 2000).

Crop	Total fertilizer N loss % of N applied	N loss by denitrification* % of N applied
Cotton	41.5	40 ± 10
Maize	39	2.7 ± 1.1
Wheat	33	3.4 ± 1.1

*In excess of the N lost by denitrification from unfertilized control plots.

In an irrigated wheat production system in Sonora, Mexico, Matson et al. (1998) found that N_2O and NO_x emissions were large, $\sim$8.5 kg N ha^{-1}, under conventional farming practices for the region. These losses were decreased to $\sim$4.4 kg N ha^{-1} by instituting alternative fertilization practices which decreased fertilizer input from 250 to 180 kg N ha^{-1} while maintaining crop yield. Total fertilizer N losses were 70 and 48 kg N ha^{-1} under conventional and improved fertilization practice, respectively.

Crill et al. (2000) employed an automated chamber system to monitor N_2O emission from a maize field in Costa Rica. They found that N_2O emissions from fertilized (122 kg N ha^{-1}/cropping season) *versus* unfertilized plots averaged 640 μg N_2O–N m^{-2} h^{-1} vs 120 μg N_2O–N m^{-2} h^{-1}. Likewise, Weitz et al. (2001) found that fertilized systems in Costa Rica had more than three times the N_2O emissions as unfertilized systems. Both authors observed that highest N_2O fluxes were associated with surface WFPS between 80 and 99% and that fertilization and soil moisture were the dominant regulators of N_2O flux.

Watanabe et al. (2000) measured N_2O emissions from maize fields at four sites in Thailand (Table VI). They found that the N_2O emission increase due to N-fertilization of 47–75 kg N ha^{-1} was small, averaging 0.1–0.4, percentage of applied N at four sites.

Khalil et al. (2000; 2002) measured N_2O emissions during an annual groundnut–maize crop rotation at the Puchong Experimental Farm of the University of Putra in Malaysia. The study was conducted within fields that had the following treatments: (1) recommended inorganic N+crop residue, (2) recommended inorganic N (($NH_4)_2SO_4$) only, and (3) half of recommended inorganic N + crop residue + chicken manure. The recommended N fertilizer rate was 30 kg N ha^{-1} for groundnut and 150 kg N ha^{-1} for maize. Over the whole groundnut-fallow–maize-fallow cropping sequence N_2O–N emissions totaled 0.6%, 0.7% and 1.0% of total N applied for treatments 1, 2 and 3, respectively. Cumulative N_2O–N emissions for each part of the cropping sequence are shown in Table VII. Emissions were generally highest when chicken manure was applied and during the maize part of the cropping sequence. Emissions were also high during fallow periods.

Nitrous oxide and CH_4 fluxes were measured weekly over a 154-day period (field preparation to post harvest) in a sorghum (*Sorghum bicolor* L.) field in a Oxisol soil near Isabela, Puerto Rico. N_2O and CH_4 fluxes averaged 57 μg N m^{-2} h^{-1} and -1.2 μg C m^{-2} h^{-1}, respectively. Fluxes measured in a nearby unfertilized pasture

TABLE VI. Estimates of N_2O emitted during the maize growing season at four locations in Thailand (Watanabe et al., 2000).

Location	Not fertilized (mg N_2O–N m^{-2})	N-fertilized (mg N_2O–N m^{-2})
Nakhon Sawan	10.0	40.3
Phra Phutthabat	12.2	35.1
Khon Kaen	10.9	28.9
Chiang Mai	11.9	19.6

TABLE VII. Cumulative N_2O emissions during groundnut-fallow–maize-fallow periods of an annual cropping sequence in Malaysia (adapted from Khalil et al., 2002).

Treatment	Period of cropping season (kg N_2O–N ha^{-1})			
	Groundnut	Fallow	Maize	Fallow
Residue + N	0.3	0.3	0.7	0.6
N only	0.3	0.2	0.8	0.3
$\frac{1}{2}$N + residue + chicken manure	0.4	1.2	1.8	0.7

that had been established more than 25 years before the study was initiated, averaged 16.3 μg N_2O–N m^{-2} h^{-1} and 7.9 μg CH_4–C m^{-2} h^{-1} (Mosier et al., 1998c).

Delmas et al. (1997) also found reduced CH_4 sink strength in agricultural fields compared to savanna soils in the Loudima region of the Congo. Flux rates in savannas averaged -0.46 μg CH_4–C m^{-2} h^{-1} while in agricultural soils they were slightly reduced in a cassava field and more reduced in a ploughed field and in a peanut field (-0.38, -0.34, and -0.28 μg CH_4–C m^{-2} h^{-1}, respectively).

3.3.3. Tree crops: Banana, cocoa, coconut, coffee, palm oil, papaya, rubber

Tropical tree-based cropping systems (e.g. banana, cocoa, coconut, coffee, palm oil, rubber) covered approximately 16 million ha in 1995 and were fertilized with about 1.04 Tg of N (IFA, 2000). Globally averaged N application rates vary from only 5 kg N ha^{-1} for coconut to 200 kg N ha^{-1} for banana. Average N-fertilization rates for coffee, palm oil, and rubber were similar to each other, ranging from 50 to 70 kg N ha^{-1} yr^{-1} (IFA, 2000). Although little trace gas emission data exist for such crops, field studies have been conducted in banana (Veldkamp and Keller, 1997) and papaya (Crill et al., 2000) plantations in Costa Rica and in a rubber plantation in Sumatra, Indonesia (Tsuruta et al., 2000).

Veldkamp and Keller (1997) measured N_2O and NO emissions from two soil types, Andosol and Inceptisol, within a 400 ha banana plantation in a humid tropical area of Costa Rica. N_2O and NO emissions averaged 314 and 556 μg N m^{-2} h^{-1} from the Andosol and 93 and 411 μg N for N_2O and NO, respectively, from the Inceptisol. These fluxes amounted to 1.3–2.9% and 5.1–5.7% as N_2O–N and NO–N, respectively, of the 360 kg N applied.

3.3.4. Fertilized grasslands

The global land area of permanent pastures was 3.21×10^9 ha in 1970 compared to 3.46×10^9 ha in 1999 (FAO, 2001). Approximately, 40% of these grazing lands are located in sub-Saharan Africa, Central and South America (Table VIII). Although the total land area in permanent pastures within subtropical and tropical regions is not readily determined, such grasslands occupy large land area and are economically important for livestock production in these regions of the world. The productive lifetime of lands converted from forest to grasslands or other agricultural uses is

TABLE VIII. Land area of permanent pastures for regions containing largely subtropical and tropical climates (FAO, 2001).

Region	Land in permanent pasture (10^6 ha)	
	1970	1999
Subsahara Africa	740	735
Central America and Caribbean	90	99
South America	450	500

relatively short in the tropics unless the land is carefully managed and supplemented with nutrients, particularly N. In this context, managed grasslands are important for livestock and dairy production. A down side of increased nutrient input is a likely concomitant increase in trace gas emissions.

An example of conversion of tropical forests to agricultural use exists on the island of Puerto Rico. Most of the forested parts of Puerto Rico were converted to intensive agriculture 100–300 years ago (Birdsey and Weaver, 1982). About 50 years ago much of the intensive agriculture was discontinued and many areas reverted to unmanaged grasslands (Lugo et al., 1986). Other areas of these grasslands are now fertilized to improve forage production for cattle. To investigate the impact of nutrient input into tropical grasslands trace gas exchange, Mosier and Delgado (1997) established field locations in three different soils (Oxisol, Ultisol and Vertisol) to quantify CH_4 and N_2O fluxes in established mixed-grasslands and to measure the impact of fertilization on the soil-atmosphere exchange of CH_4 and N_2O in these systems. The study was conducted over 28 months (1992–1995).

The mean weekly N_2O emission rates from unfertilized plots of 17 μg N m^{-2} h^{-1} were similar to those observed in the Costa Rican pastures, 10–18 μg N m^{-2} h^{-1}, and considerably lower than the 43–67 μg N m^{-2} h^{-1} from the secondary and old growth forests (Keller and Reiners, 1994) because of lower soil moisture in grasslands and more fertile soils in the forest soils. Emissions from fertilized grasslands in the study, 130, 46 and 44 μg N m^{-2} h^{-1} from Vertisol, Ultisol and Oxisol, respectively, were much lower, than from the banana plantation Andisol soils (Veldkamp and Keller, 1997). Emissions from the Ultisol and Oxisol were only about half that of the Costa Rican Inceptisol while fluxes from the Vertisol averaged roughly half those from the Costa Rican Andisol and about 1.4 times those from the Inceptisol.

Converting mean weekly flux rates to annual emissions, Mosier and Delgado (1997) estimated that the increase in N_2O emissions due to fertilizer addition represented 0.8%, 3.3% and 0.8% of fertilizer N applied in the Oxisol, Vertisol and Ultisol, respectively. The effects of N fertilization on N_2O emissions in these tropical grasslands were well within the range of N_2O emissions from fertilized temperate grasslands (Allen et al., 1996). Over the study period, CH_4 flux averaged −6.7, −4.6 and −6.3 μg CH_4–C m^{-2} h^{-1} at the Vertisol, Ultisol and Oxisol unfertilized grasslands, respectively.

In another study in Costa Rica, Veldkamp et al. (1998, 1999) measured N_2O and NO emissions from pastures under three different management systems in a volcanic soil. The management practices were: traditional (no N input from fertilizer or legume), pasture with grass and legume combined, and N-fertilized pasture (total of 300 kg N ha^{-1}yr^{-1} applied about monthly). They observed that annual N_2O and NO emission rates were 27 and 9; 49 and 13; and 250 and 53 μg N m^{-2} h^{-1} as N_2O–N and NO–N from traditional, grass-legume and N-fertilized pastures, respectively. These N_2O emission rates are about the same as those observed in Puerto Rico (Mosier and Delgado, 1997) for the unfertilized pasture but 2–6 times higher than the Puerto Rican fertilized pastures, depending on soil type. Veldkamp et al. (1998) observed that soil water content between fertilization events had a large impact on N_2O and NO emissions. Using data from Veldkamp and Keller (1997) and Matson et al. (1996) they plotted N_2O loss, as a percentage of fertilizer N applied, *versus* soil water content, expressed as water-filled-pore-space (WFPS), and found that N_2O emissions peaked at 75–80% WFPS and were relatively low at <70% WFPS and >85% WFPS.

4. Impact of land use change on trace gas exchange

4.1. BIOMASS BURNING

4.1.1. CH₄

The first step in conversion of forests to agricultural or other uses typically involves biomass burning in some form. As a result of incomplete combustion CH_4 and N-oxides are emitted. In response to declining agricultural yields and population pressures, farmers in many regions convert forests to cropland, and many of their techniques involve burning. For example, shifting cultivation requires that forests be cut, logging debris and unwanted vegetation burned, and the land farmed for several years then left fallow to re-vegetate. Savanna and rangeland biomass is often burned to improve livestock forage. Agricultural residues are also burned in the field to return nutrients to the soil or reduce shrubs on rotational fallow lands. Such agriculture-related burning may account for 50% of the biomass burned annually. Estimates indicate that 8700 Tg of biomass yr^{-1} (Andreae, 1991) and 1–5% of the world's land is burned. In addition, crop residues and animal dung are burned as fuel. Studies show that 50% of the crop residues world-wide are burned in small cooking and heating stoves (Hao et al., 1987). Delmas (1997) estimated that CH_4 emission from biomass burning, using the annual biomass burning numbers provided by Andreae (1991) and CH_4 conversion factors discussed in his paper, totaled approximately 34 Tg. Of these estimated CH_4 emissions, those from tropical forest clearing for agriculture, savanna burning, and agricultural crop residue burning are portions of total biomass burning that can be attributed to agricultural practices which sum to about 22 Tg CH_4 yr^{-1}.

4.1.2. N_2O and NO

During combustion, the N in end groups, open chains and heterocyclic rings can be converted into gaseous forms such as NH_3, NO, N_2O, N_2 and HCN (Galbally and Gillett, 1988). These compounds are liberated at all temperatures at which smoldering and combustion occur (Galbally and Gillett, 1988). Crutzen et al. (1981) estimated that 8 Tg N_2O–N yr^{-1} is emitted to the atmosphere as a result of biomass burning. However, recent work suggests that their emission factors were too high and that N_2O is probably only a minor product of biomass burning (Galbally et al., 1992; Muzio and Kramlich, 1988). Bouwman and Sombroek (1990) report that the contribution of biomass burning to N_2O emission may be less than half of the 1.5 Tg yr^{-1} assumed earlier by Seiler and Conrad (1987). Biomass burning increases the biogenic production of NO and N_2O by providing substrate to the soil (Anderson et al., 1988; Levine et al., 1988). Although NO emissions typically greatly increase following burning (Verchot et al., 1999), biogenic emissions are the dominant NO source in southern African savannas (Otter et al., 2001).

4.2. FOREST TO PASTURE CONVERSION

4.2.1. CH_4

A few studies have examined the effects of deforestation and pasture creation on CH_4 sink strength, and conclusions suggest that conversion of primary forest to pastures results in decreased net CH_4 uptake in soils and in many instances the conversion of a sink to a source (Table IX). There is great variation in the CH_4 consumption rates in forests, but as Verchot et al. (2000) pointed out, soil texture is a major determining factor. Fine texture soils in humid tropical forests consume 1.5–2.0 kg CH_4 ha^{-1} yr^{-1} (12.8–17.1 μg CH_4–C m^{-2} h^{-1}), while medium and coarse texture soils consume >4.0 kg ha^{-1} yr^{-1} (34.2 μg CH_4–C m^{-2} hr^{-1}). Pastures were for the most part sources during the wet season and sinks during the dry season. The exception to this was the young pastures in eastern Amazonia, which were sinks all year long. This may have had something to do with soil aggregation and drainage. From what we know about the mechanisms that account for the net CH_4

TABLE IX. Differences in CH_4 consumption rates in forest, old pasture, and young pasture for sites that permitted calculation of annual fluxes in the neotropics with soil texture and rainfall information for the study sites.

Site	Primary forest (kg ha^{-1} yr^{-1})	Old pasture (kg ha^{-1} yr^{-1})	Young pasture (kg ha^{-1} yr^{-1})	Rainfall	Soil texture
Guacimo, Costa Rica[1]	−4.0	−1.0 to 5.3	−1.1 to 2.8	4000	Medium
La Selva, Costa Rica[1]	−4.8 to 4.4	−2.5 to 1.5	—	4000	Medium
La Selva, Costa Rica[2]	−4.6	0.8	—	4000	Medium
Paragominas, Brazil[4]	−2.5 to 2.1	−1.3	−6.2 to 1.1	1800	Fine
Rondônia, Brazil[5]	−5.9 to −3.4	1.0 to 12.0	−0.8 to 3.4	2200	Coarse

[1] Keller et al., 1993; [2] Keller and Reiners, 1994; [4] Verchot et al., 2000; [5] Steudler et al., 1996.

flux being positive or negative, we can hypothesize that the effect of land-use change is probably governed by the interaction between soil texture and rainfall. That is, the decrease in sink strength or the source strength is governed by the amount of time when soil water exceeds field capacity.

Only two of the studies in tropical pastures derived from forests measured abandoned pastures and secondary forests. Keller and Reiners (1994) found that the abandoned pastures and secondary forests were sinks of similar magnitude to the primary forests. Verchot et al. (2000) found that degraded and abandoned pastures in eastern Amazonia were very strong sinks, consuming 50% more CH_4 than primary forest sites, and that secondary forests were weaker sinks than primary forest, consuming 50% less. The sources of these differences was in the dry season fluxes.

4.2.2. N-oxides

Conversion of tropical forest to pasture generally results in decreased rates of N cycling and a change in the nature of soil inorganic-N pools, from nitrate-dominated in forests to ammonium-dominated in pastures (Matson and Vitousek, 1990; Verchot et al., 2000). The shift from an inorganic-N pool dominated by NO_3^- to one dominated by NH_4^+ appears to be related to changes in nitrification rates with land-use change. In all of these studies where net mineralization, net nitrification or nitrification potential were measured, rates were lower than in primary forests. Results were equivocal for several pasture chronosequences in Rondônia (Neill et al., 1997), where the inorganic-N pool was dominated by NH_4^+ in pastures and some forests during the wet season, and several pastures were dominated by NO_3^- during the dry season. These shifts in N cycling have implications for N-oxide emissions from soils as NO_3^- availability is a key controlling factor for production of these gases.

Luizão et al. (1989) first raised the possibility that conversion of forests to pastures in the tropics may be responsible, at least in part, for the current increase in atmospheric N_2O burden. These authors intensively sampled N_2O fluxes in a 3-year-old pasture and a primary forest. They found that annual emissions from the pasture exceeded forest emissions by a factor of 3. They also sampled four additional pastures, representing a chronosequence from 3 to 10 years, once during the rainy season and found that two had fluxes that greatly exceeded the forest fluxes. Extrapolating the increased flux, they estimated that tropical deforestation contributed around 1 Tg of N_2O–N to the atmosphere annually, but they urged caution because their pastures were not necessarily representative of the broad diversity of soil, management and climatic conditions within the tropics. Both Keller et al. (1993) working in Costa Rica and Garcia-Montiel et al. (2001) working in Rondonia, Brazil found elevated N_2O emissions in young pastures, but significantly lower emissions in older pastures. In the Costa Rica sites, emission were elevated in pastures that were <10 years old, while in Rondonia, elevated emissions lasted only 2 years. Verchot et al. (2000) found no increase in N_2O production following land clearing in these eastern Amazonian sites; on the contrary, pastures on both sites had lower emissions than forests. They also found no consistent trend with pasture age for young pastures on their first rotation. Thus, Luizão et al. (1989) may have captured a transitory situation of elevated emissions in the 3-year-old pasture, as both Keller et al. (1993) and Garcia-Montiel et al. (2001) found lower emissions in pastures than in forests. If the trends found in these two other studies hold for this site, older pastures in this region would probably exhibit lower fluxes. Results from intensive sampling by Coolman (1994) at a site near the Luizão et al. (1989) sites showed that emissions from three abandoned pastures, that had been through a short rotation (6 years) and abandoned for 3 years, were only slightly higher that at a comparable upland primary forest site (Table X). Sufficient data to determine exactly why large site to site variation in N_2O flux have been observed are still lacking.

Data on NO fluxes from humid tropical soils are less common than data concerning N_2O. Flux estimates range between 0.7 and $1.4 \, kg \, N \, ha^{-1} \, yr^{-1}$ (an average flux of 8 and $16 \, \mu g \, N \, m^{-2} \, h^{-1}$) for these soils (Keller et al., 1993; Keller and Reiners, 1994) and the best estimate for the flux from tropical evergreen forests is $1.1 \, Tg \, yr^{-1} (12.5 \, \mu g \, N \, m^{-2} \, h^{-1})$ (Davidson and Kingerlee, 1997). Verchot et al. (2000) measured emissions of 1.46 and $1.54 \, kg \, N \, ha^{-1} \, yr^{-1}$ (an average flux of 16.6 and $17.6 \, \mu g \, N \, m^{-2} \, h^{-1}$) for primary forests in eastern Amazonia (Table X).

Unlike the increased NO emissions observed in the young pastures of Costa Rica (Keller et al., 1993), Verchot et al. (1999) and Garcia-Montiel et al. (2001)

TABLE X. Differences in N-oxide emission rates in forest, old pasture and young pasture sites in the neotropics.

Site	N_2O (kg ha^{-1} yr^{-1})			NO (kg ha^{-1} yr^{-1})			Reference
	Primary forest	Old pasture	Young pasture	Primary forest	Old pasture	Young pasture	
Guacimo, Costa Rica	6.1	1.8–10.5	34.1–51.7	4.8	0.9–1.1	3.9–8.8	Keller et al., 1993
La Selva, Costa Rica	3.5–7.9	0.9–2.6	—	0.8	1.1	0.1–0.5	Keller et al., 1993
La Selva, Costa Rica				0.9	0.2	—	Keller and Reiners, 1994
Manaus, Brazil	1.9			—	—	—	Luizão et al., 1989
Manaus, Brazil	1.4	—	1.6	—	—	—	Coolman, 1994
Rondônia, Brazil	2.5			—	—	—	Steudler et al., 1996
Paragominas, Brazil	2.6–3.3	0.1–0.3	0.5–1.7	1.5	0.5–0.7	—	Verchot et al., 1999
Luquillo, Puerto Rico	0.6–1.7	0.5–0.7	—	0.1–0.4	0.1–0.2	0.2–1.7	Erickson et al., 2000
El Verde, Puerto Rico	0.5			—	—	—	Steudler et al., 1991
Kauai, Hawaii	0.3			—	—	—	Riley and Vitousek, 1995

saw no increase in NO emissions in the young pastures. Annual fluxes in these pastures averaged $0.1–1.3\,kg\,N\,ha^{-1}$ ($1.1–14.8\,\mu g\,N\,m^{-2}\,h^{-1}$) which was similar to (Verchot et al., 1999) or considerably lower (Garcia Montiel et al., 2001) than the forest emissions. Emissions from the older pastures were 60–70% lower than from the forest and were lower than many of the young pastures in both studies.

4.3. CONVERSION OF AMAZONIAN FOREST TO INTENSIVE CROP PRODUCTION

The predominant land use systems in much of the humid tropics are not pasture or high input cropping but instead short-term cropping systems followed by the establishment of tree-based systems, including forest-fallow, tree crops, or agroforestry systems (Palm et al., 2002). Even in parts of the Amazon, pastures are not the dominant land use following forest clearing, or a large portion of land originally cleared for pastures eventually is abandoned to some sort of secondary vegetation. Few measurements of trace gas fluxes in these tree-based systems have been reported.

A long-term experiment established in 1985 in the Peruvian Amazon provided a means of making some initial comparisons of trace gas fluxes from tree-based systems with those of annual cropping systems and a secondary forest (Palm et al., 2002). The study was conducted at the Yurimaguas Experiment Station located near the town of Yurimaguas, Loreto Province in the Peruvian Amazon ($76°05'W$, $5°45'S$, 180 MASL). The area has a long-term annual average temperature of $26°C$ and average annual rainfall of 2200 mm.

The experiment provided six land management systems from which N_2O and CH_4 fluxes where compared during the period October 1997–1999. At the beginning of the experiment a 10-year-old shifting cultivation forest-fallow was slashed and burned, according to local practice, and five land management systems were established. The land management treatments included two annual cropping systems, three tree-based systems including a traditional shifting cultivation system, a multistrata agroforestry system, and a peach palm tree plantation, as well as the secondary forest-fallow control. There was no primary forest nearby to serve as the control. The shifting cultivation forest-fallow that was left undisturbed to serve as a control treatment was approximately 22 years old when the trace gases were sampled.

Palm et al. (2002) concluded that 13 years after establishment of tree-based systems following slash-and-burn agriculture, N_2O emissions were similar to that of the secondary forest while CH_4 consumption rates were slightly reduced. N_2O fluxes (Figure 3) from the tilled, limed, and fertilized high-input cropping system were three to four times that of the secondary forest while the CH_4 sink was lost, and CH_4 was emitted in this annual cropping system (Figure 3). Therefore, conversion of humid tropical forests to tree-based system will not result in a significant increase in greenhouse gas emissions, other than the large flux produced during slash-and-burn. Cropping systems, on the other hand, result in significant increases in N_2O emissions, a loss of the CH_4 sink, and net production of CH_4 from the soils, a substantial increase in the GWP compared to the forest and tree-based system.

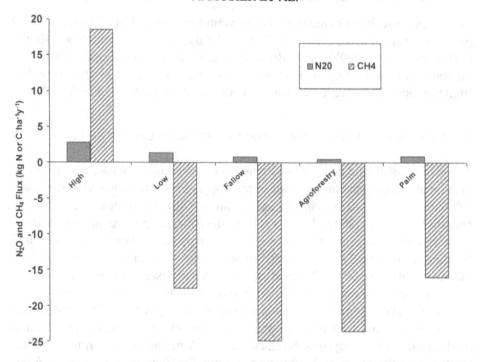

Figure 3. Annual CH$_4$ and N$_2$O fluxes in a slash-and-burn agricultural system in Sumatra (Tsuruta et al. (2000). Data are for primary forest (P1, P2), logged forest (L1, L2), burned site (B) and jungle rubber agroforest (R).

The tree-based systems in this study were not fertilized whereas many intensive tree crop systems, such as coffee and peach palm, are fertilized at rates often exceeding 300 kg N ha^{-1}yr^{-1} (Szott and Kass, 1993). Whether gaseous N losses from these fertilized tree-based systems will increase significantly compared to the forest system will depend largely on the timing of fertilizer application relative to plant demand. Perennial systems generally have high synchrony between nutrient supply and demand, so if managed properly these fertilized agroforestry systems should not be expected to have significantly large losses compared to annual cropping systems.

4.4. Land Use Change and Trace Gas Emissions: Tropical Forest to Rubber/Oil Palm Plantation

Measurement of trace gases in complex agroforests are few and have begun on a small scale in the Alternatives to Slash-and-Burn Program that spans the humid tropics. In slash-and-burn systems in southern Sumatra, forest is cleared and replaced by a multi-story rubber 'agroforest' system. This 'jungle rubber' system is characterized by relatively high density of rubber trees, with other useful fruit and timber trees interplanted. The system is established through a complex succession of production stages involving planting of crops and trees for commercial and domestic products (Williams, 2000). In a mature stage, this system has a forest-like structure.

Measurements comparing these systems with indigenous forests in Jambi province in southern Sumatra gave somewhat equivocal results for CH_4 fluxes (Tsuruta et al., 2000). A primary forest site dominated by dipterocarp (*Dipterocarpus crimtus*) and mahogany (*Shorea macrophylla*) was a strong sink for atmospheric CH_4, while a forest dominated by Mahogany and *Scaphium macropodum* was a net CH_4 source (Figure 4). Soils were a small sink during the logging and burning phase, and sink strength was only slightly stronger under the jungle rubber plantation. N_2O emissions were easier to interpret (Figure 4). In the primary forest sites, the emissions were low, were higher in the logged sites and low following burning and in the rubber plantation. While lack of replication and lack of detailed site descriptions make it difficult to draw broad generalizations about the effects of conversion of forest to tree-based agricultural systems, we do note a transient increase in N_2O emissions that was associated with increased N availability at these sites (Ishizuka, 2000). For CH_4, diffusion appeared the main factor controlling CH_4 fluxes in these soils (Ishizuka, 2000). Scaling these and other results upto the landscape scale using spatial database of land cover derived from satellite imagery,

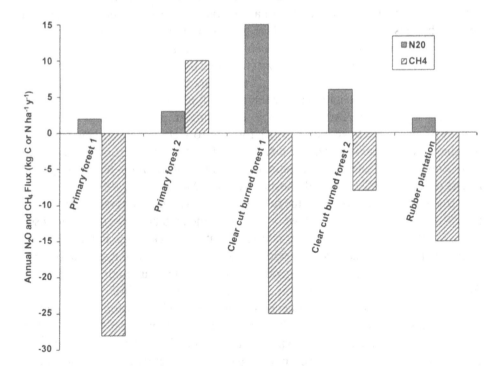

Figure 4. Annual CH_4 and N_2O fluxes in row crop and tree-based agricultural systems in Yurimaguas Peru (Palm et al., 2002). Data are for primary forest high and low input cropping systems, shade coffee and peach palm plantations and a 22-year-old tree fallow. Annual estimates are based on weighted average from three sampling periods (adapted from Palm et al., 2002).

Prasetyo et al. (2000) estimated that land-use change in Jambi province between 1986 and 1992 resulted in the loss of 50 million tons of C from aboveground biomass. Annual greenhouse gas emissions from soils increased by 4.3 million tons of CO_2, 256 tons of N_2O and the soil CH_4 sink has decreased by 183 tons yr^{-1}. These soil GHG emissions equate to 4.4 million tons of C equivalent per year, with N_2O accounting for 1.8% of this forcing and CH_4 accounting for >0.1%.

5. CH_4 and N_2O emissions from livestock

5.1. ENTERIC FERMENTATION

Although this paper does not focus on CH_4 and N_2O production in livestock systems, it is important to note that livestock and livestock waste are large sources of both gasses globally and in the tropics. Globally, approximately 80 Tg of CH_4 are emitted from livestock enteric fermentation and another 10–20 Tg CH_4 and approximately 2 Tg of N_2O–N is produced from livestock waste annually. The main source of CH_4 is from the digestive tract of ruminants. These animals (cattle, buffalo, sheep, goats and camels) have a large fore-stomach or rumen which is a particularly effective digestive system allowing utilization of a wide range of feeds, especially structural polysaccharides such as cellulose and the hemicelluloses, that are not digestible by enzymes produced directly by other animals. The process requires hydrolysis of the polysaccharides by bacteria and protozoa followed by bacterial fermentation of the released sugars to short chain fatty acids which are then utilized by the animal as an energy source. Fermentation also generates CO_2 and H_2. Methanogens, bacteria which are also present in the anaerobic environment, generate energy for their growth by using H_2 to reduce CO_2 or formate to CH_4 which is then eructated or exhaled to the atmosphere. Because the production of CH_4 from organic acids represents a loss of energy to the animals, considerable attention has been paid to means of reducing this loss in domestic livestock (Mosier et al., 1998a).

Fermentation by microflora in the rumen leads to CH_4 emissions ranging from 2% to 12% of gross feed energy intake or 5–20% of the metabolized energy (Johnson et al., 2000). CH_4 emissions from non-ruminant livestock (pigs, horses) are less than from ruminants and also vary with diet quality. Emissions range from 0.6% to 1.2% of gross energy intake for a good quality diet and are approximately doubled for a poor quality diet (Johnson et al., 1993; Crutzen et al., 1986).

Crutzen et al. (1986) developed average values for CH_4 emission per animal for all ruminants. For some animals, different emission values were provided for intensive and extensive management, typically in developed and developing countries, respectively, to reflect differences in feed quantity and quality. Crutzen et al. (1986) estimated that 55, 35, 8, 5, 1.5 kg per animal per year of CH_4 is emitted respectively from intensively managed cattle, extensively managed cattle, intensively managed sheep, extensively managed sheep and intensively managed swine. These estimates indicate that the bulk of the CH_4 emissions (~83%) are derived from cattle and buffalo, with sheep the next most important contributors (12%) (Crutzen et al. 1986).

Cattle in the USA and Europe are the most intensively managed in the world but they account for only about 17% of the total global numbers (including buffalo).The bulk of the world's cattle are under much less intensive management in the continents of Asia, S. America, and Africa (Figure 5A). CH_4 emission from cattle and buffalo are estimated using the Crutzen et al. (1986) emission estimates. Global emissions totaled 60–70 Tg CH_4 in 2000 (Figure 5B) with approximately 70% arising from developing countries, mostly from tropical regions.

5.2. ANIMAL WASTE

Significant CH_4 (10–18 Tg yr^{-1}) is also emitted from animal waste, and the amount varies with waste type and management practice. N_2O, approximately 2 Tg N yr^{-1} globally, is produced in aerobic waste handling systems and after application to soils (Mosier et al. 1998b). In general, manures from animals having a high quality diet have higher potential to generate CH_4 and N_2O than manures from animals

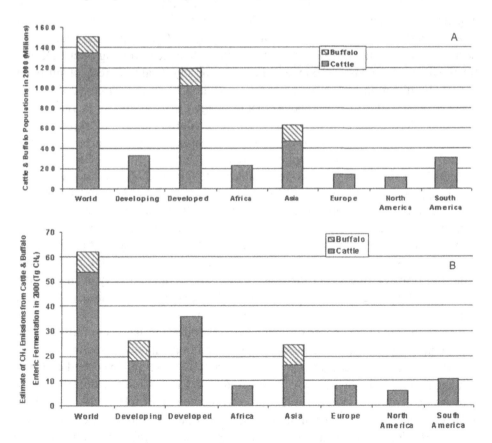

Figure 5. (A) Cattle and buffalo populations in different regions of the world in 2000 (FAO, 2001). (B) Estimate of CH_4 emissions from cattle and buffalo enteric fermentation in 2000, based on Crutzen et al. (1986) emission factors and FAO (2001) livestock population numbers.

having a low quality diet. Actual CH_4 emission values depend on the amount of manure being generated, its potential to generate CH_4, and the extent to which the CH_4 generating potential is realized (which depends on manure handling practices and climate). Not unexpectedly, the highest emissions of CH_4 from animal manures are associated with the most intensively managed animals. As a result, the majority of CH_4 emitted from livestock production is from outside the tropics. Seventy five percent of the global generation is estimated to come from cattle plus swine and 55% comes from N. America and Europe. N_2O oxide emissions from livestock waste management, on the other hand, are likely from tropical livestock production due to the much higher cattle and buffalo populations in developing countries (Figure 5A).

6. Summary and conclusions

Tropical agricultural soils are important global sources of greenhouse gases. Rice-based agriculture is a significant source of CH_4 during times when fields are flooded. The contribution of rice-based agriculture to the global CH_4 budget is not as large as generally thought a decade ago, and, due to changes in management that resulted in a decrease in organic matter returned to the field, CH_4 emissions have likely decreased during the past decade in many important rice-growing areas. During times when rice fields are flooded, they are not significant sources of nitrogen oxides. Fertilization and drainage can be coordinated so that drainage does not occur soon after fertilization and so that drainage does not greatly increase N_2O emissions during the growing season. However, during fallow and upland crop periods, N-oxide emissions from rice-based cropping systems can be similar to those observed from upland cropping systems. As a result, annual greenhouse gas budgets from specific fields need to consider both CH_4 and N_2O emissions.

It is apparent that conversion of forests to grazing lands initially induces higher N-oxide emissions than observed from the primary forest in some cases. The forest conversion appears, however, to be a much smaller pertubation to atmospheric N-oxides than early studies suggested (Verchot et al., 1999; Garcia-Montiel et al., 2001). Increased N-oxide emissions following clearing generally fall below those from the primary forest. On the other hand, CH_4 oxidation is typically greatly reduced and grazing lands may even become net CH_4 sources in situations where soil compaction from cattle traffic limits gas diffusion.

Establishment of tree-based cropping systems following slash-and-burn agriculture enhances N_2O and NO emissions during and immediately following burning. These emissions soon decline to rates similar to those observed in secondary forest while CH_4 consumption rates are slightly reduced. Conversion of tropical forest to a tree-based cropping system resulted in conversion of a net CH_4 sink to a net CH_4 source. Overall, conversion of humid tropical forests to tree-based systems probably will not result in a significant increase in greenhouse gas emissions, other than the large flux produced during slash-and-burn. Conversion to cropping systems, on the other hand, results in significant increases in N_2O emissions, a loss of the CH_4 sink, and a substantial increase in the GWP compared to the forest and tree-based systems.

High input agriculture in the tropics can be a large source of N-oxides (Veldkamp et al., 1997; Crill et al., 2000). Improved management (see discussions in Peoples et al., 1995), however, can reduce those emissions while conserving N fertilizer and other resources. Fertilization, to match crop demand, and water inputs are important considerations in optimizing crop production while limiting N-oxide emissions (Matson et al. 1998). Whether gaseous N losses from fertilized tree-based systems will increase significantly compared to the forest system will depend largely on the timing of fertilizer application relative to plant demand, soil and rainfall. Indonesian studies indicate that plant density is an important aspect in minimizing N-oxide emissions in intensive tree crop production (Tsuruta et al. 2000). Higher competition between plants and microbes in tropical systems compared to temperate systems generally results in higher plant uptake and minimal nitrogen gas losses, but efficient plant uptake depends on high plant density. Perennial systems generally have high synchrony between nutrient supply and demand, so, if managed properly, these fertilized agroforestry systems should not be expected to have significantly large losses compared to annual cropping systems.

The increasing intensification of crop production in the tropics, in which N fertilization must increase for many crops to sustain production, will most certainly increase N-oxide emissions. The increase, however, may be on the same order as that expected in temperate crop production and thus be smaller than some have predicted (Bouwman, 1998). From the studies presented above, comparison of N_2O emissions from temperate agriculture (Table II) indicates that N_2O emissions from tropical cropping systems are not, as are often quoted (e.g. Matson and Vitousek, 1990; Veldkamp, 1997; Bouwman, 1998), higher than agricultural systems in temperate zones. Even the very high N-oxide emissions noted in Veldkamp et al. (1997) fall within the range of emissions from temperate agricultural soils (Table II). The potential certainly exists for large emissions from tropical systems, but timing of rainfall and competition by plants with soil microbes for soil N appear to generally limit N-oxide emissions. As a result, appropriate management in high-input agricultural systems could limit N-oxide emissions. Such considerations would likely increase fertilizer use efficiency as well.

One must keep in mind that the generalizations in the preceding paragraphs and the status of our understanding of tropical agroecosystems are based on very limited information. Few studies exist in most tropical agroecosystems except for the measurements of CH_4 from flooded rice fields. Since trace gas fluxes vary by orders of magnitude in time and space, depending upon land use and management, soils, rainfall and nutrient input, it is clear that more information is needed that represents the major land uses and agricultural systems and encompass seasonal and annual cycles. Additionally, the fluxes of the suite of gases, CO_2, CH_4, NO_x, N_2O and possibly VOCs, need to be quantified simultaneously in ways that further our understanding of the impact of not only changes in land use changes but also of changes in agricultural management. This information can then be used to help address issues of crop production, water quality, and air quality across local and regional scales.

Acknowledgements

The authors thank the Center for Development Research (ZEF), University of Bonn for providing the forum for these discussion and for travel funding for those who were able to participate in the workshop.

References

Abao, E.B. Jr., Bronson, K.F., Wassmann, R. and Singh, U.: 2000, 'Simultaneous records of methane and nitrous oxide emissions in rice-based cropping systems under rain fed conditions', *Nutrient Cycling in Agroecosystems* **58**, 131–139.

Adhya, T.K., Bharati, K., Mohanty, S.R., Mishra, S.R., Ramakrishnan, B., Rao, V.R., Sethunathan, N. and Wassmann, R.: 2000, 'Methane emission from rice fields at Cuttack, India', *Nutrient Cycling in Agroecosystems* **58**, 95–105.

Allen, A.G., Jarvis, S.C. and Headon, D.M.: 1996, 'Nitrous oxide emissions from excreta return by livestock on grazed grasslands in the U.K.', *Soil Biology and Biochemistry*, **28**, 597–607.

Anderson, I.C., Levine, J.S., Poth, M.A. and Riggan, P.J.: 1988, 'Enhanced biogenic emissions of nitric oxide and nitrous oxide following surface biomass burning', *Journal of Geophysical Research* **93**, 3893–3898.

Andreae, M.O. and Schimel, D.S.: 1989, *Exchange of Trace Gases Between Terrestrial Ecosystems and the Atmosphere*, Chichester, John Wiley & Sons, 347 pp.

Bender, M. and Conrad, R.: 1993, 'Kinetics of methane oxidation in oxic soils', *Chemosphere*, **26**, 687–696.

Birdsey, R.A. and Weaver, F.L.: 1982, 'The forest resources of Puerto Rico', *USDA Forest Service Research Bulletin*, SO-85, 59 p.

Bouwman, A.F.: 1990, 'Exchange of greenhouse gases between terrestrial ecosystems and the atmosphere', in A.F. Bouwman (ed.), *Soils and the Greenhouse Effect*, New York, John Wiley & Sons, pp. 61–127.

Bouwman, A.F.: 1998, 'Nitrogen oxides and tropical agriculture', *Nature* **392**, 866–867.

Bouwman, A.F. and Sombroek, S.M.: 1990, 'Soils on a warmer earth', in H.W. Scharpenseel et al. (eds.), Amsterdam, Elsevier, pp. 5–30.

Bremner, J.M. and Blackmer, A.M.: 1978, 'Nitrous oxide: Emissions from soils during nitrification of fertilizer nitrogen', *Science* **199**, 295–296.

Bremner, J.M. and Blackmer, A.M.: 1982, 'Composition of soil atmospheres', in R.J. Cicerone, J.D. Shetter and C.C. Delwiche (eds.), *Methods of Soil Analysis, Part 2, Chemical and Microbiological Properties*, Agronomy Monograph No. 9, pp. 873–901.

Breuer, L., Papen, H. and Butterbach-Bahl, K.: 2000, 'N$_2$O emission from tropical forest soils of Australia', *Journal of Geophysical Research* **105**, 26353–26367.

Bronson, K.F., Neue, H.U., Singh, U. and Abao, E.B. Jr.: 1997a, 'Automated chamber measurements of methane and nitrous oxide flux in a flooded rice soil: I. Residue, nitrogen and water management', *Soil Science Society of America Journal* **61**, 981–987.

Bronson, K.F., Neue, H.U., Singh, U. and Abao, E.B. Jr.: 1997b, 'Automated chamber measurements of methane and nitrous oxide flux in a flooded rice soil: II. Fallow period emissions', *Soil Science Society of America Journal* **61**, 988–993.

Cai, Z., Xing, G., Yan, X., Xu, H, Tsuruta, H., Yagi, K. and Minami, K.: 1997, 'Methane and nitrous oxide emissions from rice paddy fields as affected by nitrogen fertilizers and water management', *Plant and Soil* **196**, 7–14.

Chameides, W.L. et al., 1994, 'Growth of continental scale metro-agro-plexes, regional ozone pollution, and world food production', *Science* **264**, 74–77.

Cicerone, R.J. and Oremland, R.S.: 1988, 'Biogeochemical aspects of atmospheric methane', *Global Biogeochemical Cycles* **2**, 299–327.

Cicerone, R.J., Shetter, J.D. and Delwiche, C.C.: 1983, 'Seasonal variation of methane flux from a California rice paddy', *Journal of Geophysical Research* **88**, 7203–7209.

CMDL.: 2001, Climate Monitoring and Diagnostic Laboratory (CMDL) of the National Oceanographic and Atmospheric Administration, Boulder, CO, USA. N$_2$O data from: http://ftp.cmdl.noaa.gov/hats/n2o/insitu Gcs/global/. Hall, B.D. et al., Halocarbons and other Atmospheric Trace Species Group, C < MDL Summary Report 1989–1999, NOAA/CMDL, in preparation 2000.

Conrad, R.: 1989, 'Control of methane production in terrestrial ecosystems', in M.O. Andreae and D.S. Schimel (eds.), *Exchange of Trace Gases Between Terrestrial Ecosystems and the Atmosphere*, Chichester, John Wiley & Sons, pp. 39–58.

Coolman, R.M.: 1994, *Nitrous Oxide Emissions from Amazonian Ecosystems*, Ph.D. dissertation, North Carolina State University.

Crill, P.M., Keller, M., Weitz, A., Grauel, B. and Veldkamp, E.: 2000, 'Intensive field measurement of nitrous oxide emissions from a tropical agricultural soil', *Global Biogeochemical Cycles* **14**, 85–95.

Crutzen, P.J.: 1981, 'Atmospheric chemical processes of the oxides of nitrogen including nitrous oxide', in C.C. Delwiche (ed.), *Denitrification, Nitrification and Atmospheric Nitrous Oxide*, New York, Wiley, pp. 17–44.

Crutzen, P.J. and Ehhalt, D.H.: 1977, 'Effects of nitrogen fertilizers and combustion on the stratospheric ozone layer', *Ambio* **6**, 112–117.

Crutzen, P.J. and Andreae, M.O.: 1990, 'Biomass burning in the tropics: Impact on atmospheric chemistry and biogeochemical cycles', *Science* **250**, 1669–1678.

Davidson, E.A.: 1991, 'Fluxes of nitrous oxide and nitric oxide from terrestrial ecosystems', in J.E. Rogers and W.B. Whitman (eds), *Microbial Production and Consumption of Greenhouse Gases: Methane, Nitrogen Oxides and Halomethanes*, Washington, D.C., American Society for Microbiology, pp. 219–235.

Davidson, E.A. and Kingerlee W.: 1997, 'A global inventory of nitric oxide emissions from soils', *Nutrient Cycling in Agroecosystems* **48**, 37–50.

Davidson, E.A. and Verchot, L.V.: 2000, 'Testing the hole-in-the-pipe model of nitric and nitrous oxide emissions from soils using the TRAGNET database', *Global Biogeochemical Cycles* **14**, 1035–1043.

Davidson, E.A., Keller, M., Erickson, H.E., Verchot, L.V. and Veldkamp, E.: 2000, 'Testing a conceptual model of soil emissions of nitrous and nitric oxides', *BioScience* **50**, 667–680.

Del Grosso, S.J., Parton, W.J., Mosier, A.R., Ojima, D.S., Potter, C.S., Borken, W., Brumme, R., Butterbasch-Bahl, K., Criss, P.M., Dobbie, K. and Smith, K.A.: 2000a, 'General CH_4 oxidation model and comparisons of CH_4 oxidation in natural and managed systems', *Global Biogeochemical Cycles* **14**, 999–1019.

Del Grosso, S.J., Parton, W.J., Mosier, A.R., Ojima, D.S., Kulmala, A.E. and Phongpan, S.: 2000b, 'General model for N_2O and N_2 gas emissions from soils due to denitrification', *Global Biogeochemical Cycles* **14**, 1045–1060.

Delmas, R., Serca, D. and Jambert, C.: 1997, 'Global inventory of NO_x sources', *Nutrient Cycling in Agroecosystems* **48**, 51–60.

Denier van der Gon, H.A.C.: 1999, 'Changes in CH_4 Emissions from Rice Fields from 1960 to the 1990s. II. The declining use of organic inputs in rice farming', *Global Biogeochemical Cycles* **13**, 1053–1062.

Denier van der Gon, H.A.C. and Neue, H.U.: 1995, 'Influence of organic matter incorporation on the methane emission from a wetland rice field', *Global Biogeochemical Cycles* **9**, 11–22.

Dias-Filho, M.B.: 1986, 'Espécies forrageiras e estabelecimento de pastagens na Amazonia', in *Pastagens na Amazônia*, FEALQ, Piricaba, Brazil, pp. 27–54.

Dias-Filho, M.B. and Serrão E.A.S.: 1987, 'Limitacões de fertilidade do solo na recuperacão de pastagen degradada de capim colonião (Panicum maximim Jacq.) em Paragominas, na Amazonia oriental', *EMBRAPA/CPATU Research Bulletin* No. 87, EMBRAPA/CPATU, Belém, Pará, Brazil.

Dlugokencky, E.: 2001, NOAA CMDL Carbon Cycle Greenhouse Gases, Global average atmospheric methane mixing ratios, NOAA CMDL cooperative air sampling network, Http://www.cmdl.noaa.gov/ccg/figures/ch4trend_global.gif.

Donahue, R.L., Miller, R.W. and Shickluna, J.C.: 1983, 'Soils', *An Introduction to Soils and Plant Growth*, 5th edn. Prentice-Hall, Inc. Englewood Cliffs, New Jersey, 665 p.

Dorr, H., Katruff, L. and Levin, I.: 1993, 'Soil texture parameterization of the methane uptake on aerated soils', *Chemosphere* **26**, 697–713.

Duxbury, J.M. and McConnaughey, P.K.: 1986, 'Effect of fertilizer source on denitrification and nitrous oxide emissions in a maize field', *Soil Science Society of America Journal* **50**, 644–648.

Ehhalt, D.H., Rohrer, F. and Wahner, A.: 1992, 'Sources and distribution of NO_x in the upper troposphere at northern mid-latitudes', *Journal of Geophysical Research* **97**, 3725–3738.

Erickson, H.E. and Keller, M.: 1997, 'Tropical land use change and soil emissions of nitrogen oxides', *Soil Use and Management* **13**, 278–287.

FAO United Nations Food and Agricultural Organization (2000). FAOSTAT: Agricultural Data, are available on the world wide web: ⟨http://www.apps.fao.org/cgi-bin/nphdb.pl?subset=agriculture⟩.

Firestone, M.K.: 1982, 'Biological denitrification', in F.J. Stevenson (ed.), *Nitrogen in Agricultural Soils*, Agronomy Monograph No. 22, ASA-CSSA-SSSA, Madison, WI, pp. 289–326.

Firestone, M.K and Davidson, E.A.: 1989, 'Microbiological basis of NO and N_2O production and consumption in soil', in M.O. Andreae and D.S. Schimel (eds.), *Exchange of Trace Gases Between Terrestrial Ecosystems and the Atmosphere*, New York, John Wiley and Sons, pp. 7–21.

Frolking, S.E., Mosier, A.R., Ojima, D.S., Li, C., Parton, W.J., Potter, C.S., Priesack, E., Stenger, R., Haberbosch, C., Dorsch, P., Flessa, H. and Smith, K.A.: 1998, 'Comparison of N_2O emissions from soils at three temperate agricultural sites: Simulations of year-round \measurements by four models', *Nutrient Cycling in Agroecosystems* **52**, 77–105.

Fung I.J., Lerner, J.J., Matthews, E., Prather, M., Steele, L.P. and Fraser, P.J.: 1991, 'Three dimensional model synthesis of the global methane cycle', *Journal of Geophysical Research* **96**, 13033–13065.

Galbally, I.E. and Gillett, R.W.: 1988, in H. Rodhe and R. Herrera (eds.), *Acidification in Tropical Countries*, Chichester, J. Wiley & Sons, pp. 73–16.

Galbally, I.E., Fraser, P.J., Meyer, C.P. and Griffith, D.W.T.: 1992, 'Biosphere-atmosphere exchange of trace gases over Australia', in R.M. Gifford and M.M. Barson (eds.), *Australia's Renewable Resources: Sustainability and Global Change*, Bureau of Rural Resources No. 14, P.J. Grills, Commonwealth Printer, Canberra, pp. 117–149.

Galloway, J.N., Schlesinger, W.H., Levy II, H., Michaels, A. and Schnoor, J.L.: 1995, 'Nitrogen fixation: Anthropogenic enhancement-environmental response' *Global Biogeochemical Cycles* **9**, 235–252.

Garcia-Mendez, G., Maass, J.M., Matson, P.A. and Vitousek, P.M.: 1991, 'Nitrogen transformations and nitrous oxide flux in a tropical deciduous forest in Mexico', *Oceologia* **88**, 362–366.

Garcia-Monteil, D.C., Steudler, P.A., Piccolo, M.C., Melillo, J.M., Neill, C. and Cerri, C.C.: 2001, 'Controls on nitrogen oxide emissions from forest and pastures in the Brazilian Amazon,' *Global Biogeochemical Cycles* **15**, 1021–1031.

Hammond, A.L.: 1990, *World Resources 1990–91, A Report by the World Resources Institute*, Oxford, UK, Oxford University Press, 383 pp.

Hao, W.M., Wofsy, M.B., McElroy, M.B., Beer, J.M. and Togan, A M.: 1987, 'Sources of atmospheric nitrous oxide from combustion', *Journal of Geophysical Research* **92**, 3098–3104.

Holland, E.A. and Lamarque, J.F.: 1997, 'Modeling bio-atmospheric coupling of the nitrogen cycle through NO_x emissions and NO_y deposition', *Nutrient Cycling in Agroecosystems* **48**, 7–24.

Holland, E.A., Braswell, B.H., Lamarque, J.F., Townsend, A., Sulzman, J.M., Muller, J.F., Dentener, F., Brasseur, G., Levy II, H., Penner, J.E. and Roelofs, G.: 1997, 'Variations in the predicted spatial distribution of atmospheric nitrogen deposition and their impact on carbon uptake by terrestrial ecosystems', *Journal Geophysical Research* **102**, 15849–15866.

Holland, E.A., Neff, J.C., Townsend, A.R. and McKeown, B.: 2000, 'Uncertainties in the temperature sensitivity of decomposition in tropical and subtropical ecosystems: Implications for models', *Global Biogeochemical Cycles* **14**, 1137–1151.

Houghton, J.T., Callander, B.A. and Varney, S.K. (eds.): 1992, Climate Change 1992. The Supplementary Report to the IPCC Scientific Assessment. Intergovenmental Panel on Climate Change. Cambridge University Press. 200 pp.

Hutchinson, G.L. and Davidson, E.A.: 1993, 'Processes for production and consumption of gaseous nitrogen oxides in soil. In Agricultural Ecosystem Effects on Trace Gases and Global Climate Change', ASA Special Publication No. 55. American Society of Agronomy, Madison WI. pp. 79–93.

IFA: 2000, International Fertilizer Industry Association, Nitrogen, Phosphate, Potash IFADATA Statistics. International Fertilizer Industry Association, Paris France (http://www.fertilizer.org).

IPCC (Intergovernmental Panel on Climate Change): 1994, Radiative Forcing of Climate Change. The 1994 Report of the Scientific Assessment Working Group of IPCC, Summary for Policymakers, Geneva, WMO/UNEP, 28 pp.

IPCC (Intergovernmental Panel on Climate Change).: 1996, in J.T. Houghton, L.G. Meira Filho, B.A. Callander, N. Harris, A. Kattenberg and K. Maskell (eds.), *Climate Change 1995*, Cambridge University Press.

Ishizuka, S., Tsuruta, H. and Murdiyarso, D.: 2000, 'Relationship between the fluxes of greenhouse gases and soil properties in a resarch site of Jambi, Sumatra', in D. Murdiyarso and H. Tsuruta (eds.), *The Impacts of Land-Use/Cover Change on Greenhouse Gas Emissions in Tropical Asia*, Global Change Impacts Centre for Southeast Asia, pp. 31–34.

Jain, M.C., Kumar, K., Wassmann, R., Mitra, S., Singh, S.D., Singh, J.P., Singh, R., Yadav, A.K. and Gupta, S.: 2000, 'Methane emissions from irrigated rice fields in Northern India (New Delhi)', *Nutrient Cycling in Agroecosystems* **58**, 75–83.

Johnson, D.E., Johnson, K.A., Ward, G.M. and Branine, M.E.: 2000, 'Ruminants and other animals, in Khalil, M.A.K. (ed.) *Atmospheric Methane*, Springer-Verlag, pp. 112–113.

Keller, M. and Reiners, W.A.: 1994, 'Soil-atmosphere exchange of nitrous oxide, nitric oxide and methane under secondary succession of pasture to forest in the Atlantic lowlands of Costa Rica', *Global Biogeochemical Cycles* **8**, 399–409.

Keller, M., Mitre, M.E. and Stallard, R.F.: 1990, 'Consumption of atmospheric methane in soils of central Panama: Effects of agricultural development', *Global Biogeochemical Cycles* **4**: 21–27.

Keller, M., Veldkamp, E., Weitz, A.M. and Reiners, W.A.: 1993, 'Effect of pasture age on soil trace-gas emissions from a deforested area of Costa Rica', *Nature* **365**, 244–246.

Khalil, M.I., Rosenani, A.B., Van Cleemput, O., Fauziah, C.I. and Shamshuddin, J.: 2000, 'Nitrous oxide emissions from a sustainable land management system in the humid tropics', *Proceedings of the International Symposium on Sustainable Land Management*, August 8–10, 2000, Kulala Lumpur, Malaysia, pp. 71–72.

Khalil, M.I., Rosenani, A.B., Van Cleemput, O., Shamshuddin, J. and Fauziah, C.I.: 2002, 'Nitrous oxide emissions from an Ultisol of the humid tropics under maize-groundnut rotation', *Journal of Environmental Quality* **31**, 1071–1078.

Knowles, R.: 1993, Methane: 'Processes of Production and Consumption', in L.A. Harper, A.R. Mosier, J.M. Duxbury and D.E. Rolston (eds.), *Agricultural Ecosystem Effects on Trace Gases and Global Climate Change*, ASA Special Pub. No. 55. Am. Soc. Agron. Madison WI, pp. 145–156.

Ko, M.K.W., Sze, N.D. and Weinstein, D.K.: 1991, 'Use of satellite data to constrain the model calculated atmospheric lifetime for N_2O: Implications for other trace gases', *Journal of Geophysical Research* **96**, 7547–7552.

Kroeze, C., Mosier, A.R. and Bouwman, L.: 1999, 'Closing the global N_2O budget: A retrospective analysis 1500–1994', *Global Biogeochemical Cycles* **13**, 1–8.

Levine, J.S.: 1988, 'The effects of fire on biogenic soil emissions of nitric oxide and nitrous oxide', *Global Biogeochemical Cycles* **2**, 445–49.

Li, C., Frolking, S. and Frolking, T.A.: 1992, 'A model of nitrous oxide evolution from soil driven by rainfall events: I. Model structure and sensitivity', *Journal of Geophysical Research* **97**, 9759–9776.

Li, C., Narayanan, V. and Harriss, R.: 1996, 'Model estimates of nitrous oxide emissions from agricultural lands in the United States', *Global Biogeochemical Cycles* **10**, 297–306.

Li, C., Aber, J., Stange, F., Butterbach-Bahl, K. and Papen, H.: 2000, 'A process-oriented model of N_2O and NO emissions from forest soils: 1. Model development', *Journal of Geophysical Research* **105**, 4369–4384.

Li, C., Zhuang, Y.H., Cao, M.Q., Crill, P.M., Dai, Z.H.., Frolking, S., Moore, B., Salas, W., Song, W.Z. and Wang XK.: 2001, 'Comparing a national inventory of N_2O emissions from arable lands in China developed with a process-based agro-ecosystem model to the IPCC methodology', *Nutrient Cycling in Agroecosystems* **60**, 159–175.

Lijinsky, W.: 1977, 'How nitrosamines cause cancer', *New Scientist* **27**, 216–217.

Linn, D.M. and Doran, J.W.: 1984, 'Effect of water-filled pore space on carbon dioxide and nitrous oxide production in tilled and non-tilled soils', *Soil Science of America Journal* **48**, 1267–1272.

Liu, S.C., Trainer, M., Carroll, M.S., Hubler, G., Montzka, D.D., Norton, R.B., Ridley, B.A., Walega, J.G., Atlas, E.L., Heides, B.G., Huebert, B.J. and Warren, W.: 1992, 'A study of the photochemistry and ozone budget during the Mauna Loa observatory photochemistry experiment', *Journal of Geophysical Research* **97**, 10463–10471.

Lu, W.F., Chen, W., Duan, B.W., Guo, W.M., Lu, Y., Lantin, R.S., Wassmann, R. and Neue H.U.: 2000, 'Methane emission and mitigation options in irrigated rice fields in Southeast China', *Nutrient Cycling in Agroecosystems* **58**, 277–284.

Lugo, A.E., Sanchez, M.J. and Brown, S.: 1986, 'Land use and organic carbon content of some tropical soils', *Plant and Soil* **96**, 185–196.

Luizao, F., Matson, P.A., Livingston, G., Luizao, R. and Vitousek, P.M.: 1989, 'Nitrous oxide flux following tropical land clearing', *Global Biogeochemical Cycles* **3**, 281–85.

Mahmood, T., Malik, K.A., Shamsi, S.R.A. and Sajjad, M.I.: 1998, 'Denitrification and total N losses from an irrigated sandy-clay loam under maize-wheat cropping systems', *Plant Soil* **199**, 239–250.

Mahmood, T., Ali, R., Sajjad, M.I., Chaudhri, M.B., Tahir, G.R. and Azam, F.: 2000, 'Denitrification and total fertilizer-N losses from an irrigated cotton field', *Biology and Fertility of Soils* **31**, 270–278.

Matthews, R.B., Wassmann, R. and Arah, J.: 2000a, 'Using a crop/soil simulation model and GIS techniques to assess methane emissions from rice fields in Asia. I. Model development', *Nutrient Cycling in Agroecosystems* **58**, 141–159.

Matthews, R.B., Wassmann, R., Knox, J. and Buendia, L.V.: 2000b, 'Using a crop/soil simulation model and GIS techniques to assess methane emissions from rice fields in Asia. IV. Upscaling of crop management scenarios to national levels', *Nutrient Cycling in Agroecosystems* **58**, 201–217.

Matson, P.A. and Vitousek, P.M.: 1990, 'Ecosystem approach to a global nitrous oxide budget', *Bioscience* **40**, 667–672.

Matson, P.A., Billow, C., Hall, S. and Zachariassen, J.: 1996, 'Fertilization practices and soil variations control nitrogen oxide emissions from tropical sugar cane', *Journal of Geophysical Research* **101**, 18533–18545.

Matson, P.A., Parton, W.J., Power, A.G. and Swift, M.J.: 1997, 'Agricultural intensification and ecosystem properties', *Science* **277**, 504–509.

Matson, P.A., Naylor, R. and Ortiz-Monasterio, I.: 1998, 'Integration of environmental, agronomic and economic aspects of fertilizer management', *Science* **280**, 112–115.

Minami, K.: 1993, 'Methane from rice production', in A.R. van Amstel (ed.), *Methane and Nitrous Oxide*, *Proceedings IPCC Workshop*, The Netherlands, Bilthoven, pp. 143–162.

Minami, K., Mosier, A.R. and Sass, R. (eds), 1994, 'CH$_4$ and N$_2$O: Global emissions and controls from rice fields and other agricultural and industrial sources', *NIAES Series 2*, Tokyo, YOKENDO Publishers, 234 p.

Mosier, A.R. and Bouwman, A.F.: 1993, 'Working group report: Nitrous oxide emissions from agricultural soils', in A.R. van Amstel (ed.), *Methane and Nitrous Oxide: Methods in National Emission Inventories and Options for Control Proceedings*, The Netherlands, National Institute of Public Health and Environmental Protection, Bilthoven, pp. 343–346.

Mosier, A.R. and Delgado, J.A.: 1997, 'Methane and nitrous oxide fluxes in grasslands in western Puerto Rico', *Chemosphere* **35**, 2059–2082.

Mosier, A.R. and Kroeze, C.: 1999, 'Contribution of agroecosystems to the global atmospheric N$_2$O budget', in R.L. Desjardins, J.C. Keng and K. Haugen-Kozyra (eds.) *Proceedings of the International Workshop on Reducing Nitrous Oxide Emissions from Agroecosystems*, Banff, Alberta, March 3–5. Agriculture and Agri-Food Canada, Research Branch; Alberta Agriculture, Food and Rural Development, Conservation and Development Branch, pp. 3–15.

Mosier, A.R., Schimel, D.S., Valentine, D.W., Bronson, K.F. and Parton, W.J.: 1991, 'Methane and nitrous oxide fluxes in native, fertilized, and cultivated grasslands', *Nature* **350**, 330–332.

Mosier, A.R., Duxbury, J.M., Freney, J.R., Heinemeyer, O. Minami, K. and Johnson, D.E.: 1998a, 'Mitigating agricultural emissions of methane', *Climatic Change* **40**, 39–80.

Mosier, A.R., Duxbury, J.M., Freney, J.R., Heinemeyer, O. and Minami, K.: 1998b, 'Assessing and mitigating N$_2$O emissions from agricultural soils', *Climatic Change* **40**, 7–38.

Mosier, A.R., Delgado, J.A. and Keller, M.: 1998c, 'Methane and nitrous oxide fluxes in an acid Oxisol in western Puerto Rico: Effects of tillage, liming and fertilization', *Soil Biology and Biochemistry* **30**, 2087–2098.

Muzio, L.J. and Kramlich, J.C.: 1988, 'An artifact in the measurement of N$_2$O from combustion sources', *Geophysical Research Letters* **15**, 1369–372.

Nelson, K.E., Turgeon, A.J. and Street, J.R.: 1980, 'Thatch influence on mobility and transformation of nitrogen carriers applied to turf', *Agronomy Journal* **72**, 487–492.

Nesbit, S.P. and Breitenbeck, G.A.: 1992, 'A laboratory study of factors influencing methane uptake by soils', *Agricultural Ecosystems and Environment*, **41**, 39–54.

Nouchi, I., Mariko, S. and Aoki, K.: 1990, 'Mechanisms of methane transport from the rhizosphere to the atmosphere through rice plant', *Plant Physiology* **94**, 59–66.

Ojima, D.S., Valentine, D.W., Mosier, A.R., Parton, W.J. and Schimel, D.S.: 1993, 'Effect of land use change on methane oxidation in temperate forest and grassland soils', *Chemosphere* **26**, 675–685.

Otter, L.B., Marufu, L. and Scholes, M.C.: 2001, 'Biogenic, biomass and biofuel sources of trace gases in southern Africa', *South African Journal of Science* **97**, 131–138.

Palm, C.A., Alegre, J.C., Arevalo, L., Mutuo, P.K., Mosier, A.R. and Coe, R.: 2002, 'Nitrous oxide and methane fluxes in six different land use systems in the Peruvian Amazon', *Global Biogeochemical Cycles* **16**, 1073.

Patrick, W.H., Jr.: 1981, 'The role of inorganic redox systems in controlling reduction in paddy soils', in *Proceedings Symp. Paddy Soil*, Science Press, Beijing, Springer Verlag, pp. 107–117.

Peoples, M.B., Mosier, A.R. and Freney, J.R.: 1995, 'Minimizing gaseous losses of nitrogen. In Bacon, P.E. (ed.) *Nitrogen Fertilization in the Environment*. Marcel Deckker Inc. pp. 565–602.

Plant, R.A. and Bouman, A.M.: 1999, 'Modeling nitrogen oxide emissions from current and alternative pastures in Costa Rica', *Journal of Environmental Quality* **28**, 866–872.

Poth, M. and Focht, D.D.: 1985, '^{15}N kinetic analysis of N$_2$O production by *Nitrosamonas europae*: An examination of nitrifier denitrification', *Applied and Environmental Microbiology* **49**, 1134–1141.

Potter, C.S., Davidson, E.A. and Verchot, L.V.: 1996, 'Estimation of global biogeochemical controls and seasonality in soil methane consumption', *Chemosphere* **32**, 2219–2246.

Potter, C.S., Davidson, E.A., Klooster, S.A., Nepstad, D.C., De Negreiros, G.H. and Brooks, V.: 1998, 'Regional application of an ecosystem production model for studies of biogeochemistry in Brazilian Amazonia', *Global Change Biology* **4**, 315–333.

Prasetyo, L.B., Saito, G. and Tsuruta, H.: 2000, 'Estimation of greenhouse gasses emissions using remote sensing and GIS techniques in Sumatra, Indonesia', in D. Murdiyarso and H. Tsuruta (eds.), *The Impacts of Land-Use/Cover Change on Greenhouse Gas Emissions in Tropical Asia*. Global Change Impacts Centre for Southeast Asia, pp. 53–59.

Robertson, G.P., Paul, E.A. and Harwood, R.R.: 2000, 'Greenhouse gases in intensive agriculture: Contributions of individual gases to radiative forcing of the atmosphere', *Science* **289**, 1922–1925.

Rudolph, J.: 1994, 'Anomalous methane', *Nature* **368**, 19–20.

Riley, H.R. and Vitousek, P.M.: 1995, 'Nutrient dynamics and nitrogen trace gas flux during ecosystem development in montane rain forest', *Ecology*, **76**, 292–304.

Sass, R.L.: 1994, 'Short summary chapter for methane', in K. Minami, A. Mosier and R. Sass (eds.), CH$_4$ and N$_2$O: *Global Emissions and Controls from Rice Fields and Other Agricultural and Industrial Sources*, Tokyo, NIAES. Yokendo Publishers, pp. 1–7.

Sass, R.L., Fisher, F.M., Turner, F.T. and Jund, M.F.: 1991, 'Methane emissions from rice fields as influenced by solar radiation, temperature, and straw incorporation', *Global Biogeochemical Cycles* **5**, 335–350.

Sass, R.L., Fisher, F.M., Wang, Y.B., Turner, F.T. and Jund, M.F.: 1992, 'Methane emission from rice fields: The effect of flood water management', *Global Biogeochemical Cycles* **6**, 249–262.

Seiler, W. and Conrad, R.: 1987, *Geophysiology of Amazonia*, Vegetation and Climate Interactions, Dickinson, R.E. (ed.), Chichester, John Wiley & Sons, pp. 133–160.

Schmidt, E.L.: 1982, 'Nitrification in Soil', in F.J. Stevenson (ed.), *Nitrogen in Agricultural Soils*, Agronomy Monograph No. 22, Madison WI, ASA-CSSA-SSSA, pp. 253–288.

Schnell, S. and King, G.M.: 1994, 'Mechanistic analysis of ammonium inhibition of atmospheric methane consumption in forest soils', *Applied and Environmental Microbiology* **60**, 3514–3521.

Schutz, H., Seiler, W. and Rennenberg,, H.: 1989, 'Soil and land use related sources and sinks of methane in the context of the global methane budget', in A.F. Bouwman (ed.), *Soils and the Greenhouse Effect*, Chichester, John Wiley and Sons, pp. 269–301.

Seiler, W. and Conrad, R.: 1987, 'Contribution of tropical ecosystems to the global budgets of trace gases, especially CH$_4$, H$_2$, CO and N$_2$O', in R.E. Dickinson (ed.), *Geophysiology of Amazonia. Vegetation and Climate Interactions*, New York, Wiley and Sons, pp. 133–160.

Seiler, W., Conrad, R. and Scharffe, D.: 1984, 'Field studies of methane emission from termite nests into the atmosphere and measurements of methane uptake by tropical soils', *Journal of Atmospheric Chemistry* **1**, 171–186.

Setyanto, P., Makarim, A.K., Fagi, A.M., Wassmann, R. and Buendia, L.V.: 2000, 'Crop management affecting methane emissions from irrigated and rainfed rice in Central Java (Indonesia)', *Nutrient Cycling in Agroecosystems* **58**, 85–93.

Shearer, M.J. and Khalil, M.A.K.: 1993, 'Rice agriculture: Emissions', in M.A.K. Khalil (ed.), *Atmospheric Methane: Sources, Sinks and Role in Global Change*, Berlin, Heidelberg, Springer-Verlag, pp. 230–253.

Smith, K.A.: 1990, 'Greenhouse gas fluxes between land surfaces and the atmosphere', *Progress in Physical Geography* **14**, 349–372.

Steudler, P.A., Bowden, R.D., Melillo, J.M. and Aber, J.D.: 1989, 'Influence of nitrogen fertilization on methane uptake in temperate forest soils', *Nature* **341**, 314–316.

Steudler, P.A., Melillo, J.M., Geigl, B.J., Neill, C., Piccolo, M.C. and Cerri, C.C.: 1996, 'Consequence of forest-to-pasture conversion on CH$_4$ fluxes in the Brazilian Amazon basin', *Journal of Geophysical Research* **101**, 18547–18554.

Steudler, P.A., Melillo, J.M., Bowden, R.D., Castro, M.S., Lugo, A.E.: 1991, 'The effect of natural and human disturbance on soil nitrogen dynamics and trace gas fluxes in a Puerto Rican wet forest,' *Biotropica* **23**, 356–363.

Szott, L.T. and Kass, D.C.: 1993, 'Fertilizers in agroforestry systems', *Agroforestry Systems* **23**, 153–176.

Takai, Y.: 1970, 'The mechanism of methane fermentation in flooded paddy soil', *Soil Science and Plant Nutrition* **16**, 238–244.

Takai, Y., Koyaman, T. and Kamura, T.: 1956, 'Microbial metabolism in reduction process of paddy soils (Part 1)', *Soil and Plant Food* **2**, 63–66.

Tsuruta, H., Ishizuka, S., Ueda, S. and Murdiyarso, D.: 2000, 'Seasonal and spatial variations of CO_2, CH_4, and N_2O fluxes from the surface soils in different forms of land-use/cover in Jambi, Sumatra', in D. Murdiyarso and H. Tsuruta (eds.), *The Impacts of Land-use/cover Change on Greenhouse Gas Emissions in Tropical Asia*, Global Change Impacts Centre for Southeast Asia and National Institute of Agro-Environmental Sciences, pp. 7–30.

Tsuruta, H., Kanda, K. and Hirose, T.: 1997, 'Nitrous oxide emission from a rice paddy field in Japan', *Nutrient Cycling in Agroecosystems* **49**, 51–58.

Vallis, I, Catchpoole, V.R., Hughes, R.M., Myers, R.J.K., Ridge, D.R. and Weier, K.L.: 1996, 'Recovery in plants and soils of 15N applied as subsurface bands of urea to sugarcane', *Australian Journal of Agricultural Research* **47**, 355–370.

Veldkamp, E. and Keller, M.: 1997, 'Nitrogen oxide emissions from a banana plantation in the humid tropics', *Journal of Geophysical Research* **102**, 15889–15898.

Veldkamp, E., Keller, M. and Nunez, M.: 1998, 'Effects of pasture management on N_2O and NO emissions from soils in the humid tropics of Costa Rica', *Global Biogeochemical Cycles* **12**, 71–79.

Veldkamp, E., Davidson, E., Erickson, H., Keller, M. and Weitz, A.: 1999, 'Soil nitrogen cycling and nitrogen oxide emissions along a pasture chronosequence in the humid tropics of Costa Rica', *Soil Biology and Biochemistry* **31**, 387–394.

Verchot, L.V., Davidson, E.A., Cattanio, J.H., Ackerman, I.L., Erickson H.E. and Keller, M.: 1999, 'Land use change and biogeochemical controls of nitrogen oxide emissions from soils in eastern Amazonia', *Global Biogeochemical Cycles*, **13**, 41–46.

Verchot, L.V., Davidson, E.A., Cattanio, J.H. and Ackerman, I.L.: 2000, 'Land-use change and biogeochemical controls of methane fluxes in soils of eastern Amazonia', *Ecosystems* **3**, 41–56.

Visscher, A. de, Boeckx, P. and van Cleemput, O.: 1998, 'Interaction between nitrous oxide formation and methane oxidation in soils: influence of cation exchange phenomena', *Journal of Environmental Quality* **27**, 679–687.

Vitousek, P.M., Aber, J., Howarth, R.W., Likens, G.E., Matson, P.A., Schindler, D.W., Schlesinger, W.H. and Tilman, D.G.: 1997, 'Human alteration of the global nitrogen cycle: Causes and consequences', *Issues in Ecology* **1**, 1–15.

Volk, C.M., Elkins, J.W., Fahey, D.W., Dutton, D.S., Gilligan, J.M., Loewestein, M., Podolske, J.R., Chan, K.R. and Gunson, M.R.: 1997, 'Evaluation of source gas lifetimes from stratospheric observations', *Journal of Geophysical Research* **102**, 25543–25564.

Wang, Z.Y., Xu, Y.C., Li, Z., Guo, Y.X., Wassmann, R., Neue, H.U., Lantin, R.S., Buendia, L.V., Ding, Y.P. and Wang, Z.Z.: 2000, 'Methane Emissions from Irrigated Rice Fields in Northern China (Beijing)', *Nutrient Cycling in Agroecosystems* **58**, 55–63.

Wassmann, R., Neue, H.U., Bueno, C., Lantin, R.S., Alberto, M.C.R., Buendia, L.V., Bronson, K., Papen, H. and Rennenberg, H.: 1998, 'Inherent properties of rice soils determining methane production potentials', *Plant and Soil* **203**, 227–237.

Wassmann, R., Lantin, R.S. and Neue, H.U. (eds.): 2000a, 'Methane emissions from major rice ecosystems in asia', *Special Issue of Nutrient Cycling in Agroecosystems* **58**(1–3), 1–398.

Wassmann, R., Lantin, R.S., Neue, H.U., Buendia, L.V., Corton ,T.M. and Lu, Y.H.: 2000b, Characterization of methane emissions from rice fields in Asia. 3. Mitigation options and future research needs', *Nutrient Cycling in Agroecosystems* **58**, 23–36.

Wassmann, R., Neue, H.U. and Lantin, R.S.: 2000c, 'Characterization of methane emissions from rice fields in Asia. 1. Comparison among field sites in five countries', *Nutrient Cycling in Agroecosystems* **58**, 1–12.

Watanabe, T., Chairoj, P., Tsuruta, H., Masarngsan, W., Wongwiwatchai, C., Wonprasaid, S., Cholitkul, W. and Minami, K.: 2000, 'Nitrous oxide emissions from fertilized upland fields in Thailand', *Nutrient Cycling in Agroecosystems* **57**, 55–65.

Weier, K.L.: 1996, 'Trace gas emissions from a trash blanketed sugarcane field in tropical Australia', in J.R. Wilson, D.M. Hogarth, J.A. Campbell and A.L. Garside (eds.), *Sugarcane: Research Towards Efficient and Sustainable Production*, Brisbane Australia, CSIRO Division of Tropical Crops and Pastures, pp. 271–272.

Weier, K.L., Rolston, D.E. and Thornburn, P.J.: 1998, 'The potential for N losses via denitrification beneath a green cane trash blanket', *Proceedings of the Australian Society of Sugar Cane Technology* **20**, 169–175.

Weitz, A.M., Keller, M., Linder, E. and Criss, P.M.: 1999, 'Spatial and temporal variability of nitrogen oxide and methane fluxes from a fertilized tree plantation in Costa Rica', *Journal of Geophysical Research*, **104**, 30097–30107.

Weitz, A.M., Linder, E., Frolking, S., Crill, P.M. and Keller, M.: 2001, 'N$_2$O emissions from humid tropical agricultural soils: Effects of soil moisture, texture and nitrogen availability', *Soil Biology and Biochemistry* **33**, 1077–1093.

Williams, S.: 1994 *Interactions between components of rugger agroforestry systems in Indonesia*, Ph.D. thesis, Bangor, University of Wales, p. 256.

Williams, E.J., Hutchinson, G.L. and Fehsenfeld, F.C.: 1992, 'NO$_x$ and N$_2$O emissions from soil', *Global Biogeochemical Cycles* **6**, 351–388.

World Development Report: 1992, *Development and the Environment*, New York, NY, Oxford University Press, 308 pp.

Yagi, K. and Minami, K.: 1990, 'Effect of organic matter application on methane emission from some Japanese paddy fields', *Soil Science and Plant Nutrition* **36**, 599–610.

Yagi, K. and Minami, K.: 1993, 'Spatial and temporal variations of methane flux from a rice paddy field', in R.S. Oremland (ed.), *Biogeochemistry of Global Change: Radiative Trace Gases*, New York, Chapmen & Hall, pp. 353–368.

Yienger, J.J. and Levy II, H.: 1995, 'Empirical model of global soil-biogenic NOx emissions', *Journal of Geophysical Research* **100**, 11447–11464.

GREENHOUSE GAS FLUXES IN TROPICAL AND TEMPERATE AGRICULTURE: THE NEED FOR A FULL-COST ACCOUNTING OF GLOBAL WARMING POTENTIALS

G. PHILIP ROBERTSON[1]* and PETER R. GRACE[2]

[1]*W.K. Kellogg Biological Station and Department of Crop and Soil Sciences, Michigan State University, Hickory Corners, MI, USA; [2]Cooperative Research Center for Greenhouse Accounting, Australian National University, Canberra, ACT Australia
(*author for correspondence, e-mail: robertson@kbs.msu.edu; fax: (269) 671-2351; tel.: (269) 671-2267)*

(Accepted in Revised form 15 January 2003)

Abstract. Agriculture's contribution to radiative forcing is principally through its historical release of carbon in soil and vegetation to the atmosphere and through its contemporary release of nitrous oxide (N_2O) and methane (CH_4). The sequestration of soil carbon in soils now depleted in soil organic matter is a well-known strategy for mitigating the buildup of CO_2 in the atmosphere. Less well-recognized are other mitigation potentials. A full-cost accounting of the effects of agriculture on greenhouse gas emissions is necessary to quantify the relative importance of all mitigation options. Such an analysis shows nitrogen fertilizer, agricultural liming, fuel use, N_2O emissions, and CH_4 fluxes to have additional significant potential for mitigation. By evaluating all sources in terms of their global warming potential it becomes possible to directly evaluate greenhouse policy options for agriculture. A comparison of temperate and tropical systems illustrates some of these options.

Key words: carbon dioxide, carbon sequestration, global warming potential, greenhouse policy, liming, methane, nitrous oxide, soil carbon, trace gas flux.

1. Introduction

Potentials for reducing the buildup of greenhouse gases in the atmosphere by sequestering carbon in soil – thereby keeping additional carbon dioxide (CO_2) out of the atmosphere – have received widespread attention in the past 5 years, and have recently led to the initiation of carbon credit markets (e.g. McCarl and Schneider, 2001; CAST, 2003). In developed regions most of the attention has focused on no-till agriculture (e.g. Lal, 1999) because of its established capacity in many cropping systems and soils to build soil carbon (C) towards levels that existed prior to agricultural conversion (Paul et al., 1997). Recently other means for sequestering soil C have been suggested, including cover cropping and natural fallows that remove land from cultivation for a period of time.

While the focus on soil C and in particular on no-till cultivation systems has been useful for stimulating policy discussions, in some respects it is short-sighted. First, there are other potentials for mitigating greenhouse gas emissions that are commonly overlooked in discussions of policy options. These other potentials can

Environment, Development and Sustainability **6**: 51–63, 2004.
© 2004 *Kluwer Academic Publishers.*

be as or more effective than soil C capture in many systems, and may be especially suitable for regions and cropping systems for which no-till agriculture is agronomically unsuitable or economically prohibitive. For example, no-till is by definition unsuitable for root crops such as potatoes and groundnuts, and where soil pathogens persist in the absence of soil disturbance. No-till is economically prohibitive where the added operational costs for herbicides and the capital costs of specialized equipment cannot be justified by better yields or are hard to finance – both of these factors are at play in many developing regions of the tropics.

Second, changes in tillage practices may have unanticipated and unwanted effects on other sources or sinks of greenhouse gases. If, for example, soil water conservation associated with no-till were to provide more moisture for nitrifying and denitrifying bacteria as well as plants, then production of the greenhouse gas N_2O might increase, offsetting some or all of the mitigation potential of carbon storage (Robertson, 1999).

Third and finally, managing systems specifically for soil C storage by boosting the production of crop residues to enhance soil organic matter inputs can be counterproductive. In particular, if greenhouse-gas generating inputs are used to stimulate residue production (if yield increases are not the primary goal), then the mitigation gained with such production can be more than offset by the greenhouse costs of that production (Schlesinger, 1999). CO_2 released during fertilizer manufacture and during the generation of power for irrigation pumps are examples of such offsetting practices (cf. Izaurralde et al., 2000).

The need to include all sources of greenhouse warming potentials in cropping systems is acute – without a complete cost-benefit analysis with respect to a cropping system's capacity to affect the radiative forcing of the atmosphere, it is difficult to judge the appropriateness of one mitigation strategy over another. It is also otherwise easy to overlook additional mitigation options that may be particularly well suited to specific cropping systems or regions, especially for those in the developing tropics.

2. Global warming potential

Global warming potential (GWP) provides a means for comparing the relative effects of one source or sink of greenhouse gas against another. By placing all fluxes in common terms, one can directly evaluate the relative cost of, for example, increased carbon storage due to residue production (GWP mitigation) against increased N_2O from additional fertilizer application (GWP source).

By convention, GWP is measured in CO_2-equivalents (IPCC, 1996a, 2001). Conversions from other gases to CO_2 are based on the effect of a particular gas on the radiative forcing of the atmosphere relative to CO_2's effect. GWP is largely a function of a molecule's ability to capture infrared radiation, its current concentration in the atmosphere, the concentration of other greenhouse gases, and its atmospheric lifetime. All else being equal, a gas molecule with a greater atmospheric lifetime

TABLE I. GWPs of greenhouse gases in agriculture (IPCC, 2001).

Greenhouse gas	Atmospheric lifetime (years)	20-year GWP	100-year GWP	500-year GWP
Carbon dioxide (CO_2)		1	1	1
Methane (CH_4)	12	62	23	7
Nitrous oxide (N_2O)	114	275	296	156

will have a higher GWP than one that cycles rapidly. For example, N_2O is long-lived relative to CH_4, the 100-year N_2O GWP (296 CO_2-equivalents) is not much different from its 20-year GWP (275 CO_2-equivalents), whereas the GWP for methane (CH_4) falls off rapidly over this period, from 62 to 23 CO_2-equivalents. Likewise, relatively novel molecules with high IR capture capacities will have higher GWPs. Sulfur hexafluoride (SF_6), for example, has a 100-year GWP that is 22 200 times that of CO_2 owing to its radiative properties, its novelty in the atmosphere, and an atmospheric lifetime of 3200 years.

In general, only three greenhouse gases are affected by agriculture: CO_2, N_2O, and CH_4. Although CH_4 and especially N_2O are at far lower atmospheric concentrations than CO_2, their GWPs are sufficiently high that small changes have a disproportionate effect on radiative forcing (Table I). Over a 20-year time horizon, the GWP of CH_4 is 62 while that of nitrous oxide (N_2O) is 275; this means that a molecule of contemporary N_2O released to the atmosphere will have 275 times the radiative impact of a molecule of CO_2 released at the same time. Thus, an agronomic activity that reduces N_2O emissions by 1 kg ha^{-1} is equivalent to an activity that sequesters 275 kg ha^{-1} CO_2 as soil C.

3. Sources of GWP in agricultural ecosystems

Sources of GWP arise from a number of agronomic practices. Some, such as soil CO_2 emission following clearing and plowing and such as CO_2 emitted by diesel farm machinery, are direct sources of CO_2. Others, such as CO_2 emitted during fertilizer and pesticide manufacture, are indirect. Still others, such as CH_4 emitted by livestock and N_2O emitted from soil bacteria following cropping, are non-CO_2 based. All must be considered when calculating the total contribution of agriculture to global warming.

Mitigation occurs when existing sources of GWP are reduced. A GWP of zero means that no net GWP is attributable to a particular cropping system or agronomic practice. A negative GWP implies mitigation, but mitigation only occurs when GWP is less than the GWP of the pre-existing cropping condition – regardless of whether the pre-existing, business-as-usual condition was net positive or net negative.

In the remainder of this section we describe the specific sources of GWP in modern cropping systems.

3.1. SOIL C CHANGE

Conversion of natural, unmanaged ecosystems to agriculture releases substantial CO_2 to the atmosphere. The release of CO_2 from cleared vegetation that is burned or left to decompose is one of the most well-documented and important sources of the atmospheric CO_2 increase (e.g. IPCC, 2002). Historically, land clearing has been a major contributor to atmospheric CO_2 loading; today it still accounts for about 25% (1.6 Gt C yr^{-1}) of the total global CO_2 loading, which includes 6.3 Gt C yr^{-1} from fossil fuel use and cement production (IPCC, 2002). Most of the contemporary flux is from land clearing in tropical regions.

Soil C is also lost upon agricultural conversion. Forests and savannahs newly cultivated tend to lose a substantial fraction of their original carbon content in the decades following initial cultivation (Figure 1). This occurs for a number of reasons: reduced plant residue inputs, tillage-induced soil disturbance, erosion, and the creation of more favorable conditions for microbial decomposition (CAST, 2003). Generally soil C contents stabilize at 40–60% of original pre-cultivation values; the new equilibrium state is a function of climate, soil physical and chemical characteristics, and agronomic management factors such as tillage, crop types and cover, and residue management (Robertson and Paul, 2000).

Because soil represents about 80% of the carbon stocks in terrestrial ecosystems (ranging from 50% in tropical forests to 95% in tundra; IPCC, 2002), the global impact of soil C loss due to agriculture is considerable. Recent estimates suggest that 50–100 Gt C (CAST, 2003) have been lost from soils in the past few hundred years, although higher estimates range to 142 Gt C (Lal et al., 1999).

Soils can also gain carbon. The soil C balance is the net difference between carbon inputs from plant roots and aboveground litter (that remaining after harvest or fire), and carbon loss from microbial respiration and erosion. In agricultural

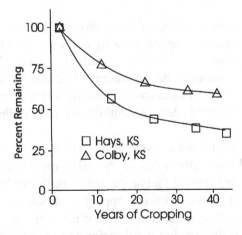

Figure 1. Soil organic matter loss following cultivation at two sites (Hays and Colby, Kansas) in the Midwestern US. Redrawn from Haas et al. (1957).

systems, manure and compost can represent additional inputs. Because erosion repositions carbon in the landscape rather than converts it to CO_2, erosion is not in itself a source of GWP. Microbial respiration, on the other hand, is a major source of GWP – where respiration is slowed, as in no-till systems, carbon can accumulate at slow but significant rates to some new equilibrium (Paustian et al., 1997).

Estimates of historic soil C loss provide a reference point for carbon sequestration potentials. Models suggest that 60–80% of the soil C lost as CO_2 could be regained under no-till conditions over a period of 50 years (IPCC, 1996a); if this is the case, then as much as 60–85 Gt C could be regained by agricultural soils at a rate of about 1.1–1.7 Gt C yr^{-1}. In soils of the US Midwest, the median rate of annual carbon gain under no-till is 30 g C m^{-2} (Franzleubbers and Steiner, 2002), which is equivalent to a GWP of -110 g CO_2-equivalents m^{-2} yr^{-1}.

3.2. Nitrogen Fertilizer

The Haber–Bosch process for producing fertilizer nitrogen results in the production of 0.375 mol of CO_2 per mole of N produced at 100% efficiency (Schlesinger, 1999); at normal efficiencies a mole of N is manufactured at a cost of about 0.58 mol of CO_2 (IPCC, 1996b). Additional CO_2 produced during the processing, transport, and application of N fertilizer pushes this value to around 1.4 mol of CO_2 released per mole of N applied (Schlesinger, 1999; Izaurralde et al., 2000).

Nitrogen fertilizer is thus a significant source of GWP in cropping systems fertilized with synthetic nitrogen. Worldwide, nitrogen fertilizer use is increasing rapidly, especially in developing regions of the tropics (with the notable exception of subsaharan Africa). Rates of nitrogen application vary widely, largely as a function of market availability, crop value, and national subsidies; typical rates in developed regions range from 50 kg N ha^{-1} for wheat to 200 kg N ha^{-1} for maize. For every 100 kg N ha^{-1} that is applied, the GWP cost is 45 g CO_2-equivalents m^{-2}.

3.3. Agricultural Lime

Agricultural lime in the form of calcium carbonate ($CaCO_3$) and dolomite ($CaMg(CO_3)_2$) are commonly applied to agricultural soils to counteract soil acidity. These carbonate minerals are mined from geologic reservoirs, ground, and applied to agricultural soils in humid regions at several-year intervals.

In regions where lime is inexpensive and readily available, lime requirements are estimated based on simple soil tests and generalized relationships, and generally strive to maintain soil pH in the range of 5.5–6.5, depending on the soil and the crop (Coleman and Thomas, 1967; van Lierop, 1990). Most varieties of alfalfa or lucerne (*Medicago sativa*), for example, perform best at pH > 6.5. Additionally, in variable charge soils typical of large areas of the tropics, soil acidity can reduce a soil's cation exchange capacity (CEC) to nil (Uehara and Gillman, 1981; Sollins et al., 1988), and raising the pH can restore CEC. In addition to raising soil pH, liming also supplies Ca^{2+} and Mg^{2+} for plant uptake. Additionally, in highly weathered soils the precipitation of reactive Al^{3+} by reaction with lime ($2Al^{3+}+3CaCO_3+6H_2O \rightarrow 3Ca^{2+} + 2Al(OH)_3 + 3H_2CO_3$) can be as or more beneficial than raising soil pH per se.

Lime applications to neutralize undesirable acidity are commonly in the range of 5–10 Mg ha^{-1}, and re-application is usually required every few years as the pH drops following fertilizer additions, nitrification, plant harvest, and leaching. As lime dissolves the fate of its carbon is difficult to predict. Carbonic acid formed in the presence of CO_2 from root and microbial respiration reacts with solid carbonates to consume H^+. For dolomite:

$$CaMg(CO_3)_2 + 2H_2CO_3 \rightarrow Ca^{2+} + Mg^{2+} + 4HCO_3^-$$

In this case, lime weathers to bicarbonate, which is then leached out of the soil profile.

If, on the other hand, carbonate comes into contact with a strong mineral acid such as nitric acid (HNO_3), the end product will be CO_2 rather than bicarbonate:

$$CaMg(CO_3)_2 + 4HNO_3 \rightarrow Ca^{2+} + Mg^{2+} + 4NO_3^- + 2CO_2 + 2H_2O$$

Nitric acid is formed by nitrifying bacteria in most soils (Robertson, 1982) including acid tropical soils (Sollins et al., 1988). Added lime thus seems likely to be a source of CO_2 where it is applied, but no data is available today to verify this.

The GWP of every metric ton of $CaCO_3$ added to soil is thus -44 to $+44$ g CO_2-equivalents m^{-2}; likewise, the GWP for $MgCO_3$ is -52 to $+52$ g CO_2-equivalents m^{-2}.

3.4. IRRIGATION

The GWP of irrigation is the result of fuel use during pumping and of carbonate reactions when calcium-saturated groundwater is sprayed on calcareous surface soils (Schlesinger, 1999). For irrigated land in the United States, the fossil fuel cost of pumping totals 22–83 g C m^{-2} yr^{-1} (Maddigan et al., 1982), equivalent to a GWP of 81–304 g CO_2-equivalents m^{-2}.

In arid regions groundwater often contains as much as 1% Ca and CO_2. When this water reaches the surface, the CO_2 (at $10\,000$ ppm$_v$) equilibrates with atmospheric CO_2 (365 ppm$_v$); $CaCO_3$ precipitates and CO_2 is released to the atmosphere:

$$Ca^{2+} + 2HCO_3 \rightarrow CaCO_3 + H_2O + CO_2$$

Schlesinger (1999) uses the average water use efficiency of arid-land plants to estimate that the net CO_2 released from the formation of soil carbonate due to irrigation would be 8.4 g C m^{-2} annually. This represents a GWP of 31 g CO_2-equivalents m^{-2} yr^{-1}.

3.5. FUEL

Diesel ($C_{16}H_{34}$) is 85% C, almost all of which is oxidized to CO_2 when burned. Agronomic activities that are mechanized, including plowing, cultivating, hoeing, spraying, planting, baling, chopping, and harvesting, exact a CO_2 cost. Every liter of fuel (at a density of $832\,g\,L^{-1}$) releases $706\,g$ C. Annual fuel use of $100\,L\,ha^{-1}$ would thus have a GWP of $26\,g$ CO_2-equivalents m^{-2}.

3.6. NITROUS OXIDE

Nitrous oxide is produced during nitrification and denitrification in agricultural soils. Nitrification is a soil microbial process in which ammonium is converted to nitrite (NO_2^-) and then to nitrate (NO_3^-) by aerobic autotrophic bacteria collectively known as nitrifiers; N_2O is a minor byproduct. Denitrification is a soil microbial process in which nitrate is converted to dinitrogen gas (N_2) by heterotrophic, facultatively anaerobic bacteria collectively known as denitrifiers; N_2O is a requisite intermediate that under some environmental conditions and for some denitrifier taxa is the end product (Cavigelli and Robertson, 2000).

Nitrification occurs whenever soil ammonium is available and environmental conditions such as temperature and moisture are favorable for nitrifier activity, which in many agronomic situations prevail most of the time (Robertson, 1982). Denitrification occurs whenever soil C and nitrate are available and oxygen is in short supply – denitrifiers can use nitrate rather than oxygen as a terminal electron acceptor if oxygen is unavailable (Robertson, 2000). This occurs in wet soils when diffusion of oxygen to microsites is slowed by saturated conditions, and inside soil aggregates in even well-drained soils. In the center of aggregates oxygen demand is often greater than can be provided by diffusion through the aggregate from the surrounding soil atmosphere (Sexstone et al., 1985). In general, controls on nitrification and denitrification in tropical soils are no different from those in temperate soils (Robertson, 1989).

Nitrous oxide can also be produced from livestock waste, though only when stored under relatively aerobic conditions such as in compost heaps. Under anaerobic conditions, as in waste lagoons, nitrification is inhibited by lack of oxygen and denitrification by the consequent lack of nitrate; further, any nitrate that is available tends to be denitrified all the way to N_2 rather than stop at N_2O (CAST, 2003) due to the low availability of electron acceptors.

Of all the sources of GWP in cropping systems, none are more poorly quantified than N_2O production. This is especially true for tropical agriculture, with the possible exception of lowland rice systems (e.g. Buresh and Austin, 1988; Bronson et al., 1997). This is mainly because of the difficulty with which N_2O fluxes are measured. Unlike for CO_2 and CH_4, N_2O flux is not suited to micrometeorological measurement (Holland et al., 1999); rather, fluxes must be measured using small chambers placed on the soil surface for 1–2 h intervals. High temporal and spatial variability means that many chambers must be deployed simultaneously at weekly or more frequent intervals in a given cropping system; sampling and analysis costs are thus high.

However, for the few cropping systems for which we have reliable N_2O fluxes, N_2O loss is frequently the major source of GWP. Robertson et al. (2000), for example, found for a 9-year measurement campaign in several annual and perennial cropping systems in the US that N_2O was the single greatest source of GWP in all four of their annual crop systems, ranging from 50 to 60 g CO_2-equivalents m^{-2} yr^{-1}. IPPC methodology assumes that 1.25% of nitrogen inputs to most cropping systems is subsequently emitted as N_2O–N; if true, then for every 100 kg N ha^{-1} applied as fertilizer, about 1.25 kg N will be emitted as N_2O, for a GWP (over a 20-year time horizon) of 54 g CO_2-equivalents m^{-2} yr^{-1}.

Soil nitrogen availability appears to be the single best predictor of N_2O flux in most terrestrial ecosystems including agricultural. Any activity or process that acts to keep available soil nitrogen low should thus lead to smaller N_2O flux. Plant demand for nitrogen is therefore one of the most important determinants of N_2O flux, and more precise application of N fertilizer – to maximize plant uptake of added N both spatially and temporally – is one of our most important means for mitigating current N_2O fluxes from agriculture.

3.7. METHANE

Methane is produced by anaerobic bacteria in soil, animal waste, and ruminant stomachs, and agricultural sources of CH_4 are thus a significant fraction of the global CH_4 budget. About 15% of the 598 Tg global CH_4 flux is from lowland rice systems, and another 15% is from enteric fermentation during livestock digestion (Hein et al., 1997; IPCC, 2001). Because methanogenesis is a strictly anaerobic process, under normal conditions upland cropping systems are not a direct source of CH_4, and CH_4 flux in paddy rice can be partly mitigated through water level and residue management and cultivar selection (Mosier et al., 1998a). Nevertheless, in rice systems CH_4 emissions can be the dominant source of GWP.

Methane is also consumed, but by a different class of soil bacteria, the methanotrophs, and CH_4 consumption in soils is a small but significant part of the global CH_4 budget, comparable in magnitude to the annual atmospheric increase in CH_4. In rice paddies and wetlands the total CH_4 flux is the net difference between methogenesis in submerged anaerobic horizons and CH_4 consumption at or above the soil–water interface. In upland soils the net flux appears to be largely a function of CH_4 consumption. Agricultural conversion tends to reduce natural rates of CH_4 consumption in soils by a factor of 5–10 (Bronson and Mosier, 1993; Smith et al., 2000), and at our current state of knowledge there is no known way to restore consumption other than allowing natural revegetation. By attenuating a natural source of mitigation, agriculture thus creates an indirect source of GWP.

Robertson et al. (2000) found for a US Midwest landscape that the GWP of CH_4 oxidation in old-growth forest was -25 g CO_2-equivalents m^{-2} yr^{-1}; for various cropping systems on the same soil type they found GWPs ranging from -4 to -6 g CO_2-equivalents m^{-2} yr^{-1}. Similar changes have been documented for a variety of soil and climates (e.g. Smith et al., 2000), including tropical (Keller and Reiners, 1994).

4. A temperate – tropical comparison

There are as yet very few cropping systems for which all significant sources of GWP have been measured. Below we contrast systems from two different regions: a temperate-region maize–soybean–wheat rotation (Robertson et al., 2000) and a tropical rice–wheat–cowpea system in the Indian Indo-Gangetic Plain (Grace et al., 2003). Except for soil C change, GWP values in the Indian systems are estimated based on IPCC methodologies; GWP values for the US systems are measured.

Table II presents GWP values for three Midwest US cropping systems and a late successional forest. In the conventional tillage system N_2O is the principal source of GWP, accounting for 52 of the system's total of 114 g CO_2-equivalents m^{-2} yr^{-1}. Contributions of N fertilizer (60 kg N ha^{-1} yr^{-1} on average) and agricultural lime were each about half of N_2O's contribution to total GWP (23–27 g CO_2-equivalents m^{-2} yr^{-1}), and fuel use accounted for about half again of this (16 g CO_2-equivalents m^{-2} yr^{-1}). Because soil C in this system was equilibrated (at about 1% C), soil C did not contribute to GWP; likewise, CH_4 oxidation contributed very little mitigation capacity, about -4 g CO_2-equivalents m^{-2} yr^{-1}.

GWP values in the no-till system were equivalent to those in the conventional system for most sources of GWP except soil C, lime, and fuel. In the no-till system soil C had accumulated at 30 g m^{-2} yr^{-1}, providing a GWP mitigation of -110 g CO_2-equivalents m^{-2} yr^{-1}. Slightly lower fuel costs (12 g CO_2-equivalents m^{-2} yr^{-1}) were offset by somewhat higher lime

TABLE II. Sources of GWP – in a maize–soybean–wheat cropping system of the US Midwest based on 9 years of measurements (Robertson et al., 2000). The system had been cropped for decades earlier, depleting soil organic matter to 1% C. N_2O and CH_4 fluxes are measured. N fertilizer was added only to maize (120 kg N ha^{-1}) and wheat (60 kg N ha^{-1}) crops. Only wheat residue was removed. GWPs for N_2O and CH_4 are based on 20-year time horizons using IPCC (1996a) values of 280 and 56, respectively (cf. Table I).

System	CO_2				N_2O	CH_4	Net GWP
	Δ Soil C	N fertilizer	Lime	Fuel			
	(g CO_2-equivalents m^{-2} yr^{-1})						
Conv. till	0	27	23	16	52	-4	114
No-till	-110	27	34	12	56	-5	14
Organic	-29	0	0	19	56	-5	41
Forest	0	0	0	0	21	-25	-4

requirements ($34 \, g \, CO_2$-equivalents $m^{-2} \, yr^{-1}$), providing a net GWP for this system of $14 \, g \, CO_2$-equivalents $m^{-2} \, yr^{-1}$, substantially lower than the $114 \, g$ CO_2-equivalents $m^{-2} \, yr^{-1}$ GWP of the conventional tillage system.

The organic system, which used legumes rather than synthetic fertilizer for nitrogen inputs, was midway between the conventional and no-till systems, having a net systemwide GWP of $41 \, g \, CO_2$-equivalent $m^{-2} \, yr^{-1}$. N_2O flux was the same as in the other systems, but there were neither N fertilizer nor lime inputs in this system, and carbon accumulated at a moderate rate ($8 \, g \, m^{-2} \, yr^{-1}$) owing to cover crop residue. The forested system, in contrast, had a net GWP of $-4 \, g$ CO_2-equivalents $m^{-2} \, yr^{-1}$ – soil C neither accumulated nor disappeared, there were no agronomic inputs, and the GWP of N_2O flux ($1.3 \, g \, N_2O$–$N \, ha^{-1} \, d^{-1}$ on average for a GWP of $21 \, g \, CO_2$-equivalents $m^{-2} \, yr^{-1}$) was counterbalanced by the GWP of CH_4 oxidation ($-9.2 \, g \, CH_4$–$C \, ha^{-1} \, d^{-1}$ on average for a GWP of $-25 \, g$ CO_2-equivalents $m^{-2} \, yr^{-1}$).

In Table III is a similar analysis for three rice–wheat–cowpea systems of the Indo-Gangetic Plain in India (Grace et al., 2002). Positive values for soil C in this example indicate loss of soil organic matter over the course of the 20-year experiment – soil C had not yet equilibrated at this site. Rates of soil C loss were similar for conventionally cultivated systems regardless of whether crop residue was burned or retained, indicating that soil microbial activity and burning are equally effective for oxidizing crop residue in these systems. But under no-till cultivation, over $100 \, g \, CO_2$-equivalents $m^{-2} \, yr^{-1}$ were lost each year. However most of the GWP cost of this system is associated with the loss of CH_4 during flooded rice cropping; some $560 \, g \, CO_2$-equivalents $m^{-2} \, yr^{-1}$ are lost during rice cultivation due to CH_4 emission; burning crop residues releases another $167 \, g$ CO_2-equivalents $m^{-2} \, yr^{-1} \, CH_4$.

It is remarkable that the total net GWP in these systems (893–$1189 \, g$ CO_2-equivalents $m^{-2} \, yr^{-1}$) is an order of magnitude greater than that for the temperate-region maize–soybean–wheat system (Table II; 14–$114 \, g$ CO_2-equivalents $m^{-2} \, yr^{-1}$). Three factors unique to the tropical system play into these differences: (1) persistent soil C loss, (2) CH_4 loss during rice cultivation

TABLE III. Sources of GWP – in a rice–wheat–cowpea cropping system of the Indian Indo-Gangetic Plain based on a 20-year trial (Ram, 2000, as reported in Grace et al., 2003). N_2O and CH_4 fluxes are estimated based on IPCC methodologies. N fertilizer ($120 \, kg \, N \, ha^{-1}$) was added only to rice and wheat crops. Crop residues were either retained or burned, as noted in table. GWPs for N_2O and CH_4 are the same as for Table II (recalculated from Grace et al., 2003).

System	CO_2				N_2O	CH_4	Net GWP
	Δ Soil C	N fertilizer	Lime	Fuel			
	GWP (g CO_2-equivalents $m^{-2} \, yr^{-1}$)						
Conv. till, residues burned	229	36	0	26	171	727	1189
Conv. till, residues retained	229	36	0	26	174	560	1025
No-till, residues retained	114	36	0	9	174	560	893

and crop residue combustion, and (3) potentially greater N_2O fluxes. The greater N_2O fluxes may be illusory-actual fluxes may be lower than those estimated here because of relatively conservative fertilizer rates. Nevertheless, even if N_2O fluxes are in fact equivalent, and after soil C equilibrates such that its annual contribution to GWP is nil, these tropical systems will continue to have higher GWP on account of CH_4 produced during rice cultivation and crop residue burning. However, even though the tropical system is not nearly as close to GWP-neutral as the temperate-region system, the 25% reduction in GWP achievable by retaining residues and implementing no-till cultivation is very significant.

This type of analysis is extremely valuable from a policy and management perspective because it shows that in both the temperate and tropical systems mitigation could achieve even greater GWP savings. Cropping systems in both regions could benefit substantially from efforts to mitigate N_2O production; these efforts could take the form of better nitrogen conservation by basing fertilizer rates on seasonal soil and plant tests, by applying fertilizers closer to the time of crop uptake, by planting cover crops to remove nitrogen from the soil solution during the non-growing season, and possibly by using nitrification inhibitors (CAST, 2003). Significant savings could also result from using nonsynthetic N fertilizers such as leguminous cover crops or manure; by reducing diesel use with either biogas production or better mechanical efficiencies; by managing soil acidity to reduce the need for lime applications; and by using cover crop and other residue management strategies in addition to no-till to increase soil C. Additionally, in the tropical system CH_4 mitigation could take the form of better water, residue, and rice cultivar management.

In all of these cases a whole-system analysis would also serve to identify the value and costs of trade-offs. Such analysis might show, for example, that soil C gained by adding manure or compost to soil might be offset by a concomitant increase in soil N_2O flux. Only a full-cost, whole-system GWP accounting can fully calculate the true net value of various cropping strategies for greenhouse gas mitigation.

Acknowledgements

This article was written while the senior author was a visiting researcher at the Australian Cooperative Research Center for Greenhouse Accounting. Support was provided by the NSF LTER program and the Michigan Agricultural Experiment Station.

References

Bronson, K.F. and Mosier, A.R.: 1993, 'Nitrous oxide emissions and methane consumption in wheat and corn-cropped systems in Northeastern Colorado,' in L.A. Harper, A.R. Mosier, J.M. Duxbury and D.E. Rolston (eds.), *Agricultural Ecosystem Effects on Trace Gases and Global Climate Change*, Madison, Wisconsin, American Society of Agronomy, pp. 133–144.

Bronson, K.F., Neue, H.U. and Abao Jr., E.B.: 1997, 'Automated chamber measurements of methane and nitrous oxide flux in a flooded rice soil: I. Residue, nitrogen, and water management', *Soil Science Society of America Journal* **61**, 981–987.

Buresh, R.J. and Austin, E.R.: 1988, 'Direct measurement of nitrogen and nitrous oxide flux in flooded rice fields', *Soil Science Society of America Journal* **52**, 681–688.

CAST: 2003, *Agriculture's Response to Global Climate Change*, Ames, Iowa, Council for Agricultural Science and Technology, in press.

Cavigelli, M.A. and Robertson, G.P.: 2000, 'The functional significance of denitrifier community composition in a terrestrial ecosystem', *Ecology* **81**, 1402–1414.

Coleman, N.T. and Thomas, G.W.: 1967, 'The basic chemistry of soil acidity', in R.W. Pearson and F. Adams (eds.), *Soil Acidity and Liming*, Madison, Wisconsin, American Society of Agronomy, pp. 1–42.

Franzluebbers, A.J. and Steiner, J.L.: 2002, 'Climatic influences on soil organic C storage with no-tillage', in J.M. Kimble and L.R. Follett (eds.), *Agricultural Practices and Policies for Carbon Sequestration in Soil*, Boca Raton, Florida, CRC Press, pp. 71–86.

Grace, P.R., Jain, M.C. and Harrington, L.: 2003, 'The long-tem sustainability of tropical and subtropical rice and wheat systems: An environmental perspective', in J.K. Ladha, J.E. Hill, J.M. Duxbury, R.K. Gupta and R.J. Buresh (eds.), *Improving the Productivity and Sustainability of Rice-Wheat Systems: Issues and Impacts*, Madison, Wisconsin, American Society of Agronomy pp. 27–43.

Haas, H.J., Evans, C.E. and Miles, E.F.: 1957, *Nitrogen and Carbon Changes in Great Plains Soils as Influenced by Cropping and Soil Treatment, USDA Technical Bulletin 1164*, Washington, DC, US Department of Agriculture.

Hein, R., Crutzen P.J. and Heinmann M.: 1997, 'An inverse modeling approach to investigate the global atmospheric methane cycle', *Global Biogeochemical Cycles* **11**, 43–76.

Holland, E.A., Robertson, G.P., Greenberg, J., Groffman, P., Boone, R. and Gosz, J.: 1999. 'Soil CO_2, N_2O, and CH_4 exchange', in G.P. Robertson, C.S. Bledsoe, D.C. Coleman and P. Sollins (eds.), *Standard Soil Methods for Long-Term Ecological Research*, New York, Oxford University Press, pp. 185–201.

IPCC: 1990. *Climate Change: The Intergovernmental Panel on Climate Change (IPCC) Scientific Assessment*, Cambridge, Cambridge University Press.

IPCC: 1996a, *Climate Change 1995: The Science of Climate Change. Contribution to the Second Assessment Report of the Intergovernmental Panel on Climate Change*, Cambridge, Cambridge University Press.

IPCC: 1996b, *Revised 1996 Guidelines for National Greenhouse Gas Inventories, Reference Manual*, Washington, DC, Organization for Economic Cooperation and Development.

IPCC: 2001, *Climate Change 2001: The Scientific Basis*, Cambridge, Cambridge University Press.

IPCC: 2002, *IPCC Special Report on Land Use, Land-Use Change and Forestry*, Cambridge, Cambridge University Press.

Izaurralde, R.C., McGill, W.B., Bryden, A., Graham, S., Ward, M. and Dickey, P.: 1998, 'Scientific challenges in developing a plan to predict and verify carbon storage in Canadian Prairie soils', in R. Lal, J. Kimble, R. Follett and B.A. Stewart (eds.), *Management of Carbon Sequestration in Soil*, Boca Raton, FL, CRC Press, pp. 433–446.

Izaurralde, R.C., McGill, W.B. and Rosenberg, N.J.: 2000, 'Carbon cost of applying nitrogen fertilizer', *Science* **288**, 809.

Keller, M. and Reiners, W.A.: 1994 'Soil-atmosphere exchange of nitrous oxide, nitric oxide, and methane under secondary succession of pasture to forest in the Atlantic lowlands of Costa Rica', *Global Biogeochemical Cycles* **8**, 399–409.

Lal, R.: 1999, 'Soil management and restoration for carbon sequestration to mitigate the accelerated greenhouse effect', *Progress in Environmental Science* **1**, 307–326.

Lal, R., Kimble J.M., Follett R.F. and Cole C.V.: 1999, *The Potential of US Cropland to Sequester Carbon and Mitigate the Greenhouse Effect*, Chelsea, Michigan, Ann Arbor Press.

Maddigan, R.J., Chern, W.S. and Rizy, C.G.: 1982, 'The irrigation demand for electricity', *American Journal of Agricultural Economics* **64**, 673.

McCarl, B.A. and Schneider, U.A.: 2001, 'Greenhouse gas mitigation in US agriculture and forestry', *Science* **294**, 2481–2482.

Mosier, A.R., Duxbury, J.M., Freney, J., Heinemeyer, O., Minami, K. and Johnson, D.E.: 1998a, 'Mitigating agricultural emissions of methane', *Climatic Change* **40**, 39–80.

Mosier, A.R., Duxbury, J.M., Freney, J.R., Heinemeyer, O. and Minami, K.: 1998b, 'Assessing and mitigating N_2O emissions from agricultural soils', *Climatic Change* **40**, 7–38.

Paul, E.A., Paustian, K., Elliot, E.T. and Cole, C.V. (eds.): 1997, *Soil Organic Matter in Temperate Ecosystems: Long-Term Experiments in North America*, Boca Raton, FL, Lewis CRC Publishers.

Paustian, K., Collins, H.P. and Paul, E.A.: 1997, 'Management controls on soil carbon', in E.A. Paul, K. Paustian, E.T. Elliott and C.V. Cole (eds.), *Soil Organic Matter in Temperate Agroecosystems: Long-Term Experiments in North America*, Boca Raton, Florida, CRC Press, pp. 15–49.

Robertson, G.P.: 1982, 'Nitrification in forested ecosystems', *Philosophical Transactions of the Royal Society of London* **296**, 445–457.

Robertson, G.P.: 1989, 'Nitrification and denitrification in humid tropical ecosystems', in J. Proctor (ed.), *Mineral Nutrients in Tropical Forest and Savanna Ecosystems*, Cambridge, Blackwell Scientific, pp. 55–70.

Robertson, G.P.: 1999, 'Keeping track of carbon', *Science* **285**, 1849.

Robertson, G.P.: 2000, 'Denitrification', in M.E. Sumner (ed.), *Handbook of Soil Science*, Boca Raton, Florida, CRC Press, pp. C181–C190.

Robertson, G.P. and Paul, E.A.: 2000, 'Decomposition and soil organic matter dynamics', in E.S. Osvaldo, R.B. Jackson, H.A. Mooney and R.W. Howarth (eds.), *Methods in Ecosystem Science*, New York, Springer Verlag, pp. 104–116.

Robertson, G.P., Paul, E.A. and Harwood R.R.: 2000, 'Greenhouse gases in intensive agriculture: Contributions of individual gases to the radiative forcing of the atmosphere', *Science* **289**, 1922–1925.

Schlesinger, W.H.: 1999, 'Carbon sequestration in soils', *Science* **284**, 2095–2097.

Sexstone, A.J., Revsbech, N.P., Parkin, T.P. and Tiedje, J.M.: 1985, 'Direct measurement of oxygen profiles and denitrification rates in soil aggregates', *Soil Science Society of America Journal* **49**, 645–651.

Smith, K.A., Dobbie, K.E., Ball, B.C., Bakken, L.R., Situala, B.K., Hansen, S. and Brumme, R.: 2000, 'Oxidation of atmospheric methane in Northern European soils, comparison with other ecosystems, and uncertainties in the global terrestrial sink', *Global Change Biology* **6**, 791–803.

Sollins, P., Robertson, G.P. and Uehara, G.: 1988, 'Nutrient mobility in variable- and permanent-charge soils', *Biogeochemistry* **6**, 181–199.

Uehara, G. and Gillman, G.P.: 1981, *The Mineralogy, Chemistry and Physics of Tropical Soils with Variable Charge Clays*, Boulder, Colorado, Westview Press.

van Lierop, W.: 1990, 'Soil pH and lime requirement determination', in *Soil Testing and Plant Analysis*, Madison, Wisconsin, American Society of Agronomy, pp. 73–126.

MITIGATING GREENHOUSE GAS EMISSIONS FROM RICE–WHEAT CROPPING SYSTEMS IN ASIA

R. WASSMANN[1]*, H.U. NEUE[2], J.K. LADHA[3] and M.S. AULAKH[4]

[1]*Institute for Meteorology and Climate Research (IMK-IFU), Forschungszentrum Karlsruhe, Garmisch-Partenkirchen, Germany; [2]Department of Soil Sciences, Umweltforschungszentrum Leipzig-Halle (UFZ), Germany; [3]International Rice Research Institute, Manila, Philippines; [4]Department of Soils, Punjab Agricultural University, Ludhiana, India (*author for correspondence; e-mail: reiner.wassmann@imk.fzk.de; fax: +49-8821-183296; tel.: +49-8821-183139)*

(Accepted in Revised form 15 January 2003)

Abstract. The rice–wheat belt comprises nearly 24–27 million ha in South and East Asia. Rice is generally grown in flooded fields whereas the ensuing wheat crop requires well-drained soil conditions. Consequently, both crops differ markedly in nature and intensity of greenhouse gas (GHG) fluxes, namely emission of (1) methane (CH_4) and (2) nitrous oxide (N_2O) as well as the sequestration of (3) carbon dioxide. Wetland rice emits large quantities of CH_4; strategies to CH_4 emissions include proper management of organic inputs, temporary (mid-season) field drainage and direct seeding. As for the wheat crop, the major GHG is N_2O that is emitted in short-term pulses after fertilization, heavy rainfall and irrigation events. However, N_2O is also emitted in larger quantities during fallow periods and during the rice crop as long as episodic irrigation or rainfall result in aerobic–anaerobic cycles. Wetland rice ensures a relatively high content of soil organic matter in the rice–wheat system as compared to permanent upland conditions. In terms of global warming potential, baseline emissions of the rice–wheat system primarily depend on the management practices during the rice crop while emissions from the wheat crop remain less sensitive to different management practices. The antagonism between CH_4 and N_2O emissions is a major impediment for devising effective mitigation strategies in rice–wheat system – measures to reduce the emission of one GHG often intensify the emission of the other GHG.

Key words: carbon dioxide, GWP, irrigation, manure, methane, nitrous oxide, organic residues, rice, sequestration, wheat.

1. Introduction

Agriculture production is an important source of greenhouse gases (GHG), namely methane (CH_4), nitrous oxide (N_2O) and carbon dioxide (CO_2). Whereas the composite effect of agricultural activities is estimated to app. one fifth of the anthropogenic greenhouse effect (IPCC, 1996; Mosier et al., this issue), individual source strengths of land use change, animal husbandry, and grain production can only be estimated in very broad ranges of uncertainty. CO_2 is primarily emitted from fossil fuel combustion; approximately $1.7\,Gt\,C\,yr^{-1}$ (corresponding to 21% of the total emission) derives from land use change (IPCC, 2001). On the other hand, terrestrial ecosystems soils can also sequester CO_2 from the atmosphere (accounting to app. $1.9\,Gt\,C\,yr^{-1}$) given favorable environment and management practice (IPCC, 2001).

Methane is emitted in substantial quantities from rice fields, domestic animals and biomass burning (Mosier et al., this issue). The total source strength of these agricultural activities accounts for 128–270 Tg CH_4 yr^{-1} corresponding to 22–46% of the global budget of CH_4 (IPCC, 1996). The N_2O fluxes generating from agriculture are assoiated with fertilizer N application, domestic animals and biomass burning. While Kroeze et al. (1999) estimated app. 17.6 Tg N_2O–N yr^{-1} as total N_2O source strength (agriculture and other sources), the different estimates on agriculture-borne emissions range from 1.4 to 18.9 Tg N_2O–N yr^{-1} (IPCC, 1996). An unknown, but probably significant, amount of GHG is generated indirectly through on and off farm activities.

The scope of this paper is to review the current state of knowledge for GHG emission from a singular, but very important production system. The rice–wheat belt in South and East Asia comprises 24–27 million ha of land cultivated under this system (Ladha et al., 2003, Jiaguo, 2000). Millions people rely on food produced under this cropping system – more than on any other individual food production system of the world. Over the past decades, the rice–wheat belt has experienced rapid development in the agricultural sector leading to very intensive land use in the vast part of the region. Moreover, as rice is generally grown in flooded fields whereas the succeeding wheat requires well-drained soil conditions, the rice–wheat system thus appears as an especially interesting object to discuss mitigation options given the strong contrast between wet and dry conditions necessitating different mitigation approaches for both crops.

2. Characterization of the rice–wheat system

In South Asia, the rice–wheat system occupies about 13.5 million ha (10 million in India, 2.2 million in Pakistan, 0.8 million in Bangladesh, and 0.5 million in Nepal), extending across the Indo-Gangetic floodplain into the Himalayan foothills (see Figure 1). Rice–wheat systems cover about 32% of the total rice area and 42% of the total wheat area in these four countries (Woodhead et al., 1994a,b). According to data given by Jiaguo (2000), China has 13 million ha of rice–wheat accounting for app. 40% of both rice and wheat cultivation area. The Chinese rice–wheat area is mainly located in the Yangtze River Valley (Figure 2) with four provinces, namely Jiangsu, Anhui, Hubei, and Sichuan, each comprising more than 1 million ha of land under this cropping system (Huke et al., 1993).

While the rice–wheat system has a long history in the region, its significance was steadily increasing as more productive crop varieties and practices for rice and wheat production became available to farmers. In the Indian states of Punjab, Haryana, and Uttar Pradesh – being traditionally wheat-growing states – the introduction of rice cultivation was mainly driven by the development of new irrigation facilities. In these areas, the abundance of water resources from rivers that bring water from Himalayas, monsoon rains and groundwater have made possible the production of rice even in coarse-textured porous soils (Aulakh and Bijay-Singh, 1997). In

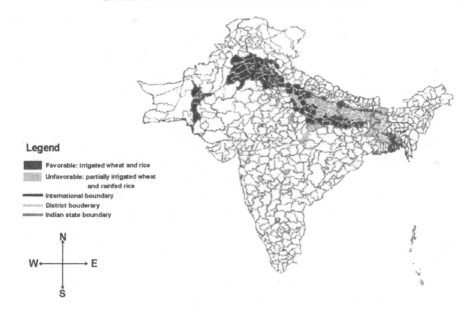

Figure 1. Agroecological analysis of rice–wheat area and productivity in the Indo-Gangetic plain (from Ladha et al., 2000).

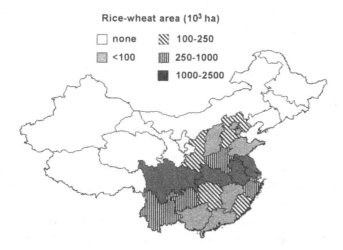

Figure 2. Geographical distribution of rice–wheat area in continental China (by province) in 1989/90; drawn with data from Huke et al. (1993).

contrast, the Indian states of Bihar and West Bengal are traditional rice-growing states that have experienced an enhancement in wheat production over the last few decades. The area of rice–wheat system in China has increased by 45% since 1952 (Huke et al., 1993).

Most of the rice–wheat cropping is fully irrigated, so that farmers are swiftly adopting modern cultivars of both cereals. Irrigated rice–wheat systems have remained the major source of the marketed surplus of food grain in South and East Asia for feeding the growing urban population and deficit areas within the country. After three decades of intensive farming, however, productivity and sustainability of this system are deteriorating (Byerlee et al., 2003; Ladha et al., 2003; Pathak et al., 2003). While the gains in food grain production have stagnated for both rice and wheat crops in recent years, there is some evidence of declining productivity (Byerlee et al., 2003). The causes for the stagnation or decline are not well understood and may include quantitative and qualitative changes in soil organic matter (SOM), nutrient imbalances due to inappropriate fertilizer applications, water scarcity as well as poor water quality (salinity), and the buildup of pests (Ladha et al., 2003). All of these symptoms indicate a threat to the resilience of the natural resource base and to the livelihoods of millions of people (Paroda et al., 1994).

Under this system, farmers grow rice in the rainy season corresponding to the summer months followed by wheat in winter. In the Indo-Gangetic plains, wheat yields are significantly lower than in any irrigated areas outside this belt (Ladha et al., 2000). One factor for the yield gap of the wheat crop could be the compaction of the soil caused by land preparation and flooding of the preceding rice crop (Sharma et al., 2003). In Northeast India and North Pakistan, farmers' prefer to grow the expensive and long-growing rice cultivars (Basmati-type), which means the prolonged rice crop delays the planting of the wheat crop against its optimum planting time. Zero tillage encompassing earlier planting of wheat seed directly in rice stubbles and the residual moisture is becoming increasingly popular in the region (Hobbs and Gupta, 2003).

3. Baseline emissions from rice–wheat system

Rice is generally grown in flooded fields whereas the ensuing wheat crop requires well-drained soil conditions. This fundamental difference in the growing conditions entails distinct properties of both crops as far as GHG emissions are concerned. While anaerobic conditions in wetland rice fields foster emission of CH_4, the upland conditions during wheat production could act as a net-sink for this gas. Conversely, N_2O fluxes are usually low during permanent flooding and are high under varying soil moisture conditions during the wheat crop and fallow periods. The third GHG to be considered for mitigation options is CO_2 that will be discussed under the section dealing with global warming potentials (GWP).

One complication in comparing emission records are the distinct units used by different authors. Emission rates are either given as average values or as cumulated emission over one season. In the following discussion, we have tried to harmonize literature values as much as possible and provide cumulated values as kg CH_4 ha^{-1} and kg N_2O–N ha^{-1}, which appeared to be the most commonly used unit for either gas. In some cases however, we have adopted daily values (in $g\,N_2O$–$N\,ha^{-1}d^{-1}$) because season lengths were not documented. Modern rice varieties are typically harvested at app. 90 to 110 d (after transplanting) and wheat varieties at app. 120–150 d (after sowing), but longer seasons also occur in the rice-wheat system depending on cropping patterns and varieties.

3.1. CH_4 EMISSIONS

The baseline emissions of CH_4 from rice fields have been reviewed thoroughly in recent publications (e.g. Neue, 1997; Aulakh et al., 2000e; Mosier et al., this issue), so that this presentation will be constrained to the specific properties of the rice–wheat system (relevant to GHG emissions) and a regional focus on the major rice–wheat belt. The regional distribution of the rice–wheat system warrants a comparative assessment of CH_4 emissions in Central China and Northern India, because these two regions collectively account for more than 75% of the global rice–wheat area. The available CH_4 emission records from rice fields in Central China showed a relatively high background level of CH_4 emissions ranging from 200 to $900\,kg$ $CH_4\,ha^{-1}$ under mineral fertilization (Zheng et al., 1997; Wassmann et al., 1993a) and up to $1100\,kg\,CH_4\,ha^{-1}$ following organic amendments (Khalil et al., 1998). Emission records from Northern India were consistently lower and did not exceed $30\,kg\,CH_4\,ha^{-1}$ under mineral fertilization (Mitra et al., 1999) and $50\,kg\,CH_4\,ha^{-1}$ under organic treatment (Debnath et al., 1996). The reasons for this pronounced difference can be attributed to a combination of soil factors and the water regimes in both regions (Wassmann et al., 2000a). In contrast to the conditions in China, the Indo-Gangetic plains have predominantly porous soils. These porous soils include sand, loamy sand and sandy loam textured classes, which distinguish themselves with high infiltration rates. Thus, floodwater cannot be retained continuously in these soils and has to be replenished frequently by the farmer. These high percolating rates imply high oxygen input into the soil which impedes CH_4 emissions even under high organic inputs (Jain et al., 2000).

Upland cultivation is known to reduce carbon (C) levels in the soils (see below), but it is unclear if this rather slow process could possibly result in a residual effect of the preceding wheat crop on CH_4 emissions during the rice crop. Methane is mainly generated from organic material that is recently formed or added during the growing season of the rice itself. Moreover, the post-wheat fallow period until the transplanting of rice crop should result in converging soil conditions. Wheat straw only slightly deviates from rice straw in its C content, so that an impact deriving from distinct residues is rather unlikely even under high doses of straw application. The composition of organic residues, however, could become a factor when rice–wheat system is compared to rice–legume rotations. In rice fields applied with residues from the preceding season, CH_4 emissions were reduced by app. 50% when cowpea was grown instead of wheat (Abao et al., 2000).

During the wheat crop and the fallow periods, the soils may either consume CH_4 (Singh et al., 1996) or act as a small source of CH_4 (Bronson et al., 1997a,b;

Abao et al., 2000). In either case, the magnitude of CH_4 net-fluxes during the wheat and fallow period is only a small share of the net-flux over the entire rice–wheat system (Singh et al., 1996; Bronson et al., 1997b). Fertilizer application generally reduces CH_4 uptake rates (Singh et al., 1996), so that intensification of wheat production may further reduce the impact of this crop against CH_4 emissions during the rice crop.

3.2. N_2O EMISSIONS

The available field data on N_2O emissions from rice fields comprise several field experiments conducted in the rice–wheat belt. Emissions recorded in Chinese rice fields were, in average, slightly higher than from upland soils (Xing and Zhu, 1997), but spread broadly in time and space. Measurements at three locations in Central China resulted in seasonal emission rates of 0.017–0.036 kg N_2O–N ha^{-1} in Yingtan (Jiangxi Province), 0.136–0.981 kg N_2O–N ha^{-1} in Nanjing (Jiangsu), and 1.780–4.720 kg N_2O–N ha^{-1} in Fengqiu (Henan Province) (Hua et al., 1997). Lower values were recorded in Northern China, i.e. annual emission of 0.025 kg N_2O–N ha^{-1} (Chen et al., 1997). In Sichuan Province, where manure application is very abundant, average N_2O emissions were relatively high with an average value of 16.8 g N_2O–N ha^{-1} d^{-1} (Khalil et al., 1998). Zheng et al. (2000) recorded N_2O emissions from rice (under rice–wheat system) in the Taihu region and found seasonal emissions of 1.95 and 5.42 kg N_2O–N ha^{-1}, respectively. These values were in the same order of magnitude as those observed in cultivated upland soils.

Available results for rice fields in Northern India indicate a level of N_2O emissions in the range of 0.034–0.06 kg N_2O–N ha^{-1} per season (Majumdar et al., 2000). In another field experiment at the same site in New Delhi, Kumar et al. (2000) recorded seasonal emission rates of 0.16 and 0.235 kg N_2O–N ha^{-1} for urea and ammonium sulfate treatment, respectively. Aulakh et al. (2001b) reported N_2O production in a well-drained sandy loam soil ranging from 15–60 g N_2O–N ha^{-1} d^{-1} during pre-rice fallow period and 15–450 g N_2O–N ha^{-1} d^{-1} during rice growing season. These values are within the lower range of most of the emission records from rice fields in China as well as outside the rice–wheat belt, e.g. 5.5 g N_2O–N ha^{-1} d^{-1} in Indonesia (Suratno et al., 1998) or 4.8 g N_2O–N ha^{-1} d^{-1} in Japan (Tsuruta et al., 1997).

While N_2O emissions are generally low during flooding periods, substantial amounts could be emitted during fallow period (Bronson et al., 1997; Abao et al., 2000) and during alternate flooding–drying cycle as in porous soils (Aulakh et al., 2001b). In Northern China, less than 3% of the total N_2O emission from rice fields is originating during flooding (Chen et al., 1997). Available data for N_2O emissions from wheat crop under the rice–wheat system is scarce. In the Taihu region of China, seasonal N_2O emissions from wheat crop were 9.79 and 13.74 kg N_2O–N ha^{-1} (Zheng et al., 2000). In NW India, N_2O emission rates during post-rice fallow and wheat crop were 20–43 and 5–33 g N_2O–N ha^{-1} d^{-1} resulting in seasonal flux of 2.6–3.4 kg N_2O–N ha^{-1} (Aulakh et al., 2001b). Field experiments in the Philippines which is outside the typical wheat production region resulted in seasonal emission of 0.64 and 0.61 kg N_2O–N ha^{-1} for two consecutive dry seasons under urea fertilization (Abao et al., 2000). These values are in the same order of magnitude as N_2O emissions during intercrop fallows as well as simultaneous emissions from unplanted/weed free plots (Abao et al., 2000).

In comparative field experiments, the wheat crop showed higher N_2O emissions than a legume crop (Abao et al., 2000) and had an intermediate rank between potato and corn (Ruser et al., 2000). Aulakh et al. (2001b) recorded 3–4 times lower seasonal N_2O emission flux in wheat fields as compared to rice fields. The cumulative N_2O production from soils during pre-rice fallow + rice crop and post-rice fallow + wheat crop periods ranged from 6.9 to 13.7 and 2.6–3.4 kg N_2O–N ha^{-1}, respectively and accounted for 3.3–6.7% of N applied in rice and 0.3–1.8% of N applied in wheat. In several earlier studies (McKenney et al., 1995; Breitenbeck et al., 1980; Aulakh et al., 1984), 3–4 fold increases in N_2O production rates have been observed from aerobic soils fertilized with ammoniacal N, representing 0.1–1% of the applied NH_4^+–N. Seasonal N losses due to denitrification and N_2O as a percentage of the N applied to irrigated corn in Nebraska, USA varied from 1% to 5% (Qian et al., 1997). The review of literature reveals that while the amount of N_2O–N produced may not be a significant concern from an agronomic efficiency point of view, irrigated rice–wheat cropping systems clearly serve as a significant source of N_2O.

Using the CENTURY-NGAS model, Mummey et al. (1998) showed that – on average – the conversion of agricultural land from conventional tillage to no-till in US agriculture entailed greater N_2O emissions per hectare. However, differences between the two tillage scenarios were strongly regional and suggest that conversion of conventionally tilled soil to no-till may have a greater effect on N_2O emissions in drier regions when higher soil moisture retained by zero-till and conservation-managed fields help support higher microbial activity (Aulakh et al., 1984; Doran et al., 1997).

In addition to management, N_2O emissions are also affected by various natural factors. Field experiments in UK clearly demonstrated overriding influence of soil moisture content upon N_2O emission rates (Yamulki et al., 1995). In turn, irrigation and/or rainfall patterns play pivotal roles in regulating N_2O emissions whereas temperature effects appear to have lower impact (Kamp et al., 1998).

3.3. GLOBAL WARMING POTENTIAL

The concept of GWP was developed to compare the ability of a gas to trap heat in the atmosphere relative to CO_2. Based on a 100-year time frame, the GWP of CH_4 is 21 times higher whereas the GWP of N_2O is 310 times higher than the reference value for CO_2 (IPCC, 1996). Thus, the GHG emissions from the rice–wheat system can be compiled in a single scale to assess the relative contribution of N_2O and CH_4, respectively. A field experiment in the Philippines (Abao et al., 2000) showed a dominance of N_2O emissions in a rainfed rice–wheat system (Figure 3a); N_2O emissions exceeded the CH_4 emissions (in terms of GWP) by a factor of 5. However, this experiment encompassed a distinct dry period during the rice crop and omitted organic inputs, i.e. resulting in a rice system with a low-emission potential for CH_4 (Wassmann et al., 2000b). Input of organic material will shift the balance towards CH_4 emissions (Figure 3b). Moreover, rice–wheat systems are typically irrigated and CH_4 emissions are on a relatively high background level in most rice–wheat systems except in highly percolating coarse-textured porous soils (see above). In a field experiment with irrigated rice (Bronson et al., 1997a), seasonal emissions were clearly dominated by CH_4 emissions while N_2O emissions became insignificant in terms of GWP (Figure 4). Organic amendments generally increase CH_4 emissions, but – at the same time – can also reduce N_2O emissions from rice fields. Tölg (1998) observed a 50% reduction of N_2O emissions in a greenhouse experiment after application of rice straw, which in part compensated for the increment in GWP (see experiment A in Figure 5). Recently, Aulakh et al. (2001b) demonstrated

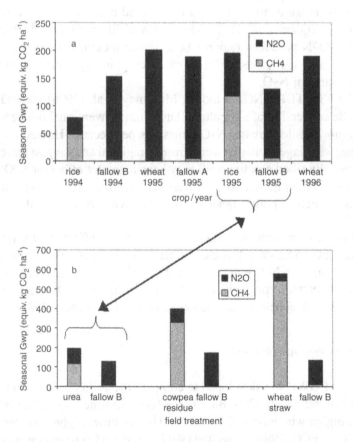

Figure 3. Emissions (in terms of GWP) from (a) rice–wheat rotation with urea fertilization from 1994 to 1996 and (b) different fertilizer treatments of the 1995 rice crop; data from a field experiment at IRRI by Abao et al. (2000).

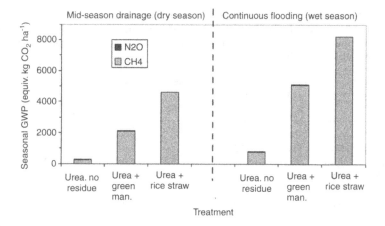

Figure 4. Emissions (in terms of GWP) from irrigated rice crop under different treatments; data from a field experiment at IRRI by Bronson et al. (1997a).

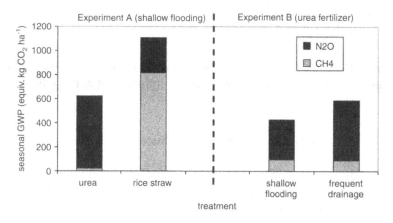

Figure 5. Emissions (in terms of GWP) from irrigated rice in two greenhouse experiments addressing impact of rice straw amendment and water management, respectively; data from Tölg (1998).

that comparable amounts of N_2O were emitted with and without crop residues and legume green manure in rice crop (Table I) suggesting that crop residue management in rice fields would not significantly alter N_2O loading of the atmosphere.

The available emission data clearly indicate that the baseline emissions of the rice–wheat system varies in a broad range, but that even low-emitting systems have an annual GWP >100 equiv. kg CO_2 ha^{-1}. Since changes in the SOM content of the soil are generally very slow, it seems unlikely that emissions of CH_4 and N_2O will be matched by CO_2 net-fluxes as long there are no drastic changes in the land use practice. The available field data also reveal that baseline emissions of the rice–wheat system primarily depend on the management practice during the rice crop while emissions from the wheat crop remained less sensitive to different

TABLE I. Effect of integrated use of fertilizer urea N (FN at 120 and 32 kg N ha^{-1}, respectively), green manure (GM at 20 t ha^{-1})[a], wheat residues (WR at 6 t ha^{-1}), and rice residues (RR at 6 t ha^{-1}) on yields, nitrate leaching, soil bulk density, C sequestration, denitrification losses, and N$_2$O emissions in rice–wheat system. (After Aulakh et al., 2000c, 2001a.)

Treatment		Rice yield[c] (t ha^{-1})	Wheat yield[c] (t ha^{-1})	Nitrate leach.[d] (kg N ha^{-1})	Soil org. C[c] (g kg^{-1})	Denitrif. losses[e] (kg N ha^{-1})	N$_2$O emiss.[e] (kg N ha^{-1})
Rice	Wheat						
Control		3.40	4.73	59	3.74	18	6.9
FN$_{120}$	FN$_{120}$	5.62	4.77	94	3.71	58	12.4
GM$_{20}$ + FN$_{32}$	FN$_{120}$	5.85	5.01	76	4.05	50	11.8
WR$_6$ + GM$_{20}$ + FN$_{32}$	FN$_{120}$	5.92	5.01	—	4.92	52	11.8
FN$_{120}$	RR$_6^b$ + FN$_{120}$	5.63	5.00	—	4.33		
LSD (0.05)		0.24	0.21	12	0.22	6	3.4

[a] Amount of 120 kg N ha^{-1} applied through 20 t GM and the balance through fertilizer N.
[b] Wheat residues (WR) and rice residues (RR) applied at 6 t ha^{-1}.
[c] 3-year (Year 2–4 of the experiment) mean yields.
[d] Measured at the end of 4-year experiment.
[e] Cumulative for pre-rice fallow and rice-growing period.

management practices. During the fallow period and the wheat crop natural factors, namely rainfall patterns, generally override the influence of management.

4. Technical options to reduce emissions

Options to reduce GHG emissions have to be seen against the specific conditions of the rice and wheat cultivation as well as the distinct mechanisms involved in the emissions of the different GHGs, namely CH$_4$, N$_2$O, and CO$_2$. Table II compiles five mitigation strategies representing a more generic level of crop management; each strategy is broken up into two to three specific measures of crop management.

4.1. Managing organic inputs (Strategy #1)

Methane emissions are generally enhanced by organic inputs into the soil such as straw or manure amendment. Changes in the management of organic inputs have been identified as the main driver for long-term changes in the CH$_4$ source strength of rice cultivation (Denier van de Gon, 2000). The increment in CH$_4$ emissions following organic inputs depends on quantity, quality and timing of the application (Yagi and Minami, 1990; Sass et al., 1991). Moreover, the mode of water management and temperature regime may reduce or amplify the incremental effect of organic inputs. Rice straw and manure are typically applied before transplanting resulting in an emission peak during the first half of the growing season. High temperatures in the weeks following the incorporation of these materials result in a

TABLE II. Possible mitigation strategies/measures in the rice–wheat rotation; the last two columns denote the targeted GHG and crop (rice or wheat).

Generic strategy	Measure for rice–wheat system	GHG	Crop
(1) Managing organic inputs	Fermentation of manure	CH_4	Rice
	Adjusting straw incorporation	CH_4	Rice; Wheat
(2) Improving nitrogen fertilization	Matching N supply with demand	N_2O	Rice; Wheat
	Selecting fertilizer type/amendment	CH_4; N_2O	Rice; Wheat
(3) Modifying irrigation patterns	Mid-season drainage	CH_4	Rice
	Alternate flooding/drying	CH_4	Rice
	Direct seeding	CH_4	Rice
(4) Improving crop cultivars	Breeding for specific traits	CH_4	Rice
	'Aerobic' rice varieties	CH_4	Rice
	Increased yield potential	CH_4; N_2O	Rice; Wheat
(5) Increasing soil organic C	Maintaining wetland conditions	CO_2	Rice
	Recycling of residues	CO_2	Wheat
	Reduced tillage	CO_2	Wheat

pronounced emission peak whereas low-temperatures during this period diminish this peak (Wassmann et al., 2000a).

Incorporation of organic material also creates a pool of readily available N and therefore often stimulates N_2O emissions (e.g. Flessa and Beese, 1995; Lemke et al., 1999; Rolston et al., 1982). On the other hand, the observed increments in N_2O emissions were not as pronounced as for CH_4 emissions. In fact, there is some evidence that N_2O emissions from rice fields could even be reduced by high straw amendments (Abao et al., 2000; Zheng et al., 2000; Aulakh et al., 2001b), which may be explained by N immobilization (Aulakh et al., 2000b, 2001a). In their field study with rice–wheat system, Aulakh et al. (2001b) observed a slight, but insignificant reduction of N_2O emission through organic amendments as compared to mere urea application (Table I).

4.1.1. Fermentation of manure

Several field studies have compared different types of organic amendments in regard to CH_4 emissions. While the differences between fresh materials, either straw, animal manure or green manure, have been relatively small, field records showed a big disparity between emissions triggered by fresh and pre-fermented material. During the fermentation process, the pool of organic matter is rapidly depleted, so that incorporation of fermented material into the soil entails a lower-emission potential.

Applying residues from a biogas generator could reduce emissions by approximately 60% as compared to fresh organic amendments and 52% as compared to the locally practiced combination of urea and organic amendments (Wassmann et al., 1993b). The combustion of biogas will also save fossil fuel consumption, so that this mitigation option could be considered a win–win solution. Biogas generators are prevalent in China, especially in areas where swine husbandry is common.

However, practical problems in operating this device, especially during the cold winter months, pose considerable constraints on its propagation as long as other sources of energy are available (Wassmann et al., 1993b).

Composting of organic inputs may offer another agricultural practice that could reduce GHG emissions (Yagi and Minami, 1990; Corton et al., 2000). Moreover, compost application results in very low N_2O emissions during the rice crop that are app. 50% lower than emissions from rice fields treated with urea (Zheng et al., 2000). However, composting is commonly done in piles dominated by anaerobic fermentation. Methane emissions during anaerobic composting process will probably counterbalance gains observed after the incorporation into the soil. These emissions during composting could possibly be reduced to a great extent through aerobic composting techniques. Organic amendments derived from aerobic composting of rice straw significantly reduced emissions as compared to fresh straw (Corton et al., 2000).

4.1.2. Adjusting straw incorporation

Crop production inevitably results in large amounts of straw residues that are typically left in the fields. Since the application of organic manure is gradually decreasing, rice soils largely rely on recycling of straw to compensate for C losses through soil cultivation and crop harvest (Verma and Bhagat, 1992; Witt et al., 1998a,b). However, straw is often burnt to prepare the field for the next cropping cycle; especially the wheat straw is burnt almost all over the Asian rice–wheat belt (Ladha et al., 2000). Removal or burning of residues insures farmers quick seedbed preparation and avoids the risk of N immobilization during decomposition of residues with wide C : N ratio (Beri et al., 1995). Incomplete combustion of carbon – which is generic to smoldering fires of harvest residues – generate substantial amounts of carbon monoxide (CO) and other pollutants and thus, have adverse effects on local air quality. Fire-borne organic compounds and nitrogen oxides lead to tropospheric ozone formation; high ozone concentrations coincide with the peak of the residue-burning season in Asia. Moreover, the burning process also releases CH_4 and N_2O into the atmosphere. Using the IPCC (1995) default values for residue to product ratio (= 1.4), C content (= 41%), C/N ratio (= 71), specific emission factors (= 0.005 for CH_4; = 0.007 for N_2O), and GWP (= 21 for CH_4, = 310 for N_2O), the emission during the burning of 4 t straw (low productivity) and 10 t straw (optimum productivity) would roughly correspond to 20 and 50 kg CH_4 ha^{-1}, respectively.

Methane emission rates become very sensitive to the mode of straw management as long as the level of C input into the soil is low. If straw is not burnt, the conventional method of straw amendment consists of ploughing the straw into the soil before transplanting. Some farming systems within the rice–wheat belt encompass a long-winter fallow, e.g. before the early rice crop in China, and therefore, allow alternative timings of straw incorporation. In a field experiment in Hangzhou, Zhejiang Province, the conventional mode of straw incorporation during spring was tested against an early straw incorporation at the start of the winter fallow (Lu et al., 2000).

This early incorporation had no effect on emissions in the first part of the early season when temperatures are low and emissions are also on a low level. However, winter fallow incorporation resulted in lower-emissions during the latter stage of this season so that seasonal emissions were reduced by 11%. Similarly, incorporation of crop residues during pre-rice 60-day fallow period did not increase N_2O emissions and also did not decrease rice yields.

Sustainable agriculture relies on proper residue management to reduce adverse impacts on the environment – at local as well as global scale. Alternative approaches of plant residue management have to be incorporated in a comprehensive decision support system that reconciles socio-economic and environmental concerns. More research is needed to develop these strategies to site-specific packages of technologies that are attractive enough for implementation.

4.2. IMPROVING NITROGEN FERTILIZATION (STRATEGY #2)

Even under the best possible fertilization practice, substantial amounts of the N applied to the field are emitted to the atmosphere. While the uptake efficiency of grain crops can range from 20% to 60% of the applied N, low-efficiency reflects improper crop management. In irrigated rice, gaseous losses of N may account for up to 48% of the N applied (Reddy and Patrick, 1980). The principal mechanisms responsible for gaseous N losses are (a) ammonia volatilization, (b) denitrification (leading to emission of N_2, nitrogen monoxide (NO), and N_2O), and (c) nitrification (leading to emission of NO and N_2O). The specific significance of these processes may vary depending on crop management as well as natural factors (Freney, 1997; Mosier et al., this issue).

Nitrification was identified in some studies as the major source of N_2O in soils (Bremner and Blackmer, 1978; Aulakh et al., 1984; Schuster and Conrad, 1993). Nitrification rates are generally high in well-aerated soils as long as the supply of ammonia (e.g. through mineralization of organic material) is ensured. However, the 'leakage rate' of N_2O during the process of nitrification is relatively low; the ratio between N_2O emission and NO_3^--formation is less than 1% (Aulakh et al., 1984; Firestone and Davidson, 1989). Actual N losses through NH_3 volatilization can vary broadly from less than 1% (Roelcke et al., 2002) to up to 32% in some calcareous soils (Zhang et al., 1992). Denitrification losses accounted for 3.1–9.4% of the fertilizer applied to wheat fields (Bronson and Fillery, 1998); the main product of denitrification in flooded rice in an Australian field experiment was N_2 that exceeded the formation N_2O by a factor of 4. Very high N losses through denitrification were observed during rice in NW India amounting to 33% of the prescribed dose of $120\,kg\,N\,ha^{-1}$ (Aulakh et al., 2001b). However, the gaseous N losses under wheat in this experiment were only 0.6–2% of the applied N and were 8–10 fold lower than those preceding rice.

With the exception of N_2O, the gaseous N compounds have no or only an indirect GWP. NO, a chemically very reactive gas, is involved in tropospheric ozone formation, which, in turn, is a very potent GHG. Atmospheric deposition of NH_3 and NO on N-limited ecosystems stimulate N turnover and subsequently, may increase N_2O emissions. Likewise, the assessment of fertilizer-borne N_2O emissions should also include off-site emissions triggered by leached nitrate transported to N-limited ecosystems as well as CO_2 emissions during fertilizer production.

Strategies to mitigate fertilizer-induced emissions of N_2O could aim at (a) reducing the proportion of N_2O produced during nitrification/denitrification or (b) limiting the intensity of microbial nitrogen turnover in the soil. The proportional N_2O emission during nitrification and denitrification primarily depends on the form of nitrogen and oxygen availability in the soil (Weier et al., 1993). Managing a rice field to maintain the reduction intensity of the flooded rice soil at an optimum range for lowest possible GHG appears difficult under actual field conditions. Nitrification inhibitors may offer some potential to reduce N_2O emissions. Emissions from urea treated rice field were significantly be reduced when dicyandiamide (DCD) was added (Majumdar et al., 2000). However farmers do not use commercial nitrification inhibitors to regulate the nitrification of fertilizer N because of the high costs involved. While comparing encapsulated calcium carbide (ECC), a commonly available and very economical material with commercial nitrification inhibitors, Aulakh et al. (2000c) observed that nitrification inhibition of applied fertilizer N in both arable crops and flooded rice systems could tremendously minimize N losses and help enhance fertilizer N-use efficiency. Their results suggested that for reducing the nitrification rate and resultant N losses in flooded soil systems, ECC is more effective than costly commercial nitrification inhibitors. Neem tree residues are used as nitrification inhibitors in parts of India (Prasad and Power, 1995). However, it remains to be seen if the will ever become a popular practice farmers beyond such regional domains.

4.2.1. Matching N supply with demand

A generic strategy to reduce N losses – and thus, to reduce N_2O emissions – is avoiding excess N in space and time. Several field experiments recorded an instant increase in N_2O emissions triggered by N fertilizer application in rice and wheat fields (Eichner, 1991; Bouwman, 1995). In contrast to the IPCC methodology that defines a default value of 1.25% (of the added N) emitted into the atmosphere, the actual fertilizer losses in the form of N_2O vary in a broad range from 0.001% to 6.8% (Freney, 1997; Aulakh et al., 2001b) have given some scope to reduce overall reduction by targeting hot spots of N_2O emissions.

Nitrogen use efficiency can significantly be enhanced through splitting of N application. Ideally, the time-dependent N demand of the crop is determined through leaf color via color chart or photometer. The consequent use of site-specific nutrient management in irrigated rice can increase yields by about 10% as compared to conventional practice as shown in eight key irrigated rice domains of Asia (Dobermann et al., 2002). N_2O emissions have not been recorded in these field experiments, but gains in emissions can be inferred from other experiments in subtropical regions.

Optimized timing of N application has reduced N_2O emissions in wheat cultivation in Mexico from 5 to 0.5 kg N_2O ha^{-1} (Matson et al., 1998). In this region, however, N application rates were very high (180–250 kg N ha^{-1}). On irrigated land in most Asian countries, farmers typically apply 100–150 kg N ha^{-1} to the dry season crop and 60–90 kg N ha^{-1} to the wet season crop. At these input levels the cost of N fertilizer roughly corresponds to 10–20% of the total costs of production.

4.2.2. Selecting fertilizer/amendment

Comparing urea and ammonium sulphate; Cai et al. (1997) showed that the latter released more N_2O than urea. Tablet urea placed into the soil entailed higher N_2O flux rates than broadcasting of granule urea (Suratno et al., 1998). High sulfate levels inhibit CH_4 formation in anaerobic systems due to the substrate competition between sulfate reducing and methanogenic bacteria (Winfrey and Zeikus, 1977). Hence, the use of sulfate-containing fertilizers as well as additional sulfate inputs has been found to reduce CH_4 emission from rice fields. Under field conditions, for example, Schütz et al. (1989) observed that CH_4 fluxes from $(NH_4)_2SO_4$ treated plots were lower as compared to urea fertilized control plots in an Italian rice field over a 3-year period. Bronson et al. (1997a) reported that seasonal CH_4 flux with $(NH_4)_2SO_4$ was one third to one fourth of the flux with urea. Cai et al. (1997) observed 42% and 60% decrease in average CH_4 emission from rice fields treated with 100 and 300 kg N ha^{-1} as $(NH_4)_2SO_4$. In other locations, however, the impact of sulfate fertilizers on CH_4 emission was marginal (Wassmann et al., 1993a) or even showed a stimulating effect (Cicerone and Shetter, 1981).

Different cost levels for sulfate-containing fertilizers affect cost-benefit calculations of this mitigation option. Provided the proper target areas are selected, the costs for mitigating CH_4 emissions through sulfate-containing fertilizer are estimated at 5–10 US dollar per ton CO_2-equivalent (Denier van der Gon et al., 2001).

Gypsum ($CaSO_4$) inputs into the soil reduced CH_4 emissions by 29–46% in field study in Louisiana (Lindau et al., 1993) and by 55–70% in the Philippines (Denier van der Gon and Neue, 1994). Gypsum application can be beneficial to neutralize the pH of alkaline soils. In the rice–wheat belt, gypsum application may offer some potential to combine the purposes of improving soil fertility and mitigating emissions. Several million hectares of the Indo-Gangetic plains are alkaline soils (Agarwal et al., 1976; Abrol et al., 1985). In China, acidity of rainwater is increasing as a consequence of higher atmospheric NO_x and SO_2 concentrations demanding for more liming in soils with low-buffer capacity.

4.3. MODIFYING IRRIGATION PATTERNS (STRATEGY #3)

Flooding of rice fields is the pre-requisite for high CH_4 emissions – the consequent alternative to reduce CH_4 emissions is therefore to grow rice under upland conditions. Upland rice comprises app. 12% of the world rice area, but yields in this ecosystem are generally much lower than in flooded paddies accounting for only 4% of the global rice production. Recent success in developing high-yielding 'aerobic rice' will be discussed in the section dealing with cultivars. Under prevailing wetland conditions in rice fields, farmers have options to reduce CH_4 emissions through distinct drainage periods in mid-season or alternate wetting and drying of the soil (Sass et al., 1992; Wassmann et al., 2000c).

However, the reductions achieved through alternative irrigation practices varies in a very broad range from 7% to 80% and imply a number of constraints due to an inverse effect on N_2O emissions. Changes in the soil moisture regime stimulate nitrification (through soil drying) and denitrification (through soil wetting) and thus, enhance emissions of N_2O (Bronson et al., 1997a,b; Zheng et al., 1997; Abao et al, 2000). Aulakh et al. (2000a) observed that the nearly-saturated soils (that are commonly encountered in porous soils under rice), being partially aerobic, support greater nitrification of applied ammoniacal fertilizer N than continuously flooded soil. In the next step, these soils showed relatively high rates of denitrification and N_2O production. Regional examples show that drainage periods are not necessarily detrimental – or can even be beneficial – to rice yields as long as timing and duration of dry intervals are prudently chosen. In any case, the drainage patterns have to be tailored to site-specific conditions to avoid water stress for the plants and water losses. These alternative practices may also save water as compared to continuous flooding, although this advantage is not applicable to soils susceptible to compaction (see below).

4.3.1. Mid-season drainage

This irrigation practice encompasses a distinct period of app. one week when irrigation is interrupted. CH_4 emission rates show a short-term peak at the beginning of soil aeration due to the release of soil-entrapped CH_4 that is followed by persistently low emissions even when the fields are reflooded. A properly timed mid-season drainage appears as a promising option to achieve net-gains in GHG emissions when the baseline of CH_4 emissions is very high (Wassmann et al., 2000c). In low CH_4-emitting rice systems, however, the net-effect of modifying water regimes may in fact become negative in terms of GWP of the gases emitted (Bronson et al., 1997b; Tölg, 1998).

Emissions are very sensitive to duration and timing of the drainage period, so that this management practice may further be improved to reduce emissions, especially in areas where the general concept of mid-season drainage is already known as a successful agronomic practice by some farmers. Mid-season drainage may increase water losses on soils that tend to form cracks when aerated. Multiple aeration during a rice cropping season may require more water. Sass et al. (1992) found that multiple aeration required 2.7 times more water than the normal floodwater treatment in the southern US. Mid-season drainage may not be feasible during the wet season and in rice fields flooded by high groundwater.

4.3.2. Alternate flooding/drying

This water regime refers to alternate flooding and aeration (drying) of the soil throughout the vegetation period. As mentioned above, irrigation practices in Northern India result in frequent flooding/drying cycles over the rice cropping season. CH_4 emissions are generally very low, but N_2O emissions from this system vary in a broad range. While the mean seasonal emission of app. 250 g N_2O ha^{-1} recorded by Kumar et al. (2000) in New Delhi is within the lower range of field records from other rice fields (see 2.1), $6.9–12.4$ kg N_2O–N ha^{-1} reported by Aulakh et al. (2001b) at Ludhiana (Punjab) are quite high. In contrast to mid-season drainage, the time intervals between wet and dry conditions appear to be too short to facilitate the shift from aerobic to anaerobic soil conditions. Frequent alternation of flooding and drying seems not to be a mitigation option because of low water use efficiency. In terms of GWP, its mitigation potential is low as long as the background level of emissions is low, which was shown in a greenhouse experiment (see experiment B in Figure 5) by Tölg (1998).

4.3.3. Direct seeding

In a field experiment in Central Philippines, direct seeding reduced CH_4 emission by app. 18% as compared to transplanting (Corton et al., 2000). The mitigation effect could be enhanced up to 50% when direct seeding was combined with mid-season drainage. On the other hand, direct seeding can increase CH_4 emissions when baseline emissions of CH_4 are very low (Wassmann et al., 2000c). The impact mechanisms of direct seeding on the CH_4 budget are not clear. Direct seeded rice has a higher root biomass (De Datta and Nantasomsaran, 1991) and thus, may introduce more organic material into the soil. The additional substrate for methanogenic bacteria is more significant in a soil environment with low-organic inputs. Under high baseline emissions, the distinct root structure of direct-seeded rice – showing a concentration of roots in the upper soil layer – may become the decisive trait determining emissions.

4.4. IMPROVING CROP CULTIVARS (STRATEGY #4)

The rice and wheat plants affect the C and N budget of this ecosystem in various pathways. Plants act as a source for the carbon budget (through photosynthesis) and represent a major source of methanogenic material through decaying plant parts and root exudation (Aulakh et al., 2001c,d). In contrast, rice and wheat plants act as a nitrogen sink in the ecosystem given the removal of biomass (grain and often straw) after harvest. However, plant material is also one of the key drivers of N cycling in the soil by providing substrate for ammonification. The ammonium produced from plant material is an important source for nitrification and thus, feeds into the denitrification chain as well.

The rice plant acts as a conduit for gases, both CH_4 and N_2O (Yu et al.,1997), *en route* from the soil to the atmosphere. This feature of the rice plant is attributed to the development of extensive aerenchyma (Aulakh et al., 2000d), which is typical for plants adapted to waterlogging. Although some wheat genotypes are also capable of developing aerenchyma (Gibberd et al., 1999), this trait will not be of any significance in well-aerated soils.

4.4.1. Breeding for specific traits

The database on rice cultivars affecting CH_4 emissions is still inconsistent. The traits that determine the CH_4 emission potential of rice cultivars are (a) root exudation (Aulakh et al., 2001c), (b) gas transfer through the aerenchyma (Butterbach-Bahl et al., 1997), and (c) duration until maturity (Setyanto et al., 2000). However, root exudation and gas transfer capacity of a given cultivar exhibit an enormous variation when grown under different greenhouse and field conditions (Wassmann and Aulakh, 2000). These variations complicate a determination of cultivar-specific emission potentials as shown by Wassmann et al. (2002) in a long-term field experiment. Therefore, maturity duration remains at present the only reliable criteria to assess cultivar-specific emission potentials. Setyanto et al. (2000) compared the emission potentials of early maturing (100 d after transplanting) and late maturing cultivars (130 d); the differences in emissions (app. 30%) basically reflected the differences in flooding periods required for the crops.

4.4.2. 'Aerobic' rice varieties

At present, upland rice cultivation is predominantly found in marginal areas where soils are never flooded for a significant time because of topography and or lack of water. The commonly used upland varieties do not achieve high yields, even when grown under more favorable upland conditions. The endeavor to develop high-yielding rice cultivars growing under upland conditions is mainly driven by the recognition that water will become a scarce commodity in future (Bouman, 2001). As of now, there are no emission records available for aerobic rice. The conversion from wetland rice to 'aerobic rice' will certainly reduce CH_4 emissions whereas the impact on N_2O emissions is unclear. It remains to be seen how successful this technique will spread in farmers' fields. 'Aerobic' rice may come at a cost unless the problems of weeds infestation and yield decline (observed at initial tests with repeatedly growing aerobic rice) can be solved through future research.

4.4.3. Increasing yield potential

Since higher yields are imperative to meet future food demand, the 'freezing' of GHG emissions at the present level could still be regarded beneficial as long as yields increase. A comparative field study on CH_4 emission included the new plant type of rice that will be disseminated to farmers' fields within the coming years (Wassmann et al., 2002). CH_4 emissions from the new plant type (characterized by fewer tillers to increase harvest index) were within the range of emissions from conventional 'high yielding varieties' as well as hybrids.

In fact, there may even be synergies between increasing yield and mitigating CH_4 emissions. Denier van der Gon et al. (2002) observed concurrence of high emission and high yields in the dry season followed by low-emissions and low-yields in the wet season of the Philippines. Unfavorable weather during the wet season resulted in inadequate allocation of assimilates for grain filling and, in turn, increased rhizodeposition (Denier van der Gon et al., 2002). However, the new rice and wheat lines have not been tested yet in terms of N_2O emissions. High yields ultimately require high N fertilization, so that proper fertilizer application is needed to 'freeze' N_2O emissions at present level.

4.5. INCREASING SOIL ORGANIC CARBON (STRATEGY #5)

While the observed increase in atmospheric CO_2 concentrations is primarily caused by fossil fuel combustion, the known source and sink strengths of the global C budget are not in balance. As recent results have shown, the 'missing sink' of $1.8 \, \text{Gt C yr}^{-1}$ may in part be attributed to C sequestration in terrestrial ecosystems (IPCC, 1996). Due to the intricate problems in curtailing emissions, recent research focused on the question if the sink strength of terrestrial ecosystems can be increased through proper management.

4.5.1. Maintaining wetland conditions

Agricultural intensification almost universally causes a decline of organic matter in tropical soils (Greenland et al., 1992). The exception from this rule is wetland rice cultivation which generally shows a stable and – in comparison to intensive upland cultivation – a relatively high level of SOM (Craswell and Lefroy, 2001). This can be illustrated in a long-term experiment at IRRI comparing rice–rice to rice–maize system (Figure 6). Under appropriate fertilization, SOM content increases in both, rice–rice and rice–maize rotation, as compared to the zero-N plots (Witt et al., 2000). The effect of a second rice crop, however, is even more pronounced; rice–rice rotation showed considerably higher SOM content as rice–maize rotation.

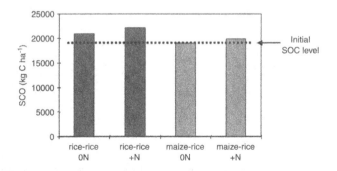

Figure 6. Changes in SOC content in a field experiment at IRRI over 5 consecutive seasons encompassing rice–rice and rice–maize rotations without (0N) and with (+N) N fertilizer; data from Witt et al. (2000).

Assuming the original SOM content as baseline, the C sequestration in this experiment corresponded to app. $3 \, t \, C \, ha^{-1}$. Since this increment was accomplished over a period of 5 seasons, however, the gains only partially compensate for the CH_4 emissions during the rice crop. Thus, maintaining wetland cropping can only be seen as mitigation option as long as this management practice is accompanied by measures to keep trace gas fluxes at acceptable levels. In the Asian lowlands, rice was originally grown on soils that were naturally flooded and had a native emission potential even before conversion – a fact that is rarely accounted for in estimating source strengths. However, increased production, the bunding of rice fields with concomitant water harvesting and artificial flooding constitute an anthropogenic CH_4 source only.

4.5.2. Recycling of residues

Constant retrieval of the autochthonous biomass ultimately results in a depletion of the organic pool as long as the nutrients or the biomass are not replenished by allochthonous inputs. A long-term experiment at four locations in India encompassed different crop rotations with three field trials, i.e. without fertilizer, with mineral fertilizer and with additional farmyard manure (Nambiar, 1994). Again the results corroborate the stimulation of SOM content by recurring rice cropping (Figure 7). Only double rice cropping resulted in enhanced SOM contents whereas rice-upland rotation generally depleted the SOM pool as compared to the original value. In all field stations the different field trials showed a clear ranking in SOM, i.e. zero-N (in average -52% of original SOM content) <mineral NPK $(+26\%)$ <farm yard manure $(+74\%)$. While farm yard manure generally increases CH_4 emission in the rice crop, the application of organic residues in the upland crop corresponds a net-mitigation of emissions due to the C sequestered in the soil.

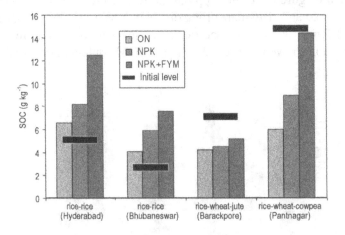

Figure 7. SOC content in long-term experiments at four locations in India encompassing different crop rotations without fertilizer (0N), with mineral fertilizer (N,P,K) and with additional farmyard manure (NPK + FYM); data from Nambiar (1994).

One recent report from a 4-year field study on rice–wheat system in NW India illustrated that soil organic carbon (SOC) concentrations remained unchanged from the start of the experiment in plots that did not receive residue or green manure (Table II). *Sesbania* legume green manuring enhanced SOC concentrations when no wheat residue was added. But the effect of wheat residue on SOC was much larger than that of green manure and masked effects of the latter. Thus incorporating green manure and crop residue in addition to urea-N applications in a rice–wheat system can result in high grain yields and have the long-term benefit of C sequestration ranging from 10 to 20% of added C (Aulakh et al., 2001a).

4.5.3. Reduced tillage
Tillage is an important determinant for oxygen fluxes into the upper soil layer and thus, for aerobic decay of SOM. Matson et al. (1997) simulated the historical SOM dynamics in the US corn belt and computed a depletion to 53% of the original value following the conversion from native vegetation to agriculture. While this depletion occurred in the first half of the 20th century, the current practice of zero-tillage has enhanced the SOC level in the US corn belt to 61% of the original value (Matson et al., 1997). Since the zero-tillage is also becoming popular in the Asian rice–wheat belt (Hobbs, 2001) this development implies mitigation potential especially in SOM depleted soil.

5. Conclusion

Mitigation options in the rice–wheat system may individually be of limited scope, but they may achieve a discernable composite effect when implemented in coordinated fashion. The 'entry point' for mitigating emissions should be national or international programs to curtail GHG emissions from the agricultural sector. Funding through 'Clean Development Mechanism' projects (introduced in Kyoto Protocol) can – at best – provide incremental incentives, but not the drivers of such projects. Mitigation programs will rely on win–win opportunities when emissions can be reduced with another concomitant benefit such as higher yields, less fertilizer, and water needs etc. Spatial heterogeneity of the rice–wheat system entails hot spots of emissions favoring mitigation programs with a regional focus. Targeting one individual gas alone seems inappropriate due to antagonistic effects in the emissions of CH_4, N_2O, and CO_2. More research is needed to combine geographic information, emission models, yield models and socio-economic information to devise site-specific packages of mitigation technologies.

References

Abao, E.B. Jr., Bronson, K.F., Wassmann, R. and Singh, U.: 2000, 'Simultaneous records of methane and nitrous oxide emissions in rice-based cropping systems under rain fed conditions', *Nutrient Cycling in Agroecosystems* **58**, 131–139.

Abrol, I. P., Bhumbla, D. R. and Meelu, O. P.: 1985, 'Influence of salinity and alkalinity on properties and management of ricelands', in *Soil Physics and Rice*, Los Banos, Philippines, International Rice Research Institute, pp. 183–198.

Agarwal, R.R., Yadav, J.S.P. and Gupta, R.N.: 1976, *'Saline and Alkali Soils of India'*, New Delhi, Indian Centre for Agricultural Research.

Ali M. and Byerlee D.: 2000, 'Productivity growth and resource degradation in Pakistan's Punjab', *Policy Research Working Paper* 2480. World Bank, Washington DC, 30 pp.

Aulakh, M.S. and Bijay-Singh.: 1997, 'Nitrogen losses and fertilizer N use efficiency in irrigated porous soils', *Nutrient Cycling in Agroecosystems* **47**, 197–212.

Aulakh, M.S., Rennie, D.A. and Paul, E.A.: 1984, 'Acetylene and N-serve effects upon N_2O production from NH_4^+ and NO_3-treated soils under aerobic and anaerobic conditions', *Soil Biology Biochemistry* **16**, 351–356.

Aulakh, M.S., Khera, T.S. and Doran, J.W.: 2000a, 'Mineralization and denitrification in upland, nearly-saturated and flooded subtropical soil. I. Effect of nitrate and ammoniacal nitrogen', *Biology and Fertility of Soils* **31**, 162–167.

Aulakh, M.S., Khera, T.S. and Doran, J.W.: 2000b, 'Mineralization and denitrification in upland, nearly-saturated and flooded subtropical soil. II. Effect of organic manures varying in N content and C : N ratio', *Biology and Feritlity of Soils* **31**, 168–174.

Aulakh, M.S., Khera T.S., Doran J.W., Kuldip-Singh and Bijay-Singh.: 2000c, 'Yields and nitrogen dynamics in a rice–wheat system using green manure and inorganic fertilizer', *Soil Science Society of America Journal* **64**, 1867–1876.

Aulakh, M.S., Wassmann, R., Rennenberg, H. and Fink, S.:2000d, 'Pattern and amount of aerenchyma relate to variable methane transport capacity of different rice cultivars', *Plant Biology* **2**, 182–194.

Aulakh, M.S., Wassmann, R. and Rennenberg, H.: 2000e, 'Methane emissions from rice fields – Quantification, mechanisms, role of management, and mitigation options', *Advances in Agronomy* **70**, 193–260.

Aulakh, M.S., Khera, T.S., Doran, J.W. and Bronson, K.F.: 2001a, 'Managing crop residue with green manure, urea, and tillage in a rice–wheat rotation', *Soil Sci Soc Am J* **65**, 820–827.

Aulakh M.S., Khera, T.S., Doran, J.W. and Bronson, K.F.: 2001b, 'Denitrification, N_2O and CO_2 fluxes in rice–wheat cropping system as affected by crop residues, fertilizer N and legume green manure', *Biology and Fertility of Soils* **34**, 375–389.

Aulakh, M.S., Wassmann, R., Bueno, C., Kreuzwieser, J. and Rennenberg, H.: 2001c, 'Characterization of root exudates at different growth stages of ten rice (Oryza sativa L.) cultivars', *Plant Biology*. **3**, 139–148.

Aulakh, M.S., Wassmann, R. and Rennenberg, H. 2001d, 'Methane emissions from rice fields-quantification, role of management, and mitigation options', *Advances in Agronomy*. **70**, 193–260.

Beri, V., Sidhu, B.S., Bahl, G.S. and Bhat, A.K.: 1995, 'Nitrogen and phosphorus transformations as affected by crop residue management practices and their influence on crop yields', *Soil Use Management* **11**, 51–54.

Bouman, B.A.M.: 2001, 'Water efficient management strategies in rice production', *International Rice Research Notes* **26**, 17–22.

Bouwman, A.F.: 1995, *'Compilation of a Global Inventory of Emissions of Nitrous oxide'*, Ph.D. thesis, Landbouwuniversiteit Wageningen.

Breitenbeck, G.A., Blackmer, A.M. and Bremner, J.M.:1980, 'Effects of different nitrogen fertilizers on emission of nitrous oxide from soil', *Geophysical Research Letters* **7**, 85–88.

Bremner, J.M. and Blackmer, A.M.: 1978, 'Nitrous oxide: Emissions from soils during nitrification of fertilizer nitrogen', *Science* **199**, 295–296.

Bronson, K.F. and Fillery, I.R.P.: 1998, 'Fate of nitrogen-15-labelled urea applied to wheat on a waterlogged texture-contrast soil', *Nutrient Cycling in Agroecosystems* **51**, 175–183.

Bronson, K.F., Neue, H.U., Singh, U. and Abao, E.B. Jr.: 1997a, 'Automated chamber measurements of methane and nitrous oxide flux in a flooded rice soil: I. Residue, nitrogen and water management', *Soil Science Society of America Journal* **61**, 981–987.

Bronson, K.F., Neue, H.U., Singh, U. and Abao, E.B. Jr.: 1997b, 'Automated chamber measurements of methane and nitrous oxide flux in a flooded rice soil: II. Fallow period emissions', *Soil Science Society of America Journal* **61**, 988–993.

Butterbach-Bahl, K., Papen, H. and Rennenberg, H.: 1997, 'Impact of gas transport through rice cultivars on methane emission from rice paddy fields', *Plant Cell Environment* **20**, 1175–1183.

Byerlee, D., Ali, M. and Siddiq, A.: 2003, 'Sustainability of the rice–wheat system in Pakistan's Punjab: How large is the problem', in J.K. Ladha in J.K. Ladha, J. Hill, R.K. Gupta, J. Duxbury, R.J. Buresh (eds.), *Improving the Productivity and Sustainability of Rice–wheat Systems: Issues and Impact*. ASA Madison, WI, ASA, Spec. Publ. **65**, 77–98.

Cai, Z., Xing, G., Yan, X., Xu, H., Tsuruta, H., Yagi, K. and Minami, K.: 1997, 'Methane and nitrous oxide emissions from rice paddy fields as affected by nitrogen fertilizers and water management', *Plant and Soil* **196**, 7–14.

Chen, G.X., Huang, G.H., Huang, B., Yu, K.W., Wu, J. and Xu, H.: 1997, 'Nitrous oxide and methane emissions from soil–plant systems', *Nutrient Cycling in Agroecosystems* **49**, 41–45.

Cicerone, R.J. and Shetter, J.D.: 1981, 'Sources of atmospheric methane: Measurements in rice paddies and a discussion', *Journal of Geophysical Research* **86**, 7203–7209.

Corton, T.M., Bajita, J., Grospe, F., Pamplona, R., Wassmann, R. and Lantin, R.S.: 2000, 'Methane emission from irrigated and intensively managed rice fields in Central Luzon (Philippines)', *Nutrient Cycling in Agroecosystems* **58**, 37–53.

Craswell, E.T. and Lefroy R.D.B.: 2001, 'The role and function of organic matter in tropical soils', *Nutrient Cycling in Agroecosystems* **61**, 7–18.

De Datta, S.K. and Nantasomsaran, P.: 1991, 'Status and prospects of direct seeded flooded rice in tropical Asia', in *Direct Seeded Flooded Rice in the Tropics*, Los Banos, Philippines, International Rice Research Institute, pp. 1–16.

Debnath, G., Jain, M.C., Kumar, S., Sarkar, K. and Sinha, S.K.: 1996, 'Methane emissions from rice fields amended with biogas slurry and farm yard manure', *Climatic Change* **33**, 97–109.

Denier van der Gon HAC (2000) Changes in CH4 emission from rice fields from 1960 to 1990s: 1. The declining use of organic inputs in rice farming. *Global Biogeochemical Cycles* **13**, 1053–1062.

Denier van der Gon H.A.C. and Neue H.U.: 1994, 'Impact of gypsum application on the methane emission from a wetland rice field', *Global Biogeochemical Cycles* **8**, 127–134.

Denier van der Gon H.A.C., van Bodegom. P.M., Wassmann, R., Lantin, R.S. and Metra-Corton T.: 2001, 'Assessing sulfate-containing soil amendments to reduce methane emissions from rice fields: Mechanisms, effectiveness and costs', *Mitigation and Adaptation Strategies for Global Change* **6**, 69–87.

Denier van der Gon, H.A.C., Kropff, M.J., van Breemen, N., Wassmann, R., Lantin, R.S., Aduna,E.A., Metra-Corton, T. and van Laar, H.H.: 2002, 'Optimising grain yields reduces CH_4 emissions from rice paddy fields', *Proceedings of the National Academy of Sciences USA* **99**, 12021–12024.

Dobermann, A., Witt, C., Dawe, D., Abdulrachman, S., Gines, H.C., Nagarajan, R., Satawathananont, S., Son, T.T., Tan, P.S., Wang, G.H., Chien, N.V., Thoa, V.T.K., Phung, C.V., Stalin, P., Muthukrishnan, P., Ravi, V., Babu, M., Chatuporn, S., Sookthongsa, J., Sun, Q., Fu, R., Simbahan, G.C., Adviento, M.A.A.: 2002, 'Site-specific nutrient management for intensive rice cropping systems in Asia', *Field Crop Research* **74**, 37–66

Doran, J.W., Aulakh, M.S., Thompson, R.L. and Thompson, S.N.: 1997, 'Soil denitrification as influenced by conservation management practices', *TERRA (Journal of Mexican Society of Soil Science)* **15**, 69–78.

Eichner, M.J.: 1991, 'Nitrous oxide emissions from fertilizer soils: Summary of available data', *Journal of Environmental Quality* **19**, 272–280.

Flessa, H. and Beese, F.: 1995, 'Effects of sugarbeet residues on soil redox potential and nitrous oxide emission', *Soil Science Society of America Journal* **59**, 1044–1051.

Firestone, M.K and Davidson, E.A.: 1989, 'Microbiological basis of NO and N_2O production and consumption in soil', in M.O. Andreae and D.S. Schimel (eds.), *Exchange of Trace Gases Between Terrestrial Ecosystems and the Atmosphere*, New York, John Wiley and Sons, pp. 7–21.

Freney, J.R.: 1997, 'Emission of nitrous oxide from soils used for agriculture', *Nutrient Cycling in Agroecosystems* **49**, 1–6.

Gibberd, M.R., Colmer, T.D., Cocks, P.S.: 1999, 'Root porosity and oxygen movement in waterlogging tolerant Trifolium tomentosum and intolerant Trifolium glomeratum', *Plant, Cell and Environment* **22**, 1161–1168.

Greenland, D.J., Wild, A., Adams, D.: 1992, 'Organic matter dynamics in soils of the tropics-from myth to complex reality', in R. Lal and P.A. Sanchez (eds.), *Myths and Science of Soils of the Tropics*, Soil Science Society of America, Inc. and American Society of Agronomy, Inc., pp. 17–33.

Hobbs P.R.: 2001, 'Tillage and Crop Establishment in South Asian Rice–Wheat Systems', *Journal of Crop Production* **4**, 1–22.

Hobbs, P. and Gupta, R.K.: 2003, Resource conserving technologies in wheat for rice–wheat systems. In J.K. Ladha, J. Hill, R.K. Gupta, J. Duxbury and R.J. Buresh (eds.), *Improving the Productivity and Sustainability of Rice–Wheat Systems: Issues and Impact*, ASA, Spec. Publ. Madison, WI **65**, 149–172.

Hua, X., Xing G., Cai Z.C. and Tsuruta, H.: 1997. 'Nitrous oxide emissions from three rice paddy fields in China', *Nutrient Cycling in Agroecosystems* **49**, 23–28.

Huke, R., Huke, E., Woodhead, T. and Huang, J.: 1993, *'Rice–Wheat Atlas of China'*, IRRI, CIMMYT, CNRRI Publication, 37 pp.

IPCC (Intergovernmental Panel on Climate Change): 1995, *'IPCC Guidelines for National Greenhouse Gas Inventories'*, Vol. II (Workbook), Bracknell (UK), Hadley Centre Meteorological Office.

IPCC (Intergovernmental Panel on Climate Change): 1996, *'Climate Change 1995'*, IPCC Third Assessment Report – Working Group I, Cambridge (UK), Cambridge University Press.

IPCC (Intergovernmental Panel on Climate Change): 2001, *'Climate Change 2001: The Scientific Basis'*, IPCC Third Assessment Report – Working Group I, Cambridge (UK), Cambridge University Press.

Jain, M.C., Kumar, K., Wassmann, R., Mitra, S., Singh, S.D., Singh, J.P., Singh, R., Yadav, A,K. and Gupta, S.: 2000, 'Methane emissions from irrigated rice fields in Northern India (New Delhi)', *Nutrient Cycling in Agroecosystems* **58**, 75–83.

Jiaguo, Z.: 2000, 'Rice–wheat cropping system in China', in P.R. Hobbs and R.K. Gupta, (eds.), *Soil and Crop Management Practices for Enhanced Productivity of the Rice–Wheat Cropping System in the Sichuan Province of China*, Rice–Wheat Consortium Paper Series 9, New Delhi, India, pp. 1–10.

Kamp, T., Steindl, H., Hantschel, R.E., Beese, F. and Munch, J.-C.: 1998, 'Nitrous oxide emissions from a fallow and wheat field as affected by increased soil temperatures', *Biology and Fertility of Soils* **27**, 307–314.

Khalil M.A.K., Rasmussen R.A., Shearer M.J., Chen Z.L., Yao H. and Yang J.: 1998, 'Emissions of methane, nitrous oxide, and other trace gases from rice fields in China', *Journal Geophysical Research – Atmosphere* **103**, 25241–25250.

Kroeze, C., Mosier, A.R. and Bouwman, L.: 1999, 'Closing the global N_2O budget: A retrospective analysis 1500–1994', *Global Biogeochemical Cycles* **13**, 1–8.

Kumar, U., Jain, M.C., Pathak, H., Kumar, S. and Majumdar, D.: 2000, 'Nitrous oxide emission from different fertilizers and its mitigation by nitrification inhibitors in irrigated rice', *Biology and Fertility of Soils* **32**, 474–478.

Ladha, J.K., Fischer, K.S., Hossain, M., Hobbs, P.R. and Hardy, B. (eds.): 2000, *'Improving the Productivity and Sustainability of Rice–Wheat Systems of the Indo-Gangetic Plains'*, Discussion Paper 2000, No. 40, IRRI, Philippines. pp. 31.

Ladha, J.K., Pathak, H., Padre, A.T., Dawe, D. and Gupta, R.K.: 2003, Productivity trends in intensive rice–wheat systems in Asia. In J.K. Ladha, J. Hill, R.K. Gupta, J. Duxbury and R.J. Buresh (eds.), *'Improving the Productivity and Sustainability of Rice–Wheat Systems: Issues and Impact'*, ASA Madison, WI, ASA, Spec. Publ. **65**, 45–76.

Lemke, R.L., Izaurralde, R.C., Nyborg, M. and Solberg, E.D.: 1999, 'Tillage and N source influence soil-emitted nitrous oxide in the Alberta Parkland region', *Canadian Journal of Soil Science* **79**, 15–24.

Lindau, C.W., Bollich, P.K., Delaune, R.D., Mosier, A.R. and Bronson, K.F.: 1993, 'Methane mitigation in flooded Louisiana rice fields', *Biology and Fertility of Soils* **15**, 174–178.

Lu, W.F., Chen, W., Duan, B.W., Guo, W.M., Lu, Y., Lantin, R.S., Wassmann, R. and Neue H.U.: 2000, 'Methane emission and mitigation options in irrigated rice fields in Southeast China', *Nutrient Cycling Agroecosystems* **58**, 277–284.

Majumdar, D., Kumar, S., Pathak, H., Jain, M.C. and Kumar, U.: 2000, 'Reducing nitrous oxide emission from an irrigated rice field of North India with nitrification inhibitors', *Agriculture, Ecosystems & Environment* **81**, 163–169.

Matson, P.A., Parton, W.J., Power, A.G. and Swift, M.J.: 1997, 'Agricultural intensification and ecosystem properties', *Science* **277**, 504–509.

Matson, P.A., Naylor, R. and Ortiz-Monasterio, I.: 1998, 'Integration of environmental, agronomic and economic aspects of fertilizer management', *Science* **280**, 112–115.

McKenney, D.S. Wang, S.W., Drury, C.F. and Findlay, W.I.: 1995, 'Denitrification, immobilization, and mineralization in nitrate limited and nonlimited residue-amended soil', *Soil Science Society of America Journal* **59**, 118–124.

Mitra S., Jain, M.C., Kumar, S., Bandyopadhyay, S.K. and Kalra, N.: 1999, 'Effect of rice cultivars on methane emission', *Agriculture Ecosystems Environment* **73**, 177–183.

Mosier, A.R., Wassmann, R., Verchot L. and Palm, C. (this issue) 'Greenhouse gas emissions from tropical agriculture: Sources, sinks and mechanisms', *Environment, Development, Sustainability.*

Mummey, D.L., Smith, L.J. and Bluhm, G.: 1998, 'Assessment of alternative soil management practices on N_2O emissions from US agriculture', *Agriculture, Ecosystems and Environment* **70**, 79–87.

Nambiar, K.K.M.: 1994, 'Yield sustainability of rice–rice and rice–wheat systems under long-term fertilizer use', in *Resource Management for Sustained Crop Productivity*, New Delhi, India, The Indian Society of Agronomy, pp. 266–269.

Neue, H.U.: 1997, 'Fluxes of methane from rice fields and potential for mitigation', *Soil Use and Management* **13**, 258–267.

Paroda, R.S., Woodhead T. and Singh R.B. (eds.): 1994, 'Sustainability of rice–wheat production systems in Asia', *RAPA Publication* 1994/11, FAO Regional Office for Asia and the Pacific, Bangkok, Thailand, pp. 209.

Pathak, H., Ladha, J.K., Aggarwal, P.K., Peng, S., Das, S., Yadvinder Singh, Bijay-Singh, Kamra, S.K., Mishra, B., Sastri, A.S.R.A.S., Aggarwal, H.P, Das, D.K. and Gupta. R.K.: 2003, 'Climatic potential and on-farm yield trends of rice and wheat in the Indo-Gangetic plains', *Field Crops Research* **80**, 223–234.

Qian, J.H., J.W. Doran, K.L. Weier, A.R. Mosier, T.A. Peterson and J.F. Power. 1997, 'Soil denitrification and nitrous oxide losses under corn irrigated with high nitrate groundwater. *Journal of Environmental Quality* **26**, 348–360.

Reddy, K.R. and Patrick, Jr., W.H.: 1980, 'Losses of applied 15NH4-N urea 15N, and organic 15N in flooded soils', *Soil Science* **130**, 326–330.

Roelcke, M., Li, S.X., Tian, X.H., Gao, Y.J. and Richter, J.: 2002, '*In situ* comparisons of ammonia volatilization from N fertilizers in Chinese loess soils', *Nutrient Cycling in Agroecosystems* **62**, 73–88.

Rolston, D.E., Sharpley, A.N., Toy, D.W. and Broadbent, F.E. 1982, 'Field Measurment of Denitrification: III. Rates During Irrigation Cycles', *Soil Science Society of America Journal* **46**, 289–296.

Ruser, R., Flessa, H., Schilling, R., Beese, F. and Munch, J.-C.: 2000, 'Effect of crop-specific field management and N fertilization on N_2O emissions from a fine-loamy soil', *Nutrient Cycling in Agroecosystems* **59**, 177–191.

Sass, R.L., Fisher, F.M., Turner, F.T. and Jund, M.F.: 1991, 'Methane emissions from rice fields as influenced by solar radiation, temperature, and straw incorporation', *Global Biogeochemical Cycles* **5**, 335–350.

Sass, R.L., Fisher, F.M., Wang, Y.B., Turner, F.T. and Jund, M.F.: 1992, 'Methane emission from rice fields: The effect of flood water management', *Global Biogeochemical Cycles* **6**, 249–262.

Schuster, M. and Conrad, R.: 1993, 'Metabolism of nitric oxide and nitrous oxide during nitrification and denitrification in soil at different incubation conditions', *Microbiology Ecology* **101**, 133–143.

Schütz, H., Holzapfel-Pschorn, A., Conrad, R., Rennenberg, H. and Seiler, W.: 1989, 'A 3-year continuous record on the influence of daytime, season and fertilizer treatment on methane emission rates from an Italian rice paddy', *Journal of Geophysical Research* **94**, 16,405–16,416

Setyanto, P., Makarim, A.K., Fagi, A.M., Wassmann, R. and Buendia, L.V.: 2000, 'Crop management affecting methane emissions from irrigated and rainfed rice in Central Java (Indonesia)', *Nutrient Cycling in Agroecosystems* **58**, 85–93.

Sharma, P.K., Ladha, J.K. and Bhushan, L.: 2003, 'Soil physical effects of puddling in rice–wheat cropping system', In J.K. Ladha, J. Hill, R.K. Gupta, J. Duxbury and R.J. Buresh (eds.), *Improving the Productivity and Sustainability of Rice–Wheat Systems: Issues and Impact*, ASA Madison, WI, ASA, Spec. Publ. **65**, 97–114.

Singh, J.S., Singh S., Raghubanshi A.S., Singh, S. and Kashyap, A.K.: 1996, 'Methane flux from rice/wheat agroecosystems as affected by crop phenology, fertilization and water level', *Plant Soil* **183**, 323–327.

Suratno, W., Murdiyarso, D., Suratmo, F.G., Anas, I., Saeni, M.S. and Rambe, A.: 1998, 'Nitrous oxide emission from irrigated rice fields in West Java', *Journal of Environmental Pollution* **102**, 159–166.

Tölg, M.: 1998, '*Einfluß von Bewirtschaftungsmaßnahmen auf Produktion und Emission von Methan im Naßreisanbau*', Ph.D. thesis (in German), Schriftenreihe des Fraunhofer-Instituts für Atmosphärische Umweltforschung, Bd. 50, Garmisch-Partenkirchen, Germany, pp. 239.

Tsuruta, H., Kanda, K. and Hirose, T.: 1997, 'Nitrous oxide emission from a rice paddy field in Japan', *Nutrient Cycling in Agroecosystems* **49**, 51–58.

Verma, T.S. and Bhagat, R.M.: 1992, 'Impact of rice straw management practices on yield, nitrogen uptake, and soil properties in a wheat-rice rotation in northern India', *Fertility Research* **33**, 97–106.

Wassmann, R. and Aulakh, M.S.: 2000, 'The role of rice plants in regulating mechanisms of methane missions', *Biology and Fertility of Soils* **31**, 20–29.

Wassmann, R., Schütz, H., Papen, H., Rennenberg, H., Seiler, W., Dai, A.G., Shen, R.X., Shangguan, X.J. and Wang, M.X.: 1993a, 'Quantification of methane emissions from Chinese rice fields (Zhejiang Province) as influenced by fertilizer treatment', *Biogeochemistry* **11**, 83–101.

Wassmann, R., Wang, M.X., Shangguan, X.J., Xie, X.L., Shen, R.X., Papen, H., Rennenberg, H., Seiler, W.: 1993b, 'First records of a field experiment on fertilizer effects on methane emission from rice fields in Hunan Province (PR China)', *Geophysical Research Letters* **20**, 2071–2074.

Wassmann, R., Neue, H.U. and Lantin, R.S.: 2000a, 'Characterization of methane emissions from rice fields in Asia. 1. Comparison among field sites in five countries', *Nutrient Cycling in Agroecosystems* **58**, 1–12.

Wassmann, R., Neue, H.U., Lantin, R.S., Makarim, K, Chareonsilp, N., Buendia, L.V. and Rennenberg, H.: 2000b, 'Characterization of methane emissions from rice fields in Asia. 2. Differences among irrigated, rainfed and deepwater rice', *Nutrient Cycling in Agroecosystems* **58**, 13–22.

Wassmann, R., Lantin, R.S., Neue, H.U., Buendia, L.V., Corton T.M. and Lu, Y.H.: 2000c, 'Characterization of methane emissions from rice fields in Asia. 3. Mitigation options and future research needs', *Nutrient Cycling in Agroecosystems* **58**, 23–36.

Wassmann, R., Aulakh, M.S., Lantin, R.S., Aduna, J.B. and Rennenberg, H.: 2002, 'Methane emission patterns from rice fields planted to several rice cultivars for nine seasons', *Nutrient Cycling in Agroecosystems* **64**, 111–124.

Weier, K.L., Doran, J.W., Power, J.F. and Walters, D.T.: 1993, 'Denitrification and the dinitrogen/nitrous oxide ratio as affected by soil water, available carbon, and nitrate', *Soil Science Society of America Journal* **57**, 66–72.

Winfrey, M.R. and Zeikus, J.G.: 1977, 'Effect of sulfate on carbon and electron flow during microbial methanogenesis in fresh water sediments', *Applied Environmental Microbiology* **33**, 275–281.

Witt, C., Cassman, K.G., Ottow, J.C.G. and Biker, U.: 1998, 'Soil microbial biomass and nitrogen supply in an irrigated lowland rice soil as affected by crop rotation and residue management', *Biology and Fertility of Soils* **28**, 71–80.

Witt, C., Cassman, K.G., Olk, D.C., Biker, U, Liboon, S.P. Samson, M.I. and Ottow, J.C.G.: 2000, 'Crop rotation and residue management effects on carbon sequestration, nitrogen cycling and productivity of irrigated rice systems', *Plant Soil* **225**, 263–278.

Woodhead, T., Huke, R. and Huke, E.: 1994a, *Rice–Wheat Atlas of Pakistan.* IRRI/CIMMYT/PARC publication, Los Baños (Philippines), pp. 1–32.

Woodhead, T., Huke, R., Huke, E. and Balababa, L.: 1994b. *Rice–Wheat Atlas of India.*: IRRI/CIMMYT/ICAR Publication, Los Baños (Philippines), pp. 1–147.

Xing, G.X. and Zhu, Z.L.: 1997, 'Preliminary studies on N2O emission fluxes from upland soils and paddy soils in China', *Nutrient Cycling in Agroecosystems* **49**, 17–22.

Yagi, K. and Minami, K.: 1990, 'Effect of organic matter application on methane emission from some Japanese paddy fields', *Soil Science and Plant Nutrition* **36**, 599–610.

Yamulki, S., Goulding, K.W.T., Webster, C.P. and Harrison, R.M.: 1995, 'Studies on NO and N2O fluxes from a wheat field', *Atmospheric Environment* **29**, 1627–1635.

Yu, K.W., Wang,Z.P. and Chen, G.X.: 1997, 'Nitrous oxide and methane transport through rice plants', *Biology and Fertility of Soils* **24**, 341–343.

Zhang, S.L., Cai, G.X., Wang, X.Z., Xu, Y.H., Zhu, Z.L. and Freney, J.R: 1992, 'Losses of urea nitrogen applied to maize grown on calcareous Fluvo–Aquic soil in North China plain', *Pedosphere* **2**, 171–178.

Zheng, X.H., Wang, M.X., Wang, Y.S., Heyer, J., Kogge, M., Papen, H., Jin, J.S. and Li, L.T.: 1997, 'N2O and CH4 emissions from rice paddies in Southeast China', *Chinese Journal of Atmospheric Sciences* **21**, 167–174.

Zheng, X.H., Wang, M.X., Wang, Y.S., Shen, R.X., Li, J., Heyer, J., Kogge, M., Papen, H., Jin, J.S. and Li, L.T.: 2000, 'Mitigation options for methane, nitrous oxide and nitric oxide emissions from agricultural ecosystems', *Advances in Atmospheric Sciences* **17**, 83–92.

IS IT POSSIBLE TO MITIGATE GREENHOUSE GAS EMISSIONS IN PASTORAL ECOSYSTEMS OF THE TROPICS?

ROBIN S. REID[1]*, PHILIP K. THORNTON[1], GRAEME J. MCCRABB[2],
RUSSELL L. KRUSKA[1], FRED ATIENO[1] and PETER G. JONES[3]

[1]*International Livestock Research Institute, P.O. Box 30709, Nairobi, Kenya;*
[2]*International Livestock Research Institute, Addis Ababa, Ethiopia;*
[3]*Centro Internacional de Agricultura Tropical, Cali, Colombia*
(*author for correspondence, e-mail: r.reid@cgiar.org; fax: (254-2)631499; tel.: (254-2)630743)

(Accepted in Revised form 15 January 2003)

Abstract. Climate change science has been discussed and synthesized by the world's best minds at unprecedented scales. Now that the Kyoto Protocol may become a reality, it is time to be realistic about the likelihood of success of mitigation activities. Pastoral lands in the tropics hold tremendous sequestration potential but also strong challenges to potential mitigation efforts. Here we present new analyses of the global distribution of pastoral systems in the tropics and the changes they will likely undergo in the next 50 years. We then briefly summarize current mitigation options for these lands. We then conclude by attempting a pragmatic look at the realities of mitigation. Mitigation activities have the greatest chance of success if they build on traditional pastoral institutions and knowledge (excellent communication, strong understanding of ecosystem goods and services) and provide pastoral people with food security benefits at the same time.

Key words: carbon sequestration, climate change, greenhouse gases, land use, mitigation, pastoral lands, tropics.

1. Introduction

Grazing lands[1] cover 32 million km^2, more than a quarter of the earth's land surface and more than twice the land area than cropland (FAO, 1999; WRI, 2000). Because these lands are so extensive, pastoralism is the most widespread human land-use system on earth (FAO, 1993). Most of the tropical pastoral lands are in Africa, where traditional pastoral peoples herd multi-species herds of livestock in common property rangelands and subsist on the products from their livestock (Blench, 2000). South American pastoral lands are also widespread; here, pastoral land-use is often on private land and can be more intensified and commercial than in Africa. Tropical pastoral lands are more restricted in Australo-Asia, limited to the Arabian Peninsula, northern Australia and New Guinea.

The vast size of these lands implies a high potential as a source and a sink for greenhouse gases. Although tropical forests have great storage capacity for carbon aboveground (212 Gt C), savannas have the greatest potential belowground in the tropics (264 Gt C; IPCC, 2000). The principal causes of carbon loss from most rangelands are land conversion and land/grazing management; of lesser importance are climate change, fertilizer and fossil fuel use, wind and water erosion, and biomass burning (IPCC, 2000; 2001a). Most pastoral lands support ruminant, grazing herbivores that emit methane (CH_4), 24.5 times more powerful a greenhouse gas than carbon dioxide (CO_2). Nitrous oxide (N_2O) emissions in pastoral lands come from fertilizer use, land conversion, increased temperature and manure (Ojima et al., 1993), although N_2O emissions are probably low in pastoral systems in developing countries that dominate the tropics. Many of these emissions have the potential to be reduced by improved technologies and strategies for livestock production, grazing and fire management and land use in pastoral lands (IPCC, 2000).

Only recently has research on greenhouse gas mitigation gone beyond what is technically possible to begin to address social issues, taking a developed world perspective (e.g., IPCC, 2001b). It is highly likely that some techniques will work better than others in the tropics because of differences in infrastructure, institutions, economics and social practices among different regions. These differences need to be recognized and addressed explicitly. In addition, few assessments address technical and social limitations of different approaches to mitigation at the same time.

Our intention in this paper is to describe the potential and limitations, from both biological and social points of view, for mitigating greenhouse gas emissions in systems dominated by livestock production (= pastoral systems) in the tropics. We do this first by developing a new pastoral systems map to describe where pastoral systems are, how large they are, and how many people live in them. We then review current and future sources and sinks of greenhouse gas emissions in these systems. To gaze into the future, we estimate how human population growth (a surrogate for land-use change) and climate change may affect emissions 50 years from today. This is followed by a brief summary of the technical options for mitigated emissions. We conclude by attempting a pragmatic view of the likelihood that pastoralists will adopt these technical alternatives in these vast lands.

2. Where are tropical pastoral systems?

We took a broad definition of pastoral systems to include all rangeland areas, including areas currently under pasture, shrublands, woodlands, grasslands at any elevation. We call them pastoral to emphasize their dominant type of land-use, but recognize that these areas include savannas that have potential to support livestock but currently do not, like the southern savannas of New Guinea (Gillison, 1983). We created a global pastoral lands map using four GIS data layers: land cover (Loveland et al., 2000; USGS EDC, 1999), length of growing period (LGP) (Fischer et al., 2000), rainfall (Jones and Thornton, 2002; IWMI, 2001), several regional human population density datasets (Deichmann, 1996a,b; Hyman et al., 2000). We then estimated future human population density in 2050 using methods from Reid et al. (2000). First, we used land cover, LGP and human population maps to establish the location of all cultivatable land (>60 growing days), all land cover currently under crops in the USGS coverage (dryland cropland and pasture, irrigated cropland and pasture, mixed dryland/irrigated cropland and pasture, cropland/grassland mosaic, and cropland/woodland mosaic), and any other areas with sufficient human population (>20 people km^{-2}) to exclude extensive rangeland use (see Reid et al., 2000; Thornton et al., 2001 for details). 'Urban' included all areas with more than 450 people/km^2. The remaining areas (not cultivatable, low human population density) were discriminated into pastoral land classes by mean annual rainfall. Areas receiving less than 50 mm of rainfall were classified as hyper-arid, those with 51–300 mm were arid, and those with 301–600 mm were semi-arid. Those areas with less than 20°C and greater than 5°C in the growing season, or less than 20°C for one month a year were classified as highland (or temperate). Any area that does not fall in one of these categories was classified as other; this category is comprised of principally rain forest. In the map we present (Figure 1) we show all tropical pastoral lands (between the Tropics of Cancer and Capricorn), plus those temperate lands that fall within developing countries. In the following analysis, we use only countries that fall at least 50% in the tropics and Australia.

About 65% of all pastoral lands in the tropics are in Africa (13 million km^2), with large areas also in South America (16%, 3.3 million km^2) and the Arabian peninsula of Asia (12%, 2.5 million km^2, Figure 1, Table I). All tropical pastoral lands are in developing countries, except the 1.5% in northern Australia. The only significant pastoral lands in tropical Australo-Asia outside the Arabian peninsula fall in Australia and New Guinea (Gillison, 1983). In Africa, 80% of pastoral lands are dry (hyper-arid, arid, and semi-arid), as are 88% of the Arabian lands. Strikingly different are the pastoral lands in South America where 75% of these lands are sub-humid/humid or in the highlands. Central American and Australo-Asian pastures are predominantly semi-arid.

Africa is dominated by the largest contiguous hyper-arid lands in the world, the Sahara (Figure 1). Ringing the Sahara are Africa's arid lands. Semi-arid lands are very widespread through all regions of the continent. Sub-humid and humid pastoral lands are restricted to central western and north central Africa. About half the countries in Africa have small highland pasture areas not yet converted to cultivation, with the majority in Ethiopia and South Africa.

The principal pastoral lands in Latin America are in the Brazilian cerrado and the Venezuelan and Colombian llanos (Sarmiento, 1983). In addition, seasonally flooded areas dominate the Gran Pantanal of Brazil, Bolivia and Paraguay, and the Llanos de Mojos of northern Bolivia. Smaller patches of pastoral lands exist in the Amazonian rainforest, in coastal zones of the Guianas, and in coastal and inland areas of Costa Rica, Panama, Nicaragua, Guatamala, Honduras, Belize, Mexico, and Cuba (Sarmiento, 1983). Native pastures in these lands have moderate to high rainfall but low productivity; substitution of improved pasture can improve both productivity and carbon sequestration in these lands strongly (Fisher et al., 1994) but probably at the cost of loss of native biodiversity.

Outside the dry areas of the Arabian Peninsula, there are many dryland areas in Asia, but we did not classify them as pastoral lands because human population is too high; these systems are now mixed agro-pastoral or farming systems. These

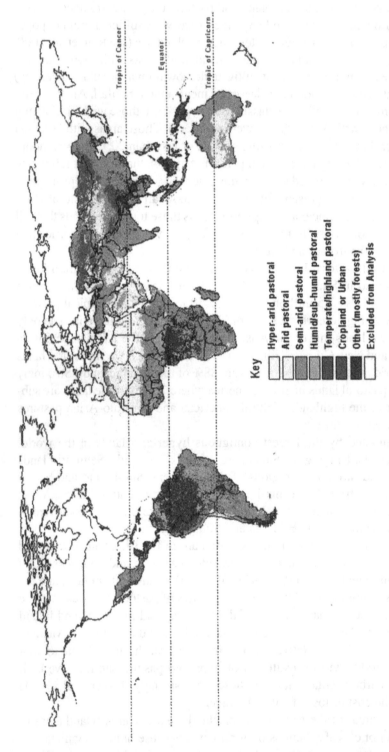

Figure 1. Distribution of tropical pastoral systems. The analysis in this paper is restricted to the areas between the tropics of Cancer and Capricorn (black horizontal lines), but we include other major pastoral areas that touch the tropics for completeness.

Key

Hyper-arid pastoral
Arid pastoral
Semi-arid pastoral
Humid/sub-humid pastoral
Temperate/highland pastoral
Cropland or Urban
Other (mostly forests)
Excluded from Analysis

Tropic of Cancer

Equator

Tropic of Capricorn

TABLE I. Areas in millions of km^2 and mean human population densities (#/km^2; in parentheses) of pastoral land and cropland/urban areas in the tropics. The tropics were defined as the area between the Tropics of Cancer and Capricorn.

	Hyper-arid pastoral	Arid pastoral	Semi-arid pastoral	Sub-humid/humid pastoral	Highland pastoral	Total pastoral area	Cropland/Urban areas
South America	0.04 (26.3)	0.14 (50.5)	0.59 (6.9)	2.12 (3.5)	0.42 (9.7)	3.31	3.50 (56.8)
Central America	0	0.02 (12.0)	0.75 (10.3)	0.03 (11.6)	0.06 (10.5)	0.86	0.97 (116.5)
Asia	0.13 (5.2)	1.73 (8.9)	0.87 (21.5)	0.12 (9.1)	0.13 (4.8)	2.98	4.45 (296.8)
Africa	2.10 (0.6)	2.08 (4.9)	6.03 (6.3)	2.43 (5.5)	0.15 (9.0)	12.98	6.05 (73.3)
Australo-Asia	0	0	0.29 (19.6)	0.04 (20.5)	0.0009 (15.5)	0.33	0.004 (50.5)
Total (mean)	2.27 (1.3)	4.15 (9.1)	8.52 (8.7)	4.73 (4.8)	0.75 (8.8)	20.42	15.21 (141.2)

Note: Central America includes Mexico.

include much of central India, parts of Pakistan, and Sri Lanka, and the central part of the southeast Asian peninsula. Central New Guinea has a highland area of savanna that is little used for livestock raising, and a southern coastal area of open to dense woodlands (Gillison, 1983).

In general, African and South American pastoral lands support fewer people (<10 people km^{-2}) than climatically similar rangelands elsewhere (Table I). Semi-arid lands on the Arabian peninsula of Asia are extensive and support twice to three times the population of similar pastoral lands on other continents.

3. What are the sources and sinks of greenhouse gases in tropical pastoral systems?

We will focus on the principal greenhouse gases: CO_2, CH_4 and N_2O. In comparison to temperate systems, tropical pastoral lands are less intensively used, more likely to convert to cropland and support lower quality forage for herbivores. Less intensive systems employ lower quantities of fossil fuel during production, and thus generate less CO_2 than more intensive systems (Subak, 1999). However, the higher conversion rates in tropical pastoral lands release more CO_2, N_2O and CH_4 (IPCC, 2001a) and the low quality fodder that grows in these lands increases CH_4 emissions per animal (Kurihara et al., 1999) compared with in temperate lands.

3.1. CARBON DIOXIDE

Because of their vast area, pastoral lands have the potential to be a significant sink for carbon. Ecosystem characteristics that affect the magnitude of carbon stores include rainfall, temperature, CO_2 concentration, productivity, species mixes, vegetation physiognomy, soil type, and rooting depth of grasses. Processes that reduce carbon sinks include: overgrazing, soil degradation, soil and wind erosion, biomass burning, land conversion to cropland; carbon can be improved by shifting species mixes, grazing and degradation management, fire management, fertilization, tree planting (agroforestry), and irrigation (Ash et al., 1996; Conant et al., 2001; Fisher et al., 1994; Ojima et al., 1993; Paustian et al., 1998).

One particularly important but controversial source (and a future potential sink) of carbon in drylands is range degradation, although the estimates of degradation vary greatly. Global assessments of drylands maintain that much of the earth's land surface is degraded (GLASOD, 1990) and that livestock are the principal global cause of desertification (Mabbutt, 1984). Analysts suggest that African pastures are 50% more degraded than those in Asia or Latin America (GLASOD, 1990). However, other analyses show that livestock numbers only exceed likely carrying capacities of arid and semi-arid rangelands in about $3–19\%^2$ of Africa at a continental scale (Ellis et al., 1999) and that there is no sustained evidence for a reduction in productivity, as measured by no change in the water-use efficiency of the Sahe-lian vegetation over 16 years (Nicholson et al., 1998). Despite these wide-ranging estimates, there is no question that some drylands are degraded and that better management could turn these sources of CO_2 into global sinks.

3.2. METHANE

The major sources of CH_4 in the tropics are enteric emissions from livestock and rice paddies, however only livestock are important for the pastoral lands. Enteric CH_4 emission from livestock and management of livestock manure are responsible for 23% and 7%, respectively, of all anthropogenic sources of CH_4 gas emissions (IPCC, 2001a), and contributes 30% of global warming potential of all agricultural emissions and about 5% of the global warming potential from all anthropogenic sources (US-EPA, 1999). Major livestock species include cattle, buffalo, sheep, goats, camels, swine, horses, and mules, however ruminant livestock (cattle, buffalo, sheep, and goats) account for 95% of global enteric CH_4 emissions (US-EPA, 1994), with cattle alone being the largest contributor (i.e., 73%). Buffalo are rare in pastoral lands and thus will be excluded from discussion here. More than half of the global cattle population is located in the tropics (McCrabb and Hunter, 1999); therefore this source is a significant global source. CH_4 emissions from manure management is significant in more intensive livestock systems, where the excreta remains wet for periods of time, such as intensive dairy cattle and pig systems. CH_4 from manure deposited on the rangelands is likely to relatively small source when compared to the enteric source.

Ruminant livestock produces CH_4 gas by microorganisms in the rumen dur-ing the process of feed digestion. The level of CH_4 gas production is closely related to the diet of the animal. Cattle consuming tropical forage diets have higher CH_4 production rates than cattle on temperate or cereal grain diets. For exam-ple, Kurihara et al. (1999) reported that CH_4 emissions of cattle fed on a medium quality tropical forage were $257\,g\,day^{-1}$ compared to $160\,g\,day^{-1}$ for a typical temperate cereal grain diet. When expressed per unit of liveweight gain (LWG), cattle fed on a tropical forage diet produced $500\,g$ methane kg^{-1} LWG compared to $127\,g\,kg^{-1}$ LWG for cattle fed on a cereal grain diet (Figure 2). Thus tropi-cal diets produce about 3.5 times more CH_4 per unit production than temperate diets.

The global distribution of cattle, sheep and goats, and thus CH_4 emissions, is not even, particularly in the tropics (Figure 3). The highest concentration of cattle is in the cerrado and llanos of South America (sometimes $>20\,km^{-2}$) and in semi-arid Africa $(10-20\,km^{-2})$, with few in Arabia. Sheep and goats are clustered in semi-arid western and eastern Africa, with few in South and Central America, and Arabia. CH_4 emissions per person are highest in South America (4–8 tropical livestock units (TLU) per person) and moderately high in the Sahel and East Africa (0.5–1 TLU per person).

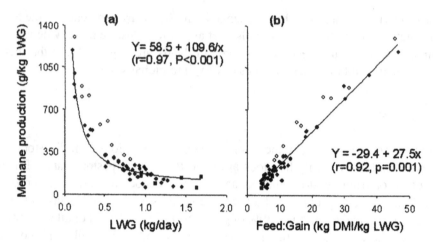

Figure 2. The relationship between (a) LWG and CH$_4$ production and (b) feed : gain ratio and CH$_4$ production for *B. taurus* (closed triangle) and *B. indicus* (closed square) cattle fed a high grain diet. Open triangles refer to cattle fed the tropical forage diet. From Kurihara et al. (1998).

3.3. NITROUS OXIDE

Another greenhouse gas, N$_2$O, with 320× the warming strength of CO$_2$, is probably of minor importance in extensive grazing systems. Sources in pastoral lands include land conversion, manure, fertilizer, and changes in temperature. Land conversion is important in these areas but both amount of manure and use of fertilizer are low in pastoral lands in the tropics (with the exception of some areas of South America).

4. How might these sources and sinks change over the next 50 years?

Currently, many of these pastoral ecosystems are under pressure to produce more livestock or to contract, to make way for more intensive land-use under crop cultivation (Niamir-Fuller, 1999; Blench, 2000). Some rangelands that used to be managed under communal land tenure are being privatized, with establishment of individual holdings; others are under state control (Galaty, 1994). This is happening first in rangelands that have more rainfall, are closer to urban centres and/or contain significant key resources that are essential for the success of crop cultivation (Galaty, 1994).

We expect these trends to continue over the next 50 years. Here, we use recent estimations of human population densities by the year 2050 by our research group (Thornton et al., 2001) to estimate shifts in land-use intensity from extensive rangeland to mixed farming systems in pastoral lands. We use 20 people km^{-2} as a threshold value between extensive systems and cropping systems, but the value is actually probably a bit lower (Reid et al., 2000) and varies from place to place. The biggest losses of pastoral lands to farming will occur in the Sahelian region of

West Africa, in Ethiopia, Mozambique and Zimbabwe, and in the southern cerrado of Brazil.

For Africa, we present here a modeling exercise by Jones and Thornton (2002) to estimate the impacts of climate change on the LGP. Throughout tropical Africa, the length of the growing season will become shorter except for a band extending about 7° north and south of the equator, where the growing season will lengthen. Thus, pastoral lands in the Sahel, in southern Africa, northern central Africa and Ethiopia will become drier. The only rangelands in Africa that will become wetter are in Kenya, northern Tanzania, parts of southern Ethiopia and southwestern Uganda. Despite improvement in areas near the equator, eastern Africa as a region will lose 20% of its land suitable for a variety of crops (Table II, LGP > 70 days), with nearly a quadrupling in the area suitable for very short season crops (LGP = 61–70 days). Southern and western Africa will lose smaller amounts of area suitable to many crops and northern Africa will almost double the area in this class. Southern, western and central Africa will see an overall drying and strong increases in arid land (LGP ≤ 40 days).

Together, human population growth and land-use change are expected to cause a contraction of the wetter pastoral lands at the same time that many of these are also drying because of climate change. On the other hand, East African pastoral lands near the equator will become wetter and will probably have strong pressures to be converted to cropland.

These changes will have profound implications for emissions of greenhouse gases and carbon sinks. When rangelands contract and are converted to cropland, 95% of the aboveground C can be lost and as much as 50% of the belowground C (Cole et al., 1989; our calculations, IPCC, 2000 figures). Pastoral lands will also lose above and belowground C when they become drier because of climate change. Within rangelands, over-grazing is the principle cause of loss of soil carbon (Ojima et al., 1993). Thus improved grazing management can have direct and substantial effects on soil carbon pools (IPCC, 2000). In addition, heavy grazing and changes in fire regimes can convert grassland systems to bushland systems. This conversion may increase C aboveground (Boutton et al., 1998) but it is not clear how carbon belowground is affected. In addition, the fragmentation and intensification of production in rangelands likely decreases the extent and severity of rangeland fires (e.g., Eva and Lambin, 2000), thus reducing emissions of CO_2, CH_4, N_2O and tropospheric ozone (Lacaux et al., 1993). More than half of the emissions from fires globally come from African pastoral lands (Crutzen and Andreae, 1990), thus the effects of these land-use changes are particularly important. In many cases, the CO_2 emitted from fires is reabsorbed during the next season's growth, however, carbon can be lost over the long term if savannas are burned too frequently (IPCC, 2000).

In addition to these changes in land use and climate, the demand for beef, mutton, pork, poultry and milk will double in the developing world in the next 20 years, particularly in East Asia, South Asia, SE Asia, West Asia and Africa (Delgado et al., 1999). In 2020, grazing systems will still provide most of the beef and milk in Africa (our calculations from de Haan et al., 1997 and Delgado et al., 1999).

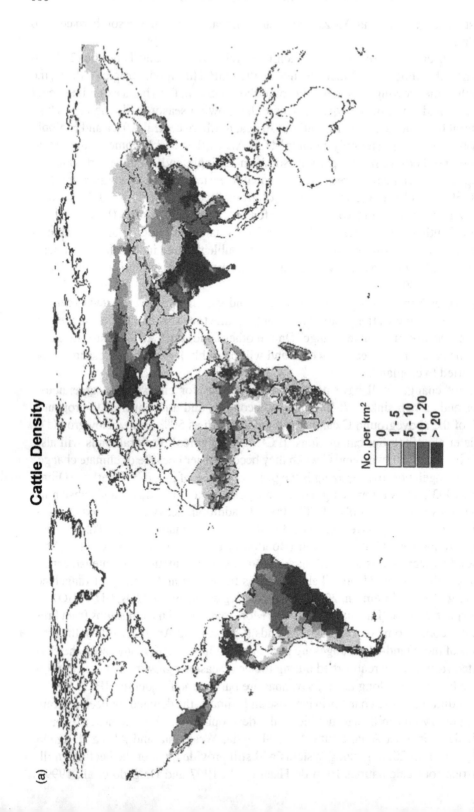

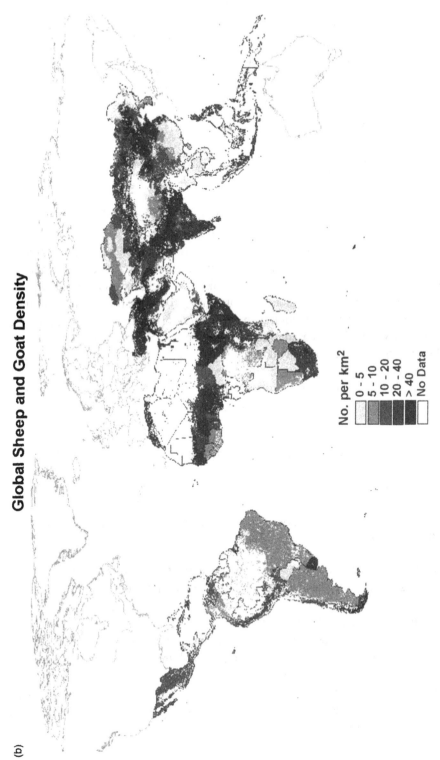

Global Sheep and Goat Density

No. per km²
0 - 5
5 - 10
10 - 20
20 - 40
> 40
No Data

(b)

Figure 3. Population densities of major ruminant species that emit CH₄ from pastoral lands in the tropics (cattle (a) and sheep/goats (b)).

TABLE II. Percentage change in the areas of land with different lengths of growing period in different regions of Africa between 2000 and 2050.

LGP (days)	Eastern	Southern	Western	Central	Northern
1–30	6.2	36.3	2.5	0.7	−1.4
31–40	−6.0	43.2	73.7	65.6	170.3
41–50	130.6	12.3	56.7	139.7	−13.7
51–60	25.4	19.8	61.4	−13.8	46.1
61–70	382.6	12.7	49.2	−12.7	189.1
>70	−20.4	−8.0	−5.3	0.02	180.2

Positive values indicate an increase in land area within a growing period, negative a decrease.

This increase in demand will be met partly by increased productivity of livestock but also will cause increases in livestock populations.

What changes will the increase in demand for livestock products bring to greenhouse gas emissions? Although this will bring increases in livestock populations, changes in technology through research will likely increase livestock production at the same time. For example, if diets and nutrition of livestock become higher quality, and levels of animal production increase, emissions per unit of animal product will reduce. Even with increased efficiency, total emissions of the larger livestock populations will still increase. In addition, if ruminant production systems become more intensive, then the management of excreta in these systems will increase the emissions of CH_4 and N_2O. Less likely, but possible, is the increased usage of N fertilizer on rangelands and consequent increases in N_2O emissions.

5. What greenhouse gas mitigation measures will work technically?

5.1. CARBON DIOXIDE MITIGATION

What are the carbon mitigation options for these vast pastoral lands? IPCC (2001) classifies three options to mitigate CO_2:

- *Protection*: active measures that maintain and preserve existing C reserves, including those in vegetation, soil organic matter, and products exported from the ecosystem (e.g., preventing the conversion of pastoral lands to croplands; reducing human use inside protected areas).
- *Sequestration*: measures, deliberately undertaken, that increase C stocks above those already present (e.g., improved grazing and fire management, reduction in degradation, improved pasture productivity).
- *Substitution*: practices that substitute renewable biological products for fossil fuels or energy-intensive products, thereby avoiding the emission of CO_2 from combustion of fossil fuels.

5.1.1. Protection

Protection is currently playing a major role for carbon sequestration in pastoral lands. In Africa, most of the protected areas are located in less productive lands and thus a disproportionate amount of pastoral lands are already protected. Even though these lands have been gazetted for conservation, the level of protection from human use varies greatly from place to place. Better management of existing protected areas would improve carbon sequestration.

More significant will be efforts to slow the conversion of rangeland into cropland. As highlighted above, this conversion can result in a 95% loss of the aboveground C and a 50% of belowground C. This is an area where payments for maintenance of currently sequestered carbon could be particularly effective. This will apply principally to the wetter savannas in Latin America and northern central Africa.

5.1.2. Sequestration

More carbon can be sequestered from improved management in grasslands than any other practice, except for adoption of agroforestry practices (Table 4-1, p. 184, IPCC, 2000). Improved management includes conversion of cropland to grassland, reductions in grazing intensity and biomass burning, improving degraded lands and reducing erosion, and changes in species mixes. Arguably, the biggest gains in carbon would result from converting wet grasslands back to woodland or forest, but the gain in forest/woodland services (ecological and economic) would have to be balanced against the loss of grassland services. Introduction of trees into pastures to gain carbon and maintain pasture services may be a compromise option. Likewise it is logical that the conversion of cropland to grassland would gain substantial carbon, but, at least in temperate semi-arid grasslands, a shift from cropland to grassland may give only small gains in carbon (Burke et al., 1995; IPCC, 2000).

Beyond these land-use changes, more subtle changes in grassland management will sequester less carbon, but still significant amounts. Fisher et al. (1994) found large gains from changing species mixes to deep-rooted grasses in wet South American grasslands, but it is unclear whether these gains could be widespread. Slightly smaller gains can be made in dry tropical grasslands by reducing grazing intensity, adjusting burning practices, and changes species mixes. For example, substantial carbon can be gained from reducing grazing intensity to spur a shift from annual to perennial grassland (Ash et al., 1996). Howden et al. (1994) found that small decreases in stocking rates in heavily grazed, savanna woodlands in Australia significantly improved carbon stocks, with additional improvements in vegetation and soils. Reduced grassland burning frequency can increase belowground carbon, aboveground litter and woody biomass (Crutzen and Andreae, 1990) but these gains can only be made in areas receiving >250–300 mm rainfall annually.

5.1.3. Substitution

Options for substitution renewable biological products for fossil fuels are principally appropriate in commercial ranch enterprises in the tropics (parts of South America,

Australia). Since these enterprises apply to only a small proportion of the pastoral people in the tropics, they will not be discussed here.

5.2. METHANE MITIGATION

Unlike other sources of anthropogenic sources of CH_4, few options exist to reduce emissions caused by enteric fermentation (US-EPA, 1999). Nevertheless some aspects of livestock management can result in lower emissions, most related to improved production efficiency. Presently available technologies that improve animal performance and reduce CH_4 emissions include: (a) improved management of livestock, and (b) production enhancing agents.

Despite the lack of easy options, most are rare 'win–win' situations for tropical pastoralists, where production gains are associated with reduced CH_4. Improved nutrition and health of animals, by reducing the number of sick and non-productive animals, will reduce emissions in all livestock production systems. Genetic improvements that improve productivity and reduce to time to market will have the same effect. Even artificial insemination will reduce CH_4 emissions by reducing the number of bulls needed to maintain reproductive herds. Improved diet quality, and hence, the level of animal productivity, will lead to reduced emission per unit of animal production. Animals with low levels of productivity, typical of many tropical pastoral systems, gain greater reductions in CH_4 production per increase in LWG than more productive animals; this applies to meat and milk (Kurihara et al., 1997; 1999). In most cases, reduced CH_4 production is associated with increased milk or meat production.

Improved grazing management can also improve livestock productivity and thus reduce CH_4 emissions (Ash et al., 1996). Steps to improve rangeland production, like improved pasture species, decreased erosion and degradation, and decreased overgrazing will improve livestock production and reduce CH_4 emissions. Grazing management that encourages cattle to consume the younger, highly digestible forages will reduce emissions and increase production because of the better quality diet.

Production enhancing agents like anabolic steroids reduce CH_4 emissions because animals reach slaughter weight in a shorter period of time than without the agents. Anabolic steroids also enhance growth rates of cattle by increasing feed conversion efficiency and lean tissue accretion. In addition, antibiotics and ionophores increase feed conversion efficiency, and hence are associated with reduced CH_4 emissions. These feed additives are expensive, and animals need to be treated with the agent frequently (e.g. daily), so farmers need to be in close proximity with cattle on a daily basis.

Scientists at CSIRO, Australia, are developing a vaccine to reduce enteric CH_4 gas emissions. This technology may be appropriate for extensive rangeland systems because animals would only have to be vaccinated infrequently (http://www.csiro.au/page.asp?type=mediaRelease&id=prmethanevax).

However, at present, the most effective CH_4 mitigation strategy is still reduction in livestock numbers. While this method will work very well technically, it is highly unlikely, unless compensated financially, that pastoralists will reduce livestock numbers in developing countries. Better solutions are urgently needed that are economically viable and can be adopted easily by pastoralists spread across vast lands.

6. But are these measures realistic socially in these lands?

Despite these technical potentials for emissions improvements, there are many infrastructural, economic, social and institutional limitations for mitigation in these vast lands. Virtually all of the tropical pastoral lands are in developing countries, and pastoral peoples are probably poorer than farmers and urban dwellers (Thornton et al., 2001). Poor people have few options to adopt new technologies or modify their management practices to improve greenhouse gas emissions, unless there are other livelihood improvements that accompany carbon sequestration. On the other hand, these lands are so vast and carbon sequestration potential is so high that they cannot be ignored in global efforts to reduce atmospheric emissions of important greenhouse gases.

Pastoral areas usually have less infrastructure than higher potential areas, partly because human populations are so low. Carbon credit schemes, as currently envisaged, will require communication between groups often distant from one another. Developing new networks for this purpose over large areas will be difficult, unless programs use mass media (e.g., radio). Pastoral people, however, have long traditions of long-distance communication, and these could be used to pass information and monitor progress.

Cultural values will be a constraint and opportunity in pastoral lands. The biggest constraints will be efforts to reduce CH_4 emissions that require reductions in herd sizes. Pastoral people live and die by their livestock and are unlikely to respond well to efforts to de-stock, particularly those sponsored by outsiders. But pastoralists are ultimately pragmatic and, if presented with a profitable alternative, will adjust their livelihood strategies. For example, if there are more productive breeds available that produce less CH_4 that can survive local disease loads, pastoralists can quickly change the composition of their herds.

Across tropical countries, there will be strong differences in the strength and ability of government institutions to implement carbon credit schemes. This will be particularly apparent when attempting to implement schemes in pastoral lands. It will be important to work initially where institutions are strong to develop the mechanisms of success with pastoral people.

These difficulties are made that much harder by the current pressures to intensify land-use and convert carbon-rich grasslands into carbon-poor croplands. At the same time, climate will be changing, putting additional stress on these lands. The first priority for pastoral lands in developing countries will be food security in risky environments, and this needs to be integrated into any schemes to sequester carbon.

These difficulties can also be advantages in carbon credit schemes for pastoral people. Each household ranges their livestock over large areas. Typical population densities in pastoral areas are 10 people km^{-2} or 1 person per 10 ha. If carbon is valued at $10 per ton and modest improvements in management can gain 0.5 t C ha^{-1} yr^{-1}, individuals might gain $50 yr^{-1}. About half of the pastoralists in Africa earn less than $1 day^{-1} or about $360 yr^{-1} (Galvin and Thornton, 2001; Kristanjanson et al., 2002). Thus modest changes in management could augment individual incomes by 15%, a substantial improvement in income. If carbon improvements are also associated with increases in production (CH$_4$ mitigation springs to mind), an elusive benefit–benefit scenario might occur. Despite the higher carbon gains that might come from agroforestry, the returns per person are likely to be lower in these systems because they principally occur in higher potential pastoral lands where human population densities are 3–10 times as much as in drier pastoral lands.

Are there other positive impacts that may be associated with carbon credits? In terms of other production benefits, unless CH$_4$ is included, probably not. As described previously, CH$_4$ mitigation has the potential for positive production and CH$_4$ benefits at the same time. In some cases, carbon sequestration will be associated with improved biodiversity conservation.

But what would be needed to make this work?

- Incentives and policies are required to realize the technical potential for mitigation in pastoral lands. These may be in the form of government regulations, taxes, and subsidies, or through economic incentives in the form of market payments for capturing and holding carbon, as suggested in the Kyoto Protocol.
- Institutional linkages between the countries involved in buying and selling carbon credits. It is hard to see that there is anyway round this – institutions and organizations have to be involved.
- Use of current pastoral networks of communication over long distances so that they know how to take advantage and can assist in monitoring credit schemes.
- Healthy extension services are needed, or some other mechanisms for reaching communities to inform them about such schemes and what sort of savanna/rangeland management is expected in return for carbon payments.
- Fully functioning and accountable community governance structures (such as for the group ranches prevalent in Kenya, for example), so that payments reach the intended beneficiaries.
- Greatly improved food security and disaster early warning systems; pastoralist welfare of course must take precedence over perceived non-compliance with terms and conditions of the scheme (i.e., in poor seasons, food security is the first priority, not carbon sequestration, and advance warning of such seasons would be a key component of efficient mitigation schemes, since it may be that C sequestration targets have to be sacrificed every so often for reasons of food security).

- Greatly improved monitoring techniques for carbon stocks, to set baselines and to monitor changes through time. Ultimately, the credibility and sustainability of carbon credit schemes are heavily dependent on appropriate monitoring schemes. For example, imperfect information will result in market failure, etc.
- Appropriate verification protocols that can be quickly and widely applied, coupled with transparent and equitable enforcement of C targets.

This is a fearsome list. But, as so often, if things are to change, then what is needed is a bottom–up approach to try this out, working on a pilot scale with local communities. There are still technical issues to be resolved, and some of the organizational problems are likely to be extremely difficult, but the potential impacts on poverty in pastoralist households could make the search for solutions to these problems very worthwhile.

7. Conclusions

These problems may appear insurmountable. It is important to remember that most of the people reading these lines did not grow up in pastoral areas, and have a settled, farming community or urban perspective. We calculate that only 1% of all people in countries that will provide the funds for carbon credit schemes are from pastoral lands. For example, Africa is most directly connected to European countries and institutions and yet only 0.2% of Europeans are from pastoral lands. Mitigation activities have the greatest chance of success if they build on traditional pastoral institutions and knowledge (excellent communication, strong understanding of ecosystem goods and services) and provide pastoral people with food security benefits at the same time.

Notes

[1] Here we define tropical grazing lands as equivalent to pastoral lands; they include all grasslands, pastures, rangeland, shrubland, and savannas, beyond the edge of the hyper-arid deserts (IPCC, 2000) between the Tropics of Cancer and Capricorn. These vary from arid to humid and include high elevation lands used for grazing.
[2] Carrying capacity is exceeded in 19% of areas receiving 0–200 mm rainfall per annum, 15% of the 200–400 mm zone, 3% of the 400–600 zone and 8.5% of the 600–800 mm zone (Ellis et al., 1999).

References

Ash, A.J., Howden, S.M., McIvor, J.G. and West, N.E.: 1996, 'Improved rangeland management and its implications for carbon sequestration', in N.E. West (ed.), *Proceedings of the Fifth International Rangeland Congress*, Denver, Society of Range Management, pp. 19–20.
Blench, R.: 2000, *'You Can't Go Home Again', Extensive Pastoral Livestock Systems: Issues and Options for the Future*, ODI/FAO Report, London, UK.

Boutton, R.W., Archer, S.R., Midwood, A.J., Zitzer, S.F. and Bol, R.: 1998, '^{13}C values of soil organic carbon and their use in documenting vegetation change in a subtropical savanna ecosystem', *Geoderma*, **82**, 5–41.

Burke, I.C., Laurenroth, W.K. and Coffin, D.P.: 1995, 'Soil organic matter recovery in semiarid grassland: implication for the Conservation Reserve Program', *Ecological Applications*, **5**, 793–801.

Cole, C.V., Stewart, J.W.B., Ojima, D.S, Parton, W.J. and Schimel, D.S.: 1989, 'Modeling land use effects on soil organic matter dynamics in the North America Great Plains', in M. Clarholm and L. Bergstrom, (eds.), *Ecology of Arable Land – Perspectives and Challenges. Developments in Plant and Soil Sciences*, Dordrecht, Netherlands, Kluwer Academic Publishers, pp. 89–98.

Conant, R.T., Paustian, K. and Elliot, E.T.: 2001, 'Grassland management and conversion to grassland: Effects on soil carbon', *Ecological Applications* **11**, 231–245.

Crutzen, P.J. and Andreae, M.O.: 1990, 'Biomass burning in the tropics: Impact on atmospheric chemistry and biogeochemical cycles', *Science* **250**, 1669–1678.

de Haan, C., Steinfeld, H. and Blackburn, H.: 1997, 'Livestock and the environment: Finding a balance', Fressingfield, UK, WRENmedia.

Deichmann, U.: 1996a, *Africa Population Database, third version*. National Center for Geographic Information and Analysis (NCGIA), University of California, Santa Barbara as a cooperative activity between NCGIA, Consultative Group on International Agricultural Research (CGIAR), United Nations Environment Programme/Global Resource Information Database (UNEP/GRID), and World Resources Institute (WRI), http://grid2.cr.usgs.gov/globalpop/africa.

Deichmann, U.: 1996b, *Asia Population Database 1996*, National Center for Geographic Information and Analysis (NCGIA), University of California, Santa Barbara as a cooperative activity between NCGIA, Consultative Group on International Agricultural Research (CGIAR), and United Nations Environment Programme/Global Resource Information Database (UNEP-GRID), Sioux Falls, http://grid2.cr.usgs.gov/globalpop/asia/intro.html.

Delgado, C., Rosegrant, M., Steinfeld, H., Ehui, S. and Courbois, C.: 1999, *Livestock to 2020: the Next Food Revolution*, Washington, D.C., IFPRI, FAO, and ILRI.

Ellis, J., Reid, R.S., Kruska, R.L. and Thornton, P.K.: 1999, 'Population growth and land use change among pastoral people: Local processes and continental patterns', in D. Eldridge and D. Feudenberger (eds.), *Proceedings of the Sixth International Rangeland Congress*, Australia, Townsville, pp. 168–169.

Eva, H. and Lambin, E.E.: 2000, 'Fires and land-cover change in the tropics: A remote sensing analysis at the landscape scale', *Journal of Biogeography* **27**, 765–776.

FAO: 1993, *1992 Production Yearbook*, Rome, Italy, Food and Agriculture Organization of the United Nations.

FAO: 1999, *1998 Production Yearbook*, Rome, Italy, Food and Agriculture Organization of the United Nations.

Fischer, G., van Velthuizen, A. and Nachtergaele, F.O.: 2000, *Global Agro-ecological Zones Assessment: Methodology and Results*. Interim Report, Laxenburg, Austria, International Institute for Applied Systems Analysis.

Fisher, M.J., Rao, I.M., Ayarza, M.A., Lascano, C.E., Sanz, J.I., Thomas, R.J. and Vera, R.R.: 1994, 'Carbon storage by introduced deep-rooted grasses in the South American savannas', *Nature* **371**, 236–238.

Galaty, J.G.:1994, 'Rangeland tenure and pastoralism in Africa', in E. Fratkin, K.A. Galvin and E.A. Roth (eds.) *African Pastoralist Systems: An Integrated Approach*, Boulder, Colorado, USA, Lynne Reiner Publishers, pp. 185–204.

Galvin, K.A. and Thornton, P.K.: 2001, 'Human ecology, economics and pastoral household modeling', in R.B. Boone and M.B. Coughenour, (eds), *A System for Integrated Management and Assessment of East African Pastoral Lands: Balancing Food Security*, Wildlife Conservation, and Ecosystem Integrity, pp 105–123, Final Report to the Global Livestock Collaborative Research Support Program, http://www.nrel.colostate.edu/projects/imas/prods/finals /GLCRSP_IMAS_2001.pdf.27

Gillison, A.: 1983, 'Tropical savannas of Australia and the southwest Pacific', in F. Bourliere, (ed.), *Tropical Savannas: Ecosystems of the World*, Amsterdam, Elsevier Scientific Publishing Company, pp. 183–243.

GLASOD (Global Assessment of Soil Degradation): 1990, International Soil Reference and Information Centre, Wageningen, Netherlands, and United Nations Environment Program, Nairobi, Kenya.

Howden, S.M., White, D.H., McKeon, G.M., Scanlan, J.C. and Carter, J.O.: 1994, 'Methods for exploring management options to reduce greenhouse gas emissions from tropical grazing systems', *Climatic Change* **27**, 49–70.

Hyman, G., Nelson, A. and Lema, G.: 2000, '*Latin America and Caribbean Population Database 2000*,' Cali, Colombia International Center for Tropical Agriculture (CIAT), as a cooperative activity between UNEP/GRID, CIAT, and WRI. http://grid2.cr.usgs.gov/ globalpop/lac/intro.html.

IPCC: 2000, *Land Use, Land-use Change, and Forestry*, Intergovernmental Panel on Climate Change, Cambridge, UK, Cambridge University Press.

IPCC: 2001a, *Climate Change 2001: The Scientific Basis*, Report of Working Group I of the Intergovernmental Panel on Climate Change, J.T. Houghton, Y. Ding, D.J. Griggs, M. Noguer, P.J. van der Linden, and D. Xiaosu (eds.), UK, Cambridge University Press, 944 pp.

IPCC: 2001b, *Climate Change 2001: Mitigation*, Report of Working Group III of the Intergovernmental Panel on Climate Change, B. Metz, O. Davidson, R. Swart, and J. Pan (eds.), UK, Cambridge University Press, p. 700.

International Water Management Institute (IWMI): 2001, *World Water and Climate Atlas*, http://www.cgiar.org/iwmi./WAtlas/atlas.htm.

Jones, P.G. and Thornton, P.K.: 2002, 'Spatial modeling of risk in natural resource management', *Conservation Ecology* 5(2), 27. [online] URL: http://www.consecol.org/vol5/iss2/art27.

Kristjanson, P., Radeny, M., Nkedianye, D., Kruska, R., Reid, R., Gichohi, H., Atieno, F. and Sanford, R.: 2002, *Valuing Alternative Land Use Options in the Kitengela Wildlife Dispersal Area of Kenya. ILRI Impact Assessment Series 10.* Nairobi, Kenya, International Livestock Research Institute.

Kurihara, M., Shibata, M., Nishada, T., Purnomoadi, A. and Terada, F.: 1997, in Onodera et al. (eds.) *Rumen Microbes and Digestive Physiology in Ruminants,* S. Karger, Basel, pp. 199–208.

Kurihara, M., Terada, F., Hunter, R.A., Nishida, T. and McCrabb, G.J.: 1998, 'The effect of diet and liveweight gain on methane production in temperate and tropical beef cattle', *Proceedings of the 8th World Conference on Animal Production,* June 28 1998, Seoul, Korea, Vol. 1, pp. 364–365.

Kurihara, M., Magner, T., Hunter, R.A. and McCrabb, G.J.: 1999, 'Methane production and energy partitioning of cattle in the tropics', *British Journal of Nutrition* 81, 263–272.

Lacaux, J.-P., Cachier, H. and Delmas, R.: 1993, 'Biomass burning in Africa: an overview of its impact on atmospheric chemistry', in P.J. Crutzen and J.G. Goldammer (eds.), *Fire in the Environment,* Chichester, UK, John Wiley and Sons, pp. 159–192.

Loveland, T.R., Reed, B.C., Brown, J.F., Ohlen, D.O., Zhu, Z., Yang, L. and Merchant, J.: 2000, 'Development of Global Land Cover Characteristics Database and IGBP DISCover from 1 km AVHRR data', *International Journal of Remote Sensing* 21(6), 1303–1330. Global Land Cover Characteristics Database Version 1.2 available online at http://edcdaac.usgs. gov/glcc/glcc.html.

Mabbutt, J.A.: 1984, 'A new global assessment of the status and trends of desertification', *Environmental Conservation* 11, 100–113.

McCrabb, G.J. and Hunter, R.A.: 1999. 'Prediction of methane emissions from beef cattle in tropical production systems', *Australian Journal of Agricultural Research* 50, 1335–1339.

Niamir-Fuller, M.: 1999, 'International aid for rangeland development: trends and challenges', in D. Eldridge and D. Freudenberger (eds.) *People, and Rangelands: Building a Future, Proceedings of the International Rangelands Congress,* Townsville, Australia, 19–23 July 1999, pp. 147–152.

Nicholson, S.E., Tucker, C.J. and Ba, M.B.: 1998, 'Desertification, drought and surface vegetation: An example from the West African Sahel', *Bulletin of the American Meteorological Society* 79(5), 815–830.

Ojima, D.S., Parton, W.J., Schimel, D.S., Scurlock, J.M.O. and Kittel, T.G.F.: 1993, 'Modeling the effects of climate and CO_2 changes on grassland storage of soil C', *Water, Air, and Soil Pollution* 70, 643–657.

Paustian, K., Cole, C.V., Sauerbeck, D. and Sampson, N.: 1998, 'CO_2 mitigation by agriculture: an overview', *Climatic Change* 40, 135–162.

Reid, R.S., Kruska, R.L., Deichmann, U., Thornton, P.K. and Leak, S.G.: 2000, 'Human population growth and extinction of the tsetse fly', *Agricultural Ecosystems and Environment* 77, 227–236.

Subak, S.: 1999, 'Environmental costs of beef production', *Ecological Economics* 30, 79–91.

Thornton, P.K., Kruska, R.L., Henninger, N., Kristjanson, P.M., Reid, R.S., Atieno, F., Odero, A. and Ndegwa, T.: 2001, *Mapping Poverty and Livestock,* Final report to DFID, August 2001.

US-EPA (Environment Protection Agency): 1994, *International Anthropogenic Methane Emission: Estimates for 1990,* EPA 230-R-93-010, Washington DC, US EPA Office of Policy, Planning and Evaluation.

US-EPA (Environmental Protection Agency): 1999, *U.S. Methane Emissions 1990–2020: Inventories, Projections, and Opportunities for Reductions,* Office of Air and Radiation, Washington DC (EPA 430-R-99-013).

USGS EDC: 1999, United States Geological Surveys – Earth Resources Observation Systems (EROS) Data Center (EDC), *1-km Land Cover Characterisation Database, with Revisions for Latin America,* Sioux Falls, South Dakota, USA.

WRI (World Resources Institute): 2000, *World Resources, 2000–2001,* Washington, DC, World Resources Institute.

IMPLICATIONS OF LAND USE CHANGE TO INTRODUCED PASTURES ON CARBON STOCKS IN THE CENTRAL LOWLANDS OF TROPICAL SOUTH AMERICA

MYLES J. FISHER* and RICHARD J. THOMAS

Centro Internacional de Agricultura Tropical CIAT, Apartado Aéreo 6713, Cali, Colombia
(*author for correspondence, e-mail: m.fisher@cgiar.org; fax: +57 (2) 445-0073; tel: +57 (2) 445-0036)

(Accepted in Revised form 15 January 2003)

Abstract. Three of the nine physiographic regions that comprise the 8.2 million km^2 (Mkm^2) of the central lowlands of tropical South America have undergone substantial conversion from the native vegetation in the last 30 years, a good deal of it to introduced pastures. The converted lands were either formerly treeless grasslands of the Brazilian Shield and the Orinoco Basin, or semi-evergreen seasonal forest mainly in the east and southwest of the Amazon Basin in Brazil. There are about 0.44 Mkm^2 of introduced *Brachiaria* pastures in the former grasslands and we estimate that there are 0.096 Mkm^2 of introduced pastures in the Amazon Basin, mostly *Brachiaria* species.

Based on extensive descriptions of the land systems of the central lowlands by Cochrane et al. (1985) we extrapolated data of carbon (C) accumulation in the soil under introduced pastures on the eastern plains of Colombia (about $3\,t\,C\,ha^{-1}\,yr^{-1}$), which are treeless grasslands of the Orinoco Basin, to estimate the probable change in C stocks as a result of conversion to pasture elsewhere. Losses of above-ground C on conversion of the former grasslands is negliglible, while in contrast the forests probably lose about 115 t C for each ha converted. We estimated the mean time since conversion started and allowed for the degradation of the pastures that commonly occurs. We concluded that introduced pastures on the former grasslands have been a net sink for about 900 million t (Mt) C, while conversion of the forest has been a net source of about 980 Mt C, leading to a net source of about 80 Mt C for the central lowlands as a whole. We identify a number of issues and possible methodologies that would improve precision of the estimates of the changes in C stocks on conversion of native vegetation to pasture.

Key words: Brachiaria, Brazil, carbon accumulation, Cerrados, Colombia, introduced pastures, savannas, tropical forests, tropical grasslands, Venezuela.

1. Introduction

The lands of the South American tropics east of the Andes, but excluding the Atlantic coast of north-east Brazil south of the Amazon, comprise the central lowlands with an area of some 8.2 million km^2 (Mkm^2) (Cochrane et al., 1985). Cochrane et al. (1985) divided the central lowlands into nine physiographic regions (Figure 1). Three of these regions, the Amazon Basin, the Brazilian Shield and the Orinoco Basin have undergone substantial change in land use in the last 30 years, a good deal of it for introduced pastures for cattle production.

In this paper, we have attempted to estimate the effects on total carbon (C) stocks of converting the former native vegetation to introduced pastures in these

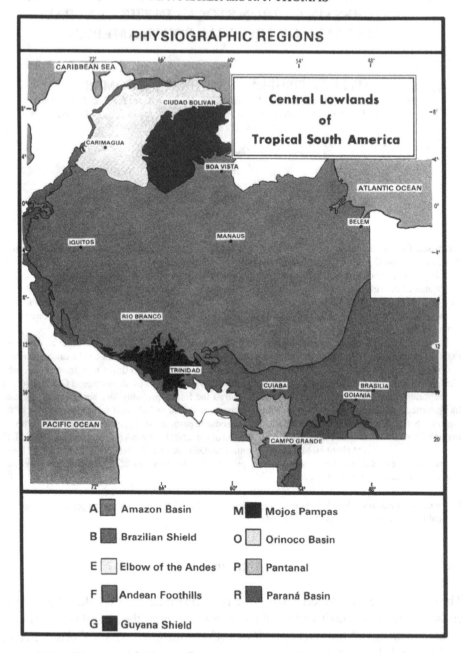

Figure 1. Physiographic regions of the central lowlands of tropical South America (from Cochrane et al., 1985).

three regions of the central lowlands. Based on Cochrane et al.'s (1985) classification of the lands of the central lowlands, we have extrapolated from published data of C stocks in the natural vegetation and the accumulation of C in the soil that occurs when the native vegetation is replaced by introduced pastures. We have also attempted to take account of the effect of the degradation that commonly occurs in introduced pastures.

2. Classification of the land in the central lowlands

In addition to the physiographic regions, Cochrane et al. (1985) also described seven climatic subregions depending on the length of the rainy season and the amount of evapotranspiration and the mean temperature during the rainy season (Table I). The combination of the physiographic regions and the climatic subregions, together with topographic classes that control drainage, broadly determine the natural vegetation classes of the central lowlands (Figure 2).

In considering the effect of changes in land use on C stocks, we focussed on the savannas and the tropical forests, estimating the changes in C stocks when each is converted to other uses. Building on this information, we then attempted to determine over what area these data could be reasonably extrapolated. The savannas have only small stocks of C above-ground, so that they have the greatest potential to accumulate C when introduced pastures replace them. In contrast, the large amounts of above-ground C lost when the forest is cleared mean that even though the soil component may come to a new, even higher equilibrium under pasture compared with the uncleared forest, areas of converted forest are likely to be net sources of C.

Using satellite images and aerial photographs supplemented by ground surveys and including published and unpublished soil survey data, Cochrane et al. (1985) divided the physiographical regions into land systems, defined as 'an area or group of areas throughout which there is a recurring pattern of climate, landscape and soils.' Each land system was given a general classification based on its topography, drainage, altitude and vegetation. For example, land system Oc201 of the Colombian eastern plains is flat, well-drained, lowland savanna, while land system Ab348 of the plains north of Manaus is flat well-drained lowland forest. Each land system consists of two or three land facets, usually only one of which makes up the major part of the land system. Cochrane et al. (1985) calculated the area of each land system and estimated the proportion of each occupied by each land facet. Within each land facet, they estimated the proportion of its area covered by one of ten vegetation types (Table II). Cochrane et al. (1985) thus synthesizes climate, vegetation, soils and phyiography, and is the only such broad survey available for the central lowlands of tropical South America.

Using these data, we calculated the area of lands classified as grasslands within each of the three physiographic regions in which they occur, subdivided on the basis of drainage, altitude and slope. We calculated that there is a total of 2.34 Mkm^2 of grasslands in the central lowlands of which 1.58 Mkm^2 are flat and well-drained, 1.32 Mkm^2 on the Brazilian Shield and 0.17 Mkm^2 in the Orinoco Basin (Table III). In general, the well-drained grasslands of the Amazon Basin occur in scattered small areas, and are of little importance. An exception is the savannas of the Macapá

TABLE 1. Areas of the nine physiographic regions and their climatic subregions in the central lowlands of tropical South America (calculated from data for the 481 land systems described by Cochrane et al., 1985, Vol. 3).

Climatic subregions

	>1300	1061–1300	900–1060	900–1060	<900	Subtropical	Other	Total
WSPE (mm)[1]	>9	8–9	6–8	6–8	<6			
No. of wet months[2]								
WSMT (°C)[3]	>23.5 a[4]	>23.5 b	>23.5 c	<23.5 d	>23.5 e	f	o	
Physiographic regions			km²					
Amazon Basin	1 634 573	2 927 221	173 619	—	—	—	—	4 735 412
Brazilian Shield	—	9 768	1 319 603	329 528	692 668	75 604	—	2 427 171
Elbow of the Andes	11 830	32 018	2 288	—	16 169	—	—	62 305
Andean Foothills	104 240	54 761	7 190	—	33 503	—	57 515	257 209
Guyana Shield	86 308	323 342	18 225	—	37 972	—	—	465 846
Mojos Pampas	—	84 919	45 522	—	—	—	—	130 441
Orinoco Basin	29 546	27 201	249 241	—	213 386	—	—	519 374
Pantanal	—	—	—	—	131 942	12 762	—	144 704
Paraná Basin	—	—	57 053	35 429	20 267	31 052	—	143 801
Total	1 866 497	3 459 229	1 872 740	364 957	1 145 907	119 418	57 515	8 886 263

[1] WSPE – Wet season potential evapotranspiration.
[2] Number of wet months each year.
[3] WSMT – Wet season mean monthly temperature.
[4] The lower-case letter is the climate subregion descriptor used as the second cipher of the land system name.

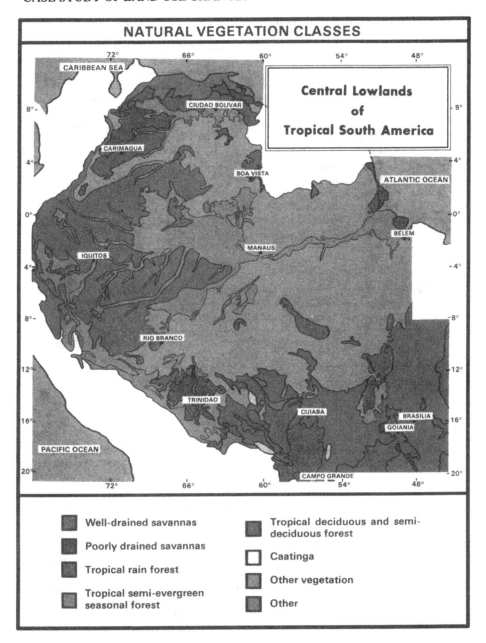

Figure 2. Natural vegetation classes of the central lowlands of tropical South America (from Cochrane et al., 1985).

cerrados at the mouth of the Amazon where the native vegetation has long been used for cattle production (Serrão and Homma, 1993).

Cochrane et al. (1985) classified 1.7 Mkm² of the central lowlands as either isohyperthermic or isothermic savannas (Table IV). Note that subsequent estimates

TABLE II. Descriptors used to classify the vegetation in each land facet and their common names (Cochrane et al., 1985).

Vegetation descriptor		Common name
Grasslands	Seasonally inundated grasslands	*Pampas*
	Grasslands on well-drained lands with occasional shrubs	*Campo limpo* (no noticeable woody vegetation) and *campo sujo* (occasional obvious shrubs)
	Open savanna	*Campo cerrado*
	Intermediate savanna	*Cerrado (sensu strictu)*
Woodlands and forest	Closed savanna	*Cerradão*
	Tropical rain forest	
	Semi-evergreen seasonal forest	
	Semi-deciduous seasonal forest	
	Scrubby xerophytic woodland	*Caatinga*
Other	Palm forest and other vegetation	

TABLE III. Areas of lands described as savannas in the three physiographic regions of the central lowlands of tropical South America that have undergone land use change in the last 30 years. Classified by altitude and topographic classes (calculated from Cochrane et al., 1985).

	Topography[3]	Amazon Basin	Brazilian Shield	Orinoco Basin	Total
Well-drained savannas			km²		
Lowland[1]	Flat	94 992	1 261 058	167 239	1 523 290
	Hilly		201 982	58 620	370 741
	Total	94 992	1 463 040	225 859	1 894 030
Upland[2]	Flat		57 561		57 561
	Hilly		41 827	9 468	51 295
	Total		99 388	9 468	108 856
Total	Flat	94 992	1 318 619	167 239	1 580 851
	Hilly		243 809	68 088	422 036
	Total	94 992	1 562 428	235 327	2 002 886
Poorly drained savannas					
Lowland	Flat	135 663	81 117	115 096	331 876
	Hilly			2 724	2 724
	Total	135 663	81 117	117 820	334 600
Total savannas					
	Flat	230 655	1 399 736	282 335	1 912 726
	Hilly		243 809	70 812	424 760
	Total	230 655	1 643 545	353 147	2 337 486

[1]Lowland, altitude <900 m.
[2]Upland, altitude >900 m.
[3]Terrain: flat, slopes <8%; hilly, slopes >8%.

put the area as substantially in excess of this, 2.5 Mkm² (Macedo, 1995), because Cochrane et al. did not include the southern parts of Goias and Mato Grosso do Sul nor the northern part of Minas Gerais. Of the 1.7 Mkm² classified by Cochrane et al. (1985), 0.68 Mkm² were on flat (slope <8%), well-drained lands classified as

TABLE IV. Area of fertility capability classification (FCC) texture classes within the savannas lands of the central lowlands of tropical South America (from Cochrane et al., 1985, Vol 1, Table VII-I).

| FCC texture class | Isohyperthermic savanna | | | | | Isothermic savanna | | | | |
| | Flat, poorly drained | Well-drained (% slope) | | | Total | Flat, poorly drained | Well-drained (% slope) | | | Total |
		<8	8–30	>30			<8	8–30	>30	
					$10^3 km^2$					
Loamy	46	198	77	60	381	1	57	71	51	180
Loamy over clayey	140	52	14	9	215	6	3	3	2	14
Clayey	80	207	48	11	346	2	158	45	3	208
Sandy	14	152	18	1	185	—	8	3	6	17
Sandy over loamy	14	11	2	—	27	1	1	1	1	4
Loamy over rock	1	4	2	3	10	—	2	2	2	6
Loamy over sandy	1	18	8	3	30	—	—	3	1	4
Sandy over clayey	44	—	—	—	44	—	—	—	—	—
Clayey over rock	—	2	8	20	30	—	—	—	—	0
Clayey over loamy	5	—	—	—	5	—	—	—	—	0
Total	345	644	177	107	1 273	10	229	128	66	433
Heavy textured[1]		457	139				218	119		
All without rock		638	167				227	126		

[1] Heavy textured soils with texture loamy or heavier and with no restriction to root growth at depth. Calculated only for soils on well-drained, flat lands (slope <8%), and for lands with intermediate slopes (8–30%).

having soil texture of loamy or heavier and no restriction to root growth in the soil profile (Table IV). It is these soils that are most likely to be similar to the soils in the Colombian llanos where Fisher et al. (1994, 1995, 1998) showed accumulation of C in the soil under pastures of introduced African grasses (see later). Sandy soils and hilly lands (slopes 8–30%) are included in Table IV, but they need to be treated with caution. Sandy soils are less likely to accumulate large amount of C because of their inherent droughtiness. Although lands with slopes greater than 8% can be managed to avoid erosion, it seems prudent to exclude them and especially those with sandy soils until their sustainable use in commercial practice is more widespread.

3. Areas sown to pasture in central lowlands

The isothermic savannas of Brazil are collectively known as the Cerrados, which is also the name of one of the main vegetation communities, called here Cerrado (*Sensu strictu*). Most of the introduced pastures sown in the Cerrados replace the grassland communities known as *Campo limpo*, which has no trees, and *Campo sujo* (see Table II), which has only a few stunted trees and shrubs. Both of these occur on infertile, acid soils and together comprise about 24% of the Cerrados or about 0.50 Mkm^2 (50 M ha) (Haridasan, 1992). The economics of cattle production on pasture cannot support the high costs of mechanical clearing the wooded

TABLE V. Relative approximate distribution of the most commonly cultivated grasses in the Cerrados region of Brazil in 1995 (Zimmer et al., 1999).

Species	Distribution (10^3 km^2)[1]	Relative distribution (%)
Brachiaria decumbens	264.0	55
B. brizantha	96.0	20
B. humidicola	43.2	9
B. ruziziensis, *B. dictyoneura*	4.8	1
Subtotal	408.0	85
Panicum maximum cv. Colonião comun	38.4	8
P. maximum cv. Tanzânia, Tobiatá, Vencedor	9.6	2
Subtotal	48.0	10
Other genera		
Andropogon, Hyparrhenia, Melinis, Cynodon	24.0	5
Total	480.0	100

[1] The multiplier given for the figures of area in Table I of Zimmer et al. (1999) is 10^3 ha, for a total area of sown pasture of only 48 000 ha. This figure conflicts with the text on page 247, where the total of sown pastures in the Cerrados is 'between 45 and 50 million hectares'. Other authors have reported figures between 35 and 50 million ha. A multiplier of 10^6 ha is used here, converted to km^2 for internal consistency within this paper.

TABLE VI. Areas of lands in the Orinoco Basin physiographic region of the central lowlands of tropical South America described as savannas. Classified by altitude and topographic classes (calculated from Cochrane et al., 1985). Savanna land system Ob639 of the Orinoco delta is excluded.

		Colombia Llanos	Venezuela			
			Western[1]	Central	Eastern	Total
		km^2				
Well-drained	Flat	58 230		57 991	51 018	109 009
	Hilly	47 855		2 619		2 619
	Total	106 085		60 610	51 018	111 628
Poorly drained	Flat	46 421	51 988			51 988
Total		152 506	51 988	60 610	51 018	163 616

[1] The well-drained llanos of western Venezuela were classified by Cochrane et al. (1985) as forested lands.

Cerrado (Sensu strictu) and *Cerradão* communities. By 1995, 0.48 Mkm^2 were sown to introduced pastures in the Brazilian Cerrados (Table V).

In the Orinoco Basin of Colombia and Venezuela, savannas, mainly treeless grasslands, are the dominant vegetation types (Table VI). However, there is such poor development of infrastructure in the eastern plains (llanos) of Colombia, coupled with increasingly severe security problems that there has been very little agricultural intensification in these lands. In 1992, only 1600 km^2 were sown to *Brachiaria* species in the Puerto López-Puerto Gaitán region of the western part of the Colombian llanos (Pizarro et al., 1996). In contrast to Colombia, much of the savannas of Venezuela have been converted to crops and pastures, although it is difficult to find data for the area that has been converted to introduced pastures. Pizarro et al. (1996) reported that between 24 000 and 40 000 km^2 were sown to *Brachiaria* species in the Venezuelan savannas, two-thirds to *B. decumbens*.

Using satellite imagery, the Brazilian national space agency, Instituto Nacional de Pesquisas Espaciais, has released figures showing the areas deforested by state in the Legal Amazon from 1978 to 1996 (Tables VII and VIII). Note that the Legal Amazon as defined by the Brazilian government is not the same as the physiographical Amazon Basin of Cochrane et al. (1985), but includes parts of lands that belong

TABLE VII. Gross area of deforestation in the states of the Legal Amazon of Brazil, 1978–1996 (Source: Instituto Nacional de Pesquisa Espaciais INPE, Brazil, http://www.dpi.inpe.br/amazonia/, accessed 8 December, 1998).

State	Original forest area[1]	Jan-78	Apr-88	Aug-89	Aug-90	Aug-91	Aug-92	Aug-94	Aug-95	Aug-96
					10^3 km^2					
Acre	154	2.5	8.9	9.8	10.3	10.7	11.1	12.1	13.3	13.7
Amapá	132	0.2	0.8	1.0	1.3	1.7	1.7	1.7	1.8	1.8
Amazonas	1561	1.7	19.7	21.7	22.2	23.2	24.0	24.7	26.6	27.4
Maranhão	155	63.9	90.8	92.3	93.4	94.1	95.2	96.0	97.8	99.3
Mato Grosso	585	20.0	71.5	79.6	83.6	86.5	91.2	103.6	112.2	119.1
Pará	1218	56.4	131.5	139.3	144.2	148.0	151.8	160.4	169.0	176.1
Rondônia	224	4.2	30.0	31.8	33.5	34.6	36.9	42.1	46.2	48.6
Roraima	188	0.1	2.7	3.6	3.8	4.2	4.5	5.0	5.1	5.4
Tocatins	58	3.2	21.6	22.3	22.9	23.4	23.8	24.5	25.1	25.5
Legal Amazon	4275	152.2	377.5	401.4	415.2	426.4	440.2	470.0	497.1	517.1

[1] Original forest area, from Serrão and Homma (1993).

TABLE VIII. Rate of increase of deforestation in the states of the Legal Amazon of Brazil (calculated from the data in Table VII.

State	78/88[1]	88/89	89/90	90/91	91/92	92/94[2]	94/95	95/96	% of the total
				km^2 yr^{-1}					
Acre	620	540	550	380	400	482	1208	433	2.7
Amapá	60	130	250	410	36	—	9	—	0.3
Amazonas	1510	1180	520	980	799	370	2114	1023	5.3
Maranhão	2450	1420	1100	670	1135	372	1745	1061	19.2
Mato Grosso	5140	5960	4020	2840	4674	6220	10391	6543	23.0
Pará	6990	5750	4890	3780	3787	4284	7845	6135	34.1
Rondônia	2340	1430	1670	1110	2265	2595	4730	2432	9.4
Roraima	290	630	150	420	281	240	220	214	1.0
Tocatins	1650	730	580	440	409	333	797	320	4.9
Legal Amazon	21 130	17 860	13 810	11 130	14 960	14 896	29 059	18 161	

[1] Mean annual rate for the decade.
[2] Mean annual rate for the two years.

to other physiographic regions, most notably parts of the Brazilian Shield. Of the total of $0.517\,Mkm^2$ deforested to August 1996 in the Legal Amazon, 85.7% is in the four states Maranhão, Mato Grosso, Pará and Rondônia, with over 57% in Mato Grosso and Pará. It is also noteworthy that the majority of the deforestation is in the vegetation class classified by Cochrane et al. (1985) as tropical semi-evergreen seasonal forest and not in the tropical rain forest. In Mato Grosso, much of the deforestation in is the vegetation class known as *Cerradão*, which forms part of the Cerrados. Although the Cerrados are commonly considered as savannas, *Cerradão* contains no continuous grass stratum. For this reason it is not strictly a savanna community but forest.

Of the $377\,500\,km^2$ estimated to have been deforested in the Brazilian Legal Amazon in 1988 (Table VII), approximately $70\,000\,km^2$, or 18.5% were estimated to have been converted to introduced pastures (Serrão and Toledo, 1990). We do not know how reliable this estimate is, nor whether the proportion of pasture has remained constant, increased or decreased in the last decade. If the estimate is valid, and if the proportion of pasture has remained constant, by 1996 the area converted to pasture was $95\,900\,km^2$. INPE forecast that the rate of clearing would decline in 1997, perhaps to as low as $11\,000\,km^2$, but anecdotal reports indicate that far from decreasing the rate increased substantially once more. Because this paper reports primarily a preliminary case study, we could not devote sufficient resources to obtain up-to-date satellite images and use them to match cleared lands sown to pasture in Amazonia with Cochrane et al.'s (1985) land systems, which would be a very large undertaking. We have assumed that the lands cleared and sown to pasture were predominantly flat, well-drained lowlands with soil texture heavier than loamy, which agrees with subjective assessments of colleagues with experience in the Amazon.

4. Variations in soil C under natural vegetation

Cochrane et al. (1985) quote data for spatial variability of soil analytical data for 18 profiles for land facet No. 1 within land system Oc201 on which the Centro Nacional de Investigación-Carimagua is situated on the eastern plains of Colombia. Among the data are soil organic matter (SOM), reported to be calculated from soil organic carbon (SOC) multiplied by 1.7. The methodology used to determine SOC is not given. Nevertheless, mean SOM over the 18 samples for the A-horizon is 3.78 with a standard deviation of 0.780. Measured values range from 1.7% to 5.3%. For the B horizon, the mean and standard deviation were 1.12 and 0.368, with extremes of 0.5 and 2.1. In no case did the sample with the highest or lowest A horizon SOM have the corresponding highest or lowest B horizon SOM. The correlation coefficient between the two data sets was only 0.69. Extrapolation from mean figures, or from data for small areas is therefore subject to considerable uncertainty. Furthermore estimation of SOC in layers deeper than those sampled seems risky.

5. C losses above-ground on conversion

The amount of standing above-ground biomass in *Campo limpo* and *Campo sujo* of the Cerrados and the treeless grasslands of the Orinoco are probably no more than $1–2\,t\,ha^{-1}$ (Fisher et al., 1992), so that losses of above-ground C when introduced pastures replace the native grasslands are trivial. Indeed, because native savannas are commonly burned as frequently as annually, in contrast to introduced pastures, which are normally only burned by accident, time-averaged above-ground C may actually increase when the lands are sown to introduced pastures (Long et al., 1992; Greenland, 1995).

In contrast, losses of C above-ground when forest lands of the Amazon are deforested are very large, although the reported data vary widely, and the amounts lost may not be as great as commonly thought. Cerri et al. (1994) quoted estimates of the above-ground biomass of Amazon forests in Brazil, ranging from 256 to $353\,t\,ha^{-1}$ near Manaus, and from $248\,t\,ha^{-1}$ at Tucuruí to $300\,t\,ha^{-1}$ at Paragominas, southwest and south respectively of Belém. They used a mean value of $290\,t\,ha^{-1}$, although clearly this mean contains considerable uncertainty. Brown and Lugo (1992) estimated the biomass of Amazon forest as '>290 Mg/ha'. Fujisaka et al. (1999) estimated that standing forest in Rondônia contained $159\,t\,C\,ha^{-1}$. Assuming that biomass is 45% C, Cerri et al. (1994) calculated the above-ground C as a little over $130\,t\,ha^{-1}$. Of this they estimated that about $13\,t\,C\,ha^{-1}$ was removed as millable timber leaving $117\,t\,C\,ha^{-1}$ as felled timber. They estimated that combustion efficiency (amount of C released to the atmosphere when the felled timber was burned) was between 20.0 and 27.6%, leading them to use a mean of 23.8% for a release of almost $28\,t\,C\,ha^{-1}$ in the burn following clearing. They assumed that the C remaining in the unburned timber decayed over the subsequent 8 year, releasing C to the atmosphere, apart from a small residue of $1.4\,t\,ha^{-1}$, presumably remaining as charcoal from the burning process. The net loss above-ground over 8 year was almost $116\,t\,C\,ha^{-1}$.

6. C accumulation in the soil under introduced pastures

Fisher et al. (1994, 1998) (Tables IX and X) compared soil carbon under introduced pastures and native savanna on land facet number 1 of land system Oc201 (Cochrane et al., 1985) on the eastern plains of Colombia. The introduced pastures accumulated C compared with the native savanna, much of it at depth in the soil greater than 20 cm. The data are for replicated experiments with plots 0.5 to 1 ha, which were grazed by cattle at normal stocking rates for introduced pastures in the region. Recent data from a study of the characteristics of the casts of the large aneic earthworm *Martiodrilis caramaguensis* show that the casts can account for the accumulation of as much as $8.6\,t\,C\,ha^{-1}\,yr^{-1}$ (Decaëns et al., 1999).

Other data, also for facet number 1 of land system Oc201, show accumulation of C in the soil under other introduced pastures. A 17-yr-old pasture of *A. gayanus*

TABLE IX. Yield, net gain of C and percentage of the net gain below the plow layer (20 cm) in introduced pastures compared with native savanna on two sites on the eastern plains of Colombia (data from Fisher et al., 1994).

Site			Matazul Farm		
Pasture	Savanna	A. gayanus/S. capitata		B. dictyoneura/C. acutifolium	
Depth	C	C	Increase	C	Increase
cm	$kg\,m^{-2}$	$kg\,m^{-2}$	$kg\,m^{-2} \pm SE$	$kg\,m^{-2}$	$kg\,m^{-2} \pm SE$
0–20	6.4	7.1	$0.7 \pm 0.20^{**}$	6.5	$0.1 \pm 0.15\,ns$
20–100	12.3	16.6	$4.4 \pm 0.97^{***}$	15.0	$2.7 \pm 0.88^{**}$
Total	18.7	23.7	$5.1 \pm 1.14^{***}$	21.5	$2.8 \pm 1.06^{*}$
% deeper than 20 cm			86.0		95.7
Site			Carimagua Research Station		
Pasture	Savanna	B. humidicola alone		B. humidicola/A. pintoi	
Depth	C	C	Increase	C	Increase
cm	$kg\,m^{-2}$	$kg\,m^{-2}$	$kg\,m^{-2} \pm SE$	$kg\,m^{-2}$	$kg\,m^{-2} \pm SE$
0–20	7.0	7.6	$0.6 \pm 0.43\,ns$	8.8	$1.8 \pm 0.42^{**}$
20–80	12.6	14.7	$2.0 \pm 0.70^{*}$	17.9	$5.3 \pm 1.17^{***}$
Total	19.7	22.3	$2.6 \pm 0.77^{**}$	26.7	$7.0 \pm 1.55^{***}$
% deeper than 20 cm			78.6		74.7

ns $P > 0.05$, $^{*}P < 0.05$, $^{**}P < 0.01$, $^{***}P < 0.001$. SE = standard error of difference between the means ($n = 14$ for Matazul Farm, $n = 12$ for Carimagua).

TABLE X. Details of the pastures in Table IX.

Site	Matazul Farm	Carimagua Research Station
Location	Eastern plains (Llanos), Puerto López, Colombia.	Eastern plains (Llanos), 200 km ENE of Puerto López, Colombia.
Latitude, longitude.	4°9′ N, 72°39′ W	4°37′ N, 71°19′ W
Altitude (m)	160	175
Mean annual rainfall (mm)	2700	2240
Soil	Oxisol	Oxisol
Texture	Clay loam	Clay loam
pH (1 : 1 water)	4.4	4.1
P (0–20 cm), Bray II (ppm)	1.8	1.5
Land system/facet number (Cochrane et al., 1985)	Oc201/1	Oc201/1
Pasture details	1989. Cropped from savanna with upland rice undersown with either mixed A. gayanus cv Carimagua 1 and S. capitata cv Capica pasture or mixed B. dictyoneura cv Llanero and C. acutifolium cv. Vichada. 1989–93. Rotationally grazed with cattle at 2 head ha^{-1}.	1984. Sown to B. humidicola cv Humidícola from savanna, with the legume Desmodium ovalifolium, which failed. 1987. Resown to B. humidicola cv Humidícola alone or with A. pintoi cv Mani Forrajero. 1988–93. Rotationally grazed with cattle at 3 head ha^{-1}.
Date soil sampled	December, 1992.	April, 1993.

subjected to severe mismanagement (burning, over- and under-grazing) at least as bad as the worst farmers' fields, accumulated C at the rate of $3\,t\,ha^{-1}\,yr^{-1}$ in the soil to a depth of 160 cm over this period (Fisher et al., 1995). A 50 ha pasture of *B. humidicola*, sown in 1979 with modest amounts of fertilizer but which received no maintenance fertilizer, had by 1997 accumulated C in the soil at the rate $2.9\,t\,ha^{-1}\,yr^{-1}$ (W. Trujillo, personal communication). This pasture was well managed as a normal farm pasture for 15 years. Recent unpublished data from Brazil, Colombia and Venezuela, obtained in a project financed by the Department for International Development of the United Kingdom government, show broadly similar accumulations in the soil under introduced pastures compared with that under native vegetation.

Neill et al. (1998) measured soil C in a number of chronosequences following forest clearing for pasture in the state of Rondônia, Brazil. They confirmed that pastures are able to contribute to the C stocks of the soil (Figure 3). Indeed, the

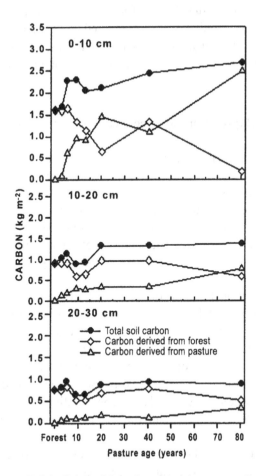

Figure 3. Changes in total soil C, in C derived from the original forest vegetation and in C derived from the sown pasture after felling of the forest. The data are from a chronosequence on the Fazenda Nova Vida, Rondônia, Brazil (Neill et al., 1997).

C concentration in the surface 10 cm increased from about $1.6 \, kg \, m^{-2}$ under the forest to almost $2.7 \, kg \, m^{-2}$ under pastures. The net increase in the 0–30 cm layer was a little over $1.6 \, kg \, m^{-2}$, equivalent to $16 \, t \, ha^{-1}$. Obviously these amounts are almost an order of magnitude less than the C stocks of the original forest, but the point is that once the forest has been cleared, introduced pastures can increase the amount of C in the soil.

Cerri et al. (1994) estimated that well-managed pastures sown after forest lose C from the soil in the first few years, but that within 8 years C in the soil has returned to an equilibrium, very close to the original level. Although Cerri et al.'s (1994) estimates from central Amazonas and Neill et al.'s (1998) data from Rondônia are consistent; they only considered C in the soil to 20 or to 30 cm, respectively. If dynamics of soil C under *Brachiaria* pastures in the forest ecosystem are similar to the data for the Colombian savannas summarized above, both Cerri et al.'s (1994) estimates and Neill et al.'s (1998) measurements could be substantially lower than the actual figures for the whole soil profile. Fujisaka et al. (1998) concluded that degraded pasture contained less C in the soil than untouched forest, but conceded that 'it also seems likely that the soil C equilibrium level [under well-managed pastures] can at least be maintained and possibly raised substantially'.

7. Mean elapsed time since conversion to introduced pasture

It is uncertain over what time scale any C accumulation in the soil under introduced pastures might have occurred. *B. decumbens* cv. Basilisk is well adapted to the infertile acid soils typical of much of the Cerrados region. Widespread conversion of the Cerrados to pastures was stimulated by its introduction from Australia in the early 1970s (Pizarro et al., 1996). We do not know the rate at which conversion occurred, but it seems likely that a broadly linear progression took place. In this case, a mean age of, say, 14 years seems a reasonable estimate.

The converted area of the Amazonian forest in Brazil grew from $0.152 \, Mkm^2$ in January 1978 to $0.517 \, Mkm^2$ in August, 1996, an increase of $0.364 \, Mkm^2$ in 18–1/2 years (Tables VI and VII). Although the rate of conversion fluctuated considerably from year to year (Table VIII), the mean rate is just under $20\,000 \, km^2 \, yr^{-1}$. If as discussed above, the proportion of the converted lands sown to pasture remained constant at the 18.5% estimated by Serrão and Toledo (1990), the area of pasture in 1996 would be $95\,900 \, km^2$. Conservatively ignoring lands sown to pasture prior to 1978, we can estimate the pastures as having a mean age of about 9 years.

There are no data for either Colombia or Venezuela of the rate of conversion to enable an estimate of mean pasture age, but since the technology was largely imported from Brazil, it seems reasonable to allow a decade to pass before conversion started, or a mean age of 9 years.

8. Pasture degradation

Since the source of any C that accumulates in the soil must be net primary productivity (NPP), anything that constrains NPP will inevitably reduce C accumulation. Pasture degradation is one such constraint, although there are few definitive data that document exactly what a degraded pasture is, nor is there a clear consensus what causes degradation (Miles et al., 1996). Compared with productive pastures, degraded pastures sustain lower stocking rate and liveweight gain per animal also falls so that liveweight production per unit area is drastically reduced (Kluthcouski et al., 1999). As a pasture degrades, individual plants die so that the sward opens up to bare soil, which either remains without plant cover or is invaded by herbaceous and woody weeds. Termite mounds become prominent and are often used as a simple indicator of failing pasture health (Boddey et al., 1996).

Pastures are thought to degrade due to one or more of N deficiency, P deficiency, over grazing, insect attack (mainly spittlebug, Homoptera:Cercopidae, Valério et al., 1996) and trampling leading to soil compaction. Although there are examples of pure grass pastures 15–20 year old that are not degraded, there is no universal recipe for their success (Fisher and Kerridge, 1996). Nor is there an unequivocally successful treatment to recuperate degraded pastures without some form of mechanical intervention, such as the Barreirão System (Kluthcouski et al., 1999) or some other form of integrated crop-pasture system (Vera et al., 1992; Thomas et al., 1995; Sanz et al., 1999; Valencia et al., 1999; Zimmer et al., 1999).

There are no data that show what form the decline in productivity takes, but for the purposes of general representation we used a declining ramp function (Figure 4). While this model is undoubtedly simplistic, we justify using it on the grounds that it is probably conservative and that it does permit the calculation of a mean 'rate of C accumulation' with minimum information, which is the current situation. The underlying assumption in the model is that a measure such as stocking rate, which has the advantage that average stocking rates are fairly well documented, is a reasonable indicator of NPP. Although other measures such as annual liveweight gain per ha may be more sensitive indicators of pasture health, and hence degradation, there are no reliable data for more than a few individual farms.

Three parameters describe the degradation relationship, maximum sustainable productivity (*Pmax*, with units of $hd\,ha^{-1}$), degraded productivity (*Pdeg*, also $hd\,ha^{-1}$) and rate of degradation (*Rdeg*, the slope of the line linking *Pmax* and *Pdeg*, with units of $hd\,ha^{-1}\,yr^{-1}$). The time in yr (*Tdeg*) for a pasture to reach *Pdeg* is given by

$$Tdeg = (Pmax - Pdeg)/Rdeg.$$

Two further parameters give a complete description of a pasture, the duration of the *Pmax* phase (*YrPmax*, with units of yr) and the cycle length between successive renovations (*YrCycle*, also yr). We assumed that the productivity of a renovated

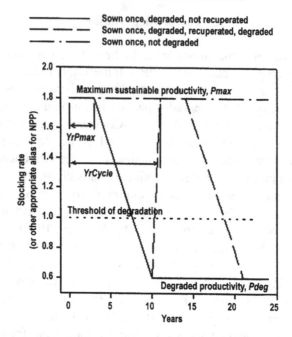

Figure 4. Conceptual model of the degradation process in pastures in terms of stocking rate.

pasture returns to *Pmax*. If these few parameters are known, we can readily calculate a mean degradation index (*Ideg*), which is (the mean productivity over any renovation cycle minus *Pdeg*) divided by (the difference between *Pmax* and *Pdeg*).

Because pastures of different species have different nutritional characteristics, the parameters of the functional relationship based on stocking rate and especially based on liveweight gain will almost certainly be species specific. There will probably be a considerable amount of site specificity as well. However, we are attempting to use some surrogate for NPP, itself an indicator for C accumulation in the soil, but none of these relations have yet been established. Given these uncertainties, it may be possible to apply a set of parameters to one group of lands with the same generalized classification within one particular climate subregion. The extent to which this is possible in practice is a clear area for research.

We calculated the influence of an arbitrary selection of scenarios for pasture degradation on the degradation index compared with a well-managed pasture, that is a pasture that does not degrade (Table XI). The main feature is that pastures that are recuperated not long after they have reached complete degradation (the 8-year and 12-year scenarios) probably accumulate as much as 50% the C in the soil compared with well-managed pastures whose productivity does not decline. It requires quite draconian mismanagement of rapid degradation and long-delayed renovation for the amount of soil C accumulated to fall below 30% of a well-managed pasture.

Fifty percent of the grass pastures of the Cerrados are said to show some degradation (Macedo, 1995). Kluthcouski et al. (1999) put the amount of pastures degraded or in the process of degradation as high as 80%. Pastures are reported to be vigorous

TABLE XI. The influence of a range of arbitrarily-chosen scenarios on the degradation index, described in the text above, compared with well-managed pasture.

Description	Symbol	Units	Scenario				
			WM[1]	8 yr	12 yr	15 yr	20 yr
Maximum sustainable productivity	Pmax	hd ha^{-1}	1.8	1.8	1.8	1.8	1.8
Degraded productivity	Pdeg	hd ha^{-1}	0.6	0.6	0.6	0.6	0.6
Rate of degradation	Rdeg	hd ha^{-1} yr^{-1}	na	0.2	0.15	0.25	0.25
Time to degrade	Tdeg	yr	na	6	8	4	4
Time at maximum productivity	YrPmax	yr	X	1	2	2	2
Time between renovation events	YrCycle	yr	X	8	12	15	20
Accumulated production:							
During maximum productivity	Maximum	hd ha^{-1} yr	1.8*X	1.8	3.6	3.6	3.6
While degrading	Degrading	hd ha^{-1} yr	0	7.2	9.6	4.8	4.8
When completely degraded	Degraded	hd ha^{-1} yr	0	0.6	1.2	5.4	8.4
Weighted mean productivity	AvProd	hd ha^{-1}	1.8	1.2	1.2	0.92	0.84
Degradation index	Ideg	%	100.0	50.0	50.0	26.7	20.0

[1] WM, well-managed pasture.

and productive during their first years, but to decline drastically in productivity after 4–10 years of use (Lopes et al., 1999). However, Gomide (1999) reported that the area of degraded pastures recuperated with the Barreirão System in 1993/94 was 4000 km^2, increasing from just 20 km^2 only three years earlier. Anecdotal reports suggest that use of the System has continued to spread rapidly. There are no estimates of the mean state of pasture health of introduced pastures in the central lowlands, that is, estimates of the mean *Ideg*. We have arbitrarily used 50%, which can easily be adjusted if and when better data become available.

9. Discussion and conclusions

There are a number of data for well-managed pastures of introduced grasses of African origin that show accumulation of C in the soil at rates close to 3 t ha^{-1} yr^{-1}. Even mismanaged *A. gayanus* accumulated C in the soil at this same rate. These data can be extrapolated to the 1600 km^2 sown to *Brachiaria* species in the Puerto López-Puerto Gaitán region of the western part of the Colombian llanos with some confidence. However, it is unclear the extent to which they might be extrapolated either to the 0.408 Mkm^2 of *Brachiaria* pastures in the Brazilian Cerrados or with greater uncertainty to the 0.10 Mkm^2 of pastures in the Amazon, largely *Brachiaria* spp.

All the data of C accumulation in the soil come from land facet number 1 of land system Oc201 (Cochrane et al., 1985), described as well-drained, flat, lowland savannas in climate subregion 'c'. Climate subregion 'c', the third data column of Table I, has wet season potential evapotranspiration 900–1060 mm, 6–8 months growing season and wet season mean temperature >23.5°C. Reference to Table I will show that of a total of 2.43 Mkm^2 in the Brazilian Shield physiographic region, 1.32 Mkm^2 or 54% have the same climate descriptor. Well-drained, flat lowland savannas on the Brazilian Shield occupy 1.26 Mkm^2 or 52% of the total area. It is not possible to say with certainty that the 0.41 Mkm^2 of *Brachiaria* pastures in the Cerrados all lie in climate subregion 'c', or are on lands that are well-drained, flat, lowland savannas, but on balance it is likely that most of them do. Because of this, while extrapolation to them is not without risk, it does not seem unduly heroic.

Most of the lands converted from forest in Amazonia are in the tropical semi-evergreen seasonal forest and semi-deciduous seasonal forest vegetation classes (Figure 2) in eastern, central and southwestern Amazonia. These areas lie mainly in climate subregion 'b' (Table I), which has a longer growing season than subregion 'c', in which the data of C accumulation in the soil under introduced pastures were measured. It does not seem unreasonable to suggest that the NPP of well-managed introduced pastures in subregion 'b' should be higher than the NPP of similar pastures in subregion 'c' with a shorter growing season. Since the source of any C that accumulates in the soil must be NPP, it is conservative to assume that subregion 'b' accumulates C in the soil at the same rate as in the drier subregion 'c'.

We can now summarize the main areas converted to pastures in the central lowlands of tropical South America and their possible accumulation of C in the soil (Tables XII and XIII). The data indicate that the lands sown to pastures in the savannas are strong sinks for C, potentially accumulating as much as 1800 million t (Mt) C in the soil to a depth of 100 cm. Even discounting this by 50% to account for pasture degradation still gives an accumulation of 900 Mt C. In contrast to use-conversion in the savannas, the forest areas sown to pasture, which account for less than 20% of the total forest area converted, have been a powerful source of atmospheric C,

TABLE XII. Estimates of possible C accumulation to a depth of 100 cm in soils under pastures in the central lowlands of tropical South America without considering degradation.

	Total area (see Tables II and III)	Area of *Brachiaria* pasture	Mean time since conversion	Area yr	Degradation discount	Potential accumulation[2]
Savannas	10^3 km^2	10^3 km^2	yr	10^3 km^2 yr	%	10^6 t C
Brazilian Shield	1261.1	408.0	14	5712	0	1713
Orinoco Basin	167.2					
Colombia	58.2	1.6	9	14	0	4
Venezuela	109.0	24.0–40.0	9	288	0	86
	Total cleared area	Area of *Brachiaria* pasture			Amount	
Forest	10^3 km^2	10^3 km^2	yr	10^3 km^2 yr	t ha^{-1}	10^6 t C
Clearing losses		95.9		96	115.6	−1109
Pasture gains		95.9	9	863	0%	259
Balance						−850

[1] Flat (slopes <8%), well-drained, lowland savannas.
[2] Assuming a mean rate of C accumulation in the soil of 3 t ha^{-1} yr^{-1}.

TABLE XIII. Estimates of possible C accumulation to a depth of 100 cm in soils under pastures in the central lowlands of tropical South America allowing 50% discount due to degradation.

	Total area (see Tables II and III)	Area of Brachiaria pasture	Mean time since conversion	Area yr	Degradation discount	Potential accumulation[2]
	10^3 km^2	10^3 km^2	yr	10^3 km^2 yr %		10^6 t C
Savannas						
Brazilian Shield	1261.1	408.0	14	5712	50	857
Orinoco Basin	167.2					
Colombia	58.2	1.6	9	14	50	2
Venezuela	109.0	24.0–40.0	9	288	50	43
	Total cleared area	Area of Brachiaria pasture			Amount	
	10^3 km^2	10^3 km^2	yr	10^3 km^2 yr t ha^{-1}		10^6 t C
Forest				96	115.6	−1109
Clearing losses		95.9				
Pasture gains		95.9	9	863	50%	130
Balance						−979

[1]Flat (slopes <8%), well-drained, lowland savannas.
[2]Assuming a mean rate of C accumulation in the soil of 3 t ha^{-1} yr^{-1}.

as much as 980 Mt assuming a 50% discount for degradation. At the level of 50% discount for degradation in both the forest and the savannas converted to pasture, the net scenario is a source of about 80 Mt C.

Of course there are uncertainties in these figures, which we have been at pains to identify in the foregoing text. In all cases, we have tried to make our estimates conservative. Clearly, there are pressing needs for better data as well as further research to enable more confident estimates. We have attempted to identify these needs below.

10. Research and data needs

- Any attempt to inventory soil C over the central lowlands of tropical South America is severely hampered by the lack of data. There is a clear need for many more data of soil C in this region.
- There is a need to extend Cochrane et al.'s (1985) land system classification to include the whole of the Brazilian Cerrados.
- The distribution of pastures by land system, preferably including pasture species and some estimate of pasture condition are needed to enable extrapolation from point sources of data to broader areas with greater confidence.
- Can remote sensing technologies be developed to provide these data? For example, can pastures be identified compared with crops? If so, can pastures be identified by species? Then by level of productivity?
- What are the characteristics of degradation in more precise terms than are currently available? What is the form of the functional relations of pasture degradation in terms of NPP? Animal performance? Stocking rate?

- If the area of sown pasture within a more or less well–defined area such as a municipality can be determined, could data of animal turnoff from this area be sufficiently sensitive to act as a proxy for pasture state, itself a proxy for NPP and hence for C accumulation in the soil?
- Could well-designed rural censuses provide improved estimates of land classes sown to pasture and their level of productivity?
- To what extent and with what degree of certainty can land systems with similar general descriptions within the same climatic subregion be treated as the same for the purposes of extrapolation?
- There are a number of errors in the land system and land facet descriptions of Cochrane et al. (1985) that require the publication of an erratum document.

References

Boddey, R.M., Alves, B.J.R. and Urquiaga, S.: 1996, 'Nitrogen cycling and sustainability of improved pastures in the Brazilian Cerrados', in R.C. Pereira and L.C. Nasser (eds.), *8o. Simposio Sobre o Cerrado. 1st International Symposium on Tropical Savannas*. Brasilia, DF, EMBRAPA-CPAC, pp. 33–38.

Brown, S. and Lugo, A.E.: 1992, 'Above-ground biomass estimates for tropical moist forests of the Brazilian Amazon', *Interciencia* **17**, 8–18.

Cerri, C.C., Bernoux, M. and Blair, G.J.: 1994, 'Carbon pools and fluxes in Brazilian natural and agricultural systems and the implications for the global CO_2 balance', *Transactions of the 15th World Congress of Soil Science*, Acapulco, Mexico, 10–16 July, 1994, Volume 5a, pp. 309–406.

Cochrane, T.T, Sánchez, L.G., de Azevedo, L.G., Porras, J.A. and Carver, C.L.: 1985, '*Land in Tropical America*', Cali, Colombia, Centro Internacional de Agricultura Tropical and Planaltina, Brazil, Empresa Brasilera de Pesquisa Agropecuária, Centro de Pesquisa Agropecuária dos Cerrados, Three volumes.

Decaëns, T., Rangel, A.F., Asakawa, N. and Thomas, R.J.: 1999, 'Carbon and nitrogen dynamics in ageing earthworm casts in grasslands of the eastern plains of Colombia', *Biology and Fertility of Soils* **30**, 20–28.

Fisher, M.J. and Kerridge, P.C.: 1996, 'The agronomy and physiology of *Brachiaria* species', in J.W. Miles, B.L. Maass and C.B. do Valle (eds.), *Brachiaria: Biology, Agronomy, and Improvement*, Cali, Colombia, Centro Internacional de Agricultura Tropical and Campo Grande, Brazil, Empresa Brasileira de Pesquisa Agropecuária/Centro Nacional de Pesquisa de Gado de Corte, pp. 43–52.

Fisher, M.J., Lascano, C.E., Vera, R.R. and Rippstein, G.: 1992, 'Integrating the native savanna resource with improved pastures', in: *Pastures for the Tropical Lowlands: CIAT's contribution*, Cali, Colombia, CIAT, pp. 75–99.

Fisher, M.J., Rao, I.M., Ayarza, M.A., Lascano, C.E., Sanz, J.I., Thomas, R.J. and Vera, R.R.: 1994, 'Carbon storage by introduced deep-rooted grasses in the South American savannas', *Nature* (London) **371**, 236–238.

Fisher, M.J., Rao, I.M., Lascano, C.E., Sanz, J.I., Thomas, R.J., Vera, R.R. and Ayarza, M.A.: 1995, 'Pasture soils as carbon sink', *Nature* (London) **376**, 472–473.

Fisher, M.J., Thomas, R.J. and Rao, I.M.: 1998, 'Management of tropical pastures in acid-soil savannas of South America for carbon sequestration in the soil', in R. Lal, J.M. Kimble, R.F. Follett and B.A. Stewart (eds.), *Management of Carbon Sequestration in Soil*, Advances in Soil Science Series, Boca Raton, Florida, CRC Press, pp. 405–420.

Fujisaka, S., Castilla, C., Escobar, G., Rodrigues, V., Veneklaas, E.J., Thomas, R. and Fisher, M.: 1999, 'The effects of forest conversion on annual crops and pastures: Estimates of carbon emissions and plant species loss in a Brazilian Amazon colony' *Agriculture, Ecosystems and Environment* **69**, 17–26.

Gomide, J. de C.: 1999, 'Transferencia de tecnología: el caso del Sistema Barreirão en Goiás, Brasil', in E.P. Guimarães, J.I. Sanz, I.M. Rao, M.C. Amézquita and E. Amézquita (eds.), *Sistemas Agropastoriles en Sabanas Tropicales de América Latina*, Cali, Colombia, CIAT and Brasilia, Brazil, EMBRAPA, pp. 107–114.

Greenland, D.J.: 1995, 'Land use and soil carbon in different agroecological zones', in R. Lal, J. Kimble, E. Levine and B.A. Stewart (eds.), *Soil Management and Greenhouse Effect*, Advances in Soil Science Series, Boca Raton, Florida, CRC Press, pp. 9–24.

Haridasan, M.: 1992, 'Observations on soils, foliar nutrient concentrations and floristic composition of cerrado sensu stricto and cerradão communities in central Brazil', in P.A. Furley, J. Proctor and J.A. Ratter (eds.), *Nature and Dynamics of Forest-Savanna Boundaries*, London, UK, Chapman & Hall.

Kluthcouski, J., de Oliveira, I.P., Yokoyama, L.P., Dutra, L.G., Portes, T. de A., da Silva, A.E., Pinheiro, B. da S., Ferreira, E., de Castro, E. da M., Guimarães, C.M., Gomide, J. de C. and Balbino, L.C.: 1999, 'Sistema Barreirão: recuperación/renovación de pasturas degradadas utilizando cultivos anuales', in E.P. Guimarães, J.I. Sanz, I.M. Rao, M.C. Amézquita and E. Amézquita (eds.), *Sistemas Agropasto-riles en Sabanas Tropicales de América Latina*, Cali, Colombia, CIAT and Brasilia, Brazil, EMBRAPA, pp. 195–231.

Long, S.P., Jones, M.B. and Roberts, M.J.: 1992, *Primary Productivity of Grass Ecosystems of the Tropics and Sub-tropics*, New York, Chapman and Hall.

Lopes, A., Ayarza, M. and Thomas, R.: 1999, 'Sistemas agropastoriles de las sabanas de América Latina tropical: Lecciones del desarrollo agrícola de los Cerrados de Brasil', in E.P. Guimarães, J.I. Sanz, I.M. Rao, M.C. Amézquita and E. Amézquita (eds.), *Sistemas Agropastoriles en Sabanas Tropicales de América Latina*, Cali, Colombia, CIAT and Brasilia, Brazil, EMBRAPA, pp. 9–30.

Macedo, J.: 1995, *Prospectives for the Rational Use of the Brazilian Cerrados for Food Production*, Brasilia, EMBRAPA/CPAC.

Miles, J.W., Maass, B.L. and do Valle, C.B.: 1996, 'Report of working groups', in J.W. Miles, B.L. Maass and C.B. do Valle (eds.), *Brachiaria: Biology, Agronomy, and Improvement*, Cali, Colombia, Centro Internacional de Agricultura Tropical and Campo Grande, Brazil, Empresa Brasileira de Pesquisa Agropecuária/Centro Nacional de Pesquisa de Gado de Corte, pp. 272–277.

Neill, C., Cerri, C., Melillo, J.M., Feigl, B.J., Steudler, P.A., Moraes, J.F.L. and Piccolo, M.C.: 1998, 'Stocks and dynamics of soil carbon following deforestation for pasture in Rondônia', in R. Lal, J.M. Kimble, R.F. Follett and B.A. Stewart (eds.), *Soil Processes and the Carbon Cycle*, Advances in Soil Science Series, Boca Raton, Florida, CRC Press, pp. 9–28.

Pizarro, E.A., do Valle, C.B., Keller-Grein, G., Schultze-Kraft, R. and Zimmer, A.H.: 1996, 'Regional experience with *Brachiaria*: tropical America-savannas', in J.W. Miles, B.L. Maass and C.B. do Valle (eds.), *Brachiaria: Biology, Agronomy, and Improvement*, Cali, Colombia, Centro Internacional de Agricultura Tropical and Campo Grande, Brazil, Empresa Brasileira de Pesquisa Agropecuária/Centro Nacional de Pesquisa de Gado de Corte, pp. 225–246.

Sanz, J.I., Zeigler, R.S., Sarkarung, S., Molina, D.L. and Rivera, M.: 1999, 'Sistemas mejoradas arroz-pasturas para sabana nativa y pasturas degradadas en suelos acidos de América Sur', in E.P. Guimarães, J.I. Sanz, I.M. Rao, M.C. Amézquita and E. Amézquita (eds.), *Sistemas Agropastoriles en Sabanas Tropicales de América Latina*, Cali, Colombia, CIAT and Brasilia, Brazil, EMBRAPA, pp. 232–244.

Serrão, E.A.S. and Homma, A.K.O.: 1993, 'Brazil', in *Sustainable Agriculture and the Environment in the Humid Tropics*, National Research Council, National Academy Press, Washington DC, pp. 265–351.

Serrão, E.A.S. and Toledo, J.M.: 1990, 'The search for sustainability in Amazonian pastures', in A.B. Anderson (ed.), *Alternatives to Deforestation: Steps Toward Sustainable Use of the Amazon Rain Forest*, New York, Columbia University Press, pp. 195–214.

Thomas, R.J., Fisher, M.J., Ayarza, M.A. and Sanz, J.I.: 1995, 'The role of forage grasses and legumes in maintaining the productivity of acid soils in Latin America', in R. Lal and B.A. Stewart (eds.), *Soil Management: Experimental Basis for Sustainability and Environmental Quality*, Advances in Soil Science Series, Boca Raton, USA, CRC Lewis Pubs., pp. 61–83.

Valencia R., R.Q., Salamanca, C.R., Navas R., G.E., Baquero P., J.E., Rincón, A. and Delgado, H.: 1999, 'Evaluación de sistemas agropastoriles en la altillanura de la Orinoquia Colombiana', in E.P. Guimarães, J.I. Sanz, I.M. Rao, M.C. Amézquita and E. Amézquita, (eds.), *Sistemas Agropastoriles en Sabanas Tropicales de América Latina*, Cali, Colombia, CIAT and Brasilia, Brazil, EMBRAPA, pp. 284–300.

Valério, J.R., Lapointe, S.L., Kelemu, S., Fernandes, C.D. and Morales, F.J.: 1996, 'Pests and diseases of *Brachiaria* species', in J.W. Miles, B.L. Maass and C.B. do Valle (eds.), *Brachiaria: Biology, Agronomy, and Improvement*, Cali, Colombia, Centro Internacional de Agricultura Tropical and Campo Grande, Brazil, Empresa Brasileira de Pesquisa Agropecuária/Centro Nacional de Pesquisa de Gado de Corte, pp. 87–105.

Vera, R.R., Thomas, R., Sanint, L. and Sanz, J.I.: 1992, 'Development of sustainable ley-farming systems for the acid-soil savanas of tropical America', *An Acad Bras Ci* **64** (Suppl 1), 105–125.

Zimmer, A.H., Macedo, M.C.M., Kichel, A.N. and Euclides, V.P.B.: 1999, 'Sistemas integrados de producción agropastoril', in E.P. Guimarães, J.I. Sanz, I.M. Rao, M.C. Amézquita and E. Amézquita (eds.), *Sistemas Agropastoriles en Sabanas Tropicales de América Latina*, Cali, Colombia CIAT and Brasilia, Brazil, EMBRAPA, pp. 245–283.

ESTIMATION OF SOIL CARBON GAINS UPON IMPROVED MANAGEMENT WITHIN CROPLANDS AND GRASSLANDS OF AFRICA

NIELS H. BATJES

International Soil Reference and Information Centre (ISRIC)/World Data Centre for Soils, P.O. Box 353, 6700 AJ Wageningen, The Netherlands (e-mail: batjes@isric.nl)

(Accepted in Revised form 15 January 2003)

Abstract. Many agro(eco)systems in Africa have been degraded as a result of past disturbances, including deforestation, overgrazing, and over exploitation. These systems can be managed to reduce carbon emissions and increase carbon sinks in vegetation and soil. The scope for soil organic carbon gains from improved management and restoration within degraded and non-degraded croplands and grasslands in Africa is estimated at 20–43 Tg C year^{-1}, assuming that 'best' management practices can be introduced on 20% of croplands and 10% of grasslands. Under the assumption that new steady state levels will be reached after 25 years of sustained management, this would correspond with a mitigation potential of 4–9% of annual CO_2 emissions in Africa. The mechanisms that are being put in place to implement the Kyoto Protocol – through C emission trading – and prevailing agricultural policies will largely determine whether farmers can engage in activities that enhance C sequestration in Africa. Mitigation of climate change by increased carbon sequestration in the soil appears particularly useful when addressed in combination with other pressing regional challenges that affect the livelihood of the people, such as combating land degradation and ensuring food security, while at the same time curtailing global anthropogenic emissions.

Key words: Africa, carbon sequestration, croplands, grasslands, soil degradation, soil management, soil organic matter

1. Introduction

Africa, the second largest continent ($\sim 30 \times 10^8$ ha), has a wide diversity of climates, ecosystems and soil conditions (FAO, 1993; UNEP, 1992, pp. 27–39). Since the 1960s, the continent has been experiencing serious economic and environmental problems. The growth of its population, the highest in the world, has placed pressure on many ecosystems (UNEP, 1997). In addition, there has been political and social turmoil in many countries.

Farming systems in Africa have been evolving towards land-use intensification in response to population growth and the scarcity of land suitable for long-fallow shifting cultivation. In many agro(eco)systems the soil has been degraded, due mainly to deforestation, overgrazing, agricultural mismanagement, and over exploitation (Oldeman et al., 1991). These degraded systems can be managed to reduce carbon emissions and increase carbon sinks in vegetation and soil, thus contributing to global climate change mitigation, an option that has been considered in article 3.4 of the Kyoto Protocol (WBBGU, 1998, pp. 56–67). In tropical and subtropical regions, the greatest potential to increase current soil carbon stocks is probably through improved management of the agricultural land, particularly degraded croplands and grasslands, as well as conservation of marginal lands and wetlands and peat lands.

The soil is an important natural sink for carbon released into the atmosphere by fossil fuel combustion and land use changes and this is due to the relatively large size and long residence time of soil-organic carbon (SOC) in humus. The content of organic carbon can be increased in three ways: (a) by improved management within a land use system, (b) by conversion of one land use to another land use with higher carbon stocks, and (c) by increasing carbon content in harvested products (Sampson and Scholes, 2000). Primary production and soil microbial activity are the main biological processes controlling organic matter dynamics in the soil. Input rates and the quality of vegetation-carbon are largely dependent on climate, especially temperature and precipitation, vegetation type and landscape, soil type, and management practices (Jastrow and Miller, 1997). Humification, aggregation, and translocation to the subsoil are the main processes of carbon sequestration in soils, while erosion, decomposition and leaching are important processes causing carbon concentrations to decrease in the soil (Lal et al., 1998). The intensity of many of these processes is affected by changes in land use systems and hence by human activity. Options for carbon sequestration thus must be chosen on the basis of knowledge of the nature and likely magnitude of carbon pools, whether organic or inorganic, in the soils of a given biome or major agro-ecological region and the responses of these soils to different land uses and management systems.

In the short-term, the main factors controlling carbon levels in managed soils are the type of land use, and the specific soil and crop management practices. Increased levels of soil organic matter will favourably affect the physical, chemical, and thermal properties of the soil as well as biological activity, and they are important in sustaining soil productivity and biodiversity.

Recommended best management practices to build-up carbon stocks in the soil are those that increase the input of organic matter to the soil, and/or decrease the rates of soil organic matter decomposition. These practices often include a combination of the following: tillage methods and residue management; soil fertility and nutrient management; erosion control; water management; and crop selection and rotation. Suitable strategies for organic carbon management of croplands, grasslands, and forestlands, as well as for land use conversion have been documented elsewhere (Lal et al., 1998; Paustian et al., 1998; Sampson and Scholes, 2000).

Potential rates of carbon sequestration in response to improved management and conservation or restoration practices vary widely as a function of land use, climate, soil, and other non-biophysical factors. Available data do not yet allow definitive evaluation of potential rates of carbon gain for all agro-ecological regions and management options (Batjes, 1999; Lal, 1999; Sampson and Scholes, 2000).

About 65% of the agricultural land and 31% of the permanent pastureland in Africa are affected by human-induced soil degradation (Oldeman, 1994). The objective of this exploratory study is to estimate the eco-technological scope for increased soil carbon sequestration within grasslands and croplands of Africa, where a large portion of the rural population lives, subsequent on improved management or the restoration of degraded soils. In addition, background data are provided on SOC stocks, by broad agro-ecological region, to provide an indication of possible 'historic' levels of soil carbon sequestration, in the broader context of climate change mitigation at a world level.

2. Materials and methods

The semi-quantitative approach used considers differences in SOC-sequestration potential by broad land use type, agro-ecological zone (AEZ), soil type, and severity of soil degradation, as defined on available 0.5° by 0.5° data sets for Africa (Figure 1). Information about the type and distribution of croplands and grasslands is derived from land cover data of IMAGE2.1 (Alcamo et al., 1998). The division into AEZs is according to criteria of FAO-IIASA (de Pauw et al., 1996; FAO, 1996). Information about the degree and extent of degraded soils is derived from the 'World Map of the Status of Human-Induced Soil Degradation', GLASOD (Oldeman et al., 1991). The corresponding database expresses the severity of soil degradation in terms of declining agricultural productivity and the degree to which the original biotic functions are impaired. Reduction factors are introduced into the model to reflect that it becomes increasingly difficult and expensive to restore SOC levels through improved management and restoration as the degree of soil degradation increases from low to very high.

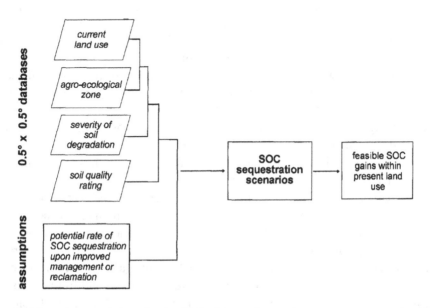

Figure 1. Schematic approach for estimating SOC gains upon improved management or restoration within current land use systems.

TABLE I. Potential rates of carbon sequestration upon introduction of improved management within or restoration of croplands and grasslands, by broad agro-ecological zone.

Management practices	Carbon sequestration potential by agro-ecological zone[a] ($Mg C ha^{-1} yr^{-1}$)[b]			
	Arid	Sub-tropics	Tropics	Temperate (Mediterranean)
Improved management of Croplands[c]	0–0.10	0.1–0.3	0.3–0.6	0.2–0.5
Improved management of Grasslands[c]	0–0.05	0.05–0.1	0.1–0.3	0.1–0.2
Restoration of eroded/ degraded soils	0–0.1	0.2–0.4	0.4–0.8	0.3–0.6

[a]For definitions of agro-ecological zones see de Pauw et al. (1996).
[b]Indicative rates, based on data of Bruce et al. (1999), Lal et al. (1998; 1999), and Sampson and Scholes (2000, p. 199).
[c]Categories aggregated from land cover data of IMAGE (Alcamo et al., 1998).

Table I lists default estimates of carbon sequestration, upon improved cropland and grassland management and restoration of degraded soils, by broad agro-ecological zone. Generally, the rate of carbon increase in soil will follow a sigmoid curve, and be at its maximum between 10 and 15 years after adoption of appropriate management or restorative measures. The net SOC accumulation usually becomes small after 20–50 years – depending on land use, management practices, climatic conditions, and soil properties – as the system approaches a new steady state (Lal et al., 1999; Sampson and Scholes, 2001). A time span of 25 years is considered appropriate in the context of this exploratory study.

The scope for increasing humus levels, upon improved management of a soil, will vary with soil quality (e.g. there is a higher scope for medium textured soils than for coarse textured soils within a given agro-ecological zone), and many soils in Africa have a low inherent quality (FAO, 1993). Therefore, a system for rating the overall potential of a soil for sequestering carbon is considered in the analyses. The scheme considers soil pH, cation exchange capacity, organic matter content, rootable soil depth, and available water capacity to rate each soil type into three broad soil quality classes (low, medium, high). Median values for the selected parameters are taken from the WISE database (Batjes, 1997). The class limits for rating and the procedure for combining the various soil attributes into an overall soil quality rating is after Batjes and Bouwman (1989). Arbitrarily, the reduction factors have been set at 1.0 for areas with soils with a high-inferred soil quality, at 0.8 for those with a medium rating, and at 0.6 for the remaining soils.

Only a part of the croplands and grasslands in Africa can be subjected to improved management due to a range of environmental, socio-economic, and policy constraints. A figure of 20% for croplands and 10% for grasslands seems appropriate for most developing countries (Sampson and Scholes, 2000). However, so as to present so-called 'windows-of-opportunities' for the achievable gains in soil carbon levels, the assumption is that from 10% to 30% of current croplands and from 5% to 15% of current grasslands can either be improved or restored. The eco-technological potential *per se* would correspond with 100% use of improved management and/or conservation measures. In practice, however, such estimates of SOC sequestration potential may well provide too bright a picture of realistically achievable levels of carbon sequestration, and as such they will not be presented here.

3. Results and discussion

3.1. CONTEMPORARY SOC STOCKS

The total stock of organic carbon for Africa is 90–96 Pg C ($1\,Pg = 10^3 \times Tg = 10^{15}\,g$) for the upper 0.3 m of soil and 170–179 Pg C for the first 1 m of soil. This corresponds with an area-weighted mean content of about 3 kg C m^{-2} in the top 0.3 m, and 6 kg C m^{-2} to 1 m depth. Soil carbon content increases from about 3 kg C m^{-2} to 1 m depth in the Arid zone to about 9 kg C m^{-2} in the Humid Tropics (Table II). Table III shows the area-weighted SOC content by length of growing period, with the highest content being observed in the '>365 days' category and the lowest in the '<30 days' category. On the whole, the potential for increased soil carbon sequestration in Africa, given improved management or restoration, thus will increase with increasing precipitation and with decreasing temperature for any particular level of precipitation. Due to regional differences in overall soil quality, agro-ecological conditions and land use practices, as described in FAO (1993), there are clear regional differences in area-weighted SOC stocks between (and within) North, West, Southern, East, and Middle Africa (Table IV).

3.2. ACHIEVABLE SOC GAINS

Table V shows different 'windows-of-opportunity' for achievable gains in soil carbon stocks upon improved management and/or restoration within grassland and croplands in Africa. In the 'reference' scenario, which assumes that 20% of the

TABLE II. Area-weighted stocks of soil organic carbon by broad agro-ecological zone in Africa[a].

Agro-ecological zone	Total soil organic carbon (kg C m^{-2})	
	0–0.3 m	0–1 m
Arid	1.6–1.9	2.9–3.3
Subtropics, summer rains	3.1–3.4	6.1–6.4
Subtropics, winter rains	2.7–3.0	5.2–5.6
Temperate (Mediterranean)	2.8–3.7	4.6–5.5
Tropics, warm humid	4.9–5.0	9.2–9.4
Tropics, warm seasonally dry	3.3–3.5	6.4–6.7
Tropics, cool	4.3–4.4	8.3–8.6

[a]Based on analyses of WISE database, using the calculation procedure described in Batjes (1996).

TABLE III. Area-weighted stocks of soil organic carbon by length of growing period in Africa.

Length of growing period (in days)	Total soil organic carbon (kg C m^{-2})	
	0–0.3 m	0–1 m
<30	1.6–1.9	2.9–3.3
30–60	2.2–2.4	4.4–4.7
60–90	2.4–2.6	4.7–5.0
90–120	2.8–3.0	5.7–6.0
120–150	3.3–3.6	6.8–7.1
150–210	4.1–4.3	7.9–8.2
210–240	4.0–4.2	7.6–7.8
240–270	4.2–4.4	7.8–8.1
250–300	4.6–4.8	8.7–8.9
300–330	5.3–5.4	10.1–10.3
>330	5.6–5.7	10.9–11.0

Length of growing period according to definitions and criteria of FAO-IIASA (de Pauw et al., 1996).

TABLE IV. Area-weighted stocks of soil organic carbon for Africa by region.

Region	Total soil organic carbon (kg C m^{-2})	
	0–0.3 m	0–1 m
North Africa	2.1–2.4	4.0–4.4
West Africa	2.3–2.6	4.2–4.5
Southern Africa	2.7–2.9	5.3–5.5
East Africa	3.7–3.9	7.2–7.5
Middle Africa	4.2–4.4	8.1–8.3
Africa	3.0–3.2	5.7–6.0

TABLE V. Technologically achievable gains in soil carbon stocks upon improved management or restoration within croplands and grasslands in Africa.

Land use system			
Croplands		Grasslands	
Area improved (%)	SOC gain (Tg C year^{-1})	Area improved (%)	SOC gain (Tg C year^{-1})
30	11–22	15	20–42
20	7–15	10	13–28
10	4–8	5	7–14

croplands and 10% of the grasslands can be realistically subjected to improved management or restoration, between 20 and 43 Tg C year^{-1} can be sequestered. This would correspond with a mitigation potential of 0.5–1.1 Pg C, under the assumption that 25 years of sustained best management practices will be needed to reach new steady state levels (see 2). About one third of this total would be achieved by improved soil management and restoration within croplands. When looking at the whole 'window-of-opportunities', the scope for SOC mitigation is 4–22 Tg C year^{-1} for croplands, and 7–42 Tg C year^{-1} for grasslands.

Africa's CO_2 emissions are low both in absolute and per capita terms (Boden et al., 1994). Between 1995 and 2000, the continent contributed to about 5–7% of global CO_2 emissions (IMAGE-team, 2001). The present reference scenario thus could help mitigating an estimated 4–9% of (current) annual African CO_2 emissions, until new steady state conditions are reached.

3.3. EVALUATION

Validation of the results is difficult. A simple evaluation can be made, however, with reference to soil degradation data published by Oldeman (1994). According to that study, about 65% of agricultural lands (121 Mha) and 31% of permanent pastures (243 Mha) in Africa is degraded. The remainder is considered to be non-degraded (i.e. 66 Mha for agricultural land and 550 Mha for permanent pasture). Using the same assumptions for the area that can be 'realistically' improved or reclaimed within agricultural lands (20%) and permanent pastures (10%), and broad ranges for SOC sequestration (i.e. 0.2–0.4 Mg C ha^{-1} year^{-1} for improved agricultural land, 0.1–0.3 Mg C ha^{-1} year^{-1} for improved permanent pasture, and 0.4–0.6 Mg C ha^{-1} year^{-1} for all degraded lands; see Table I) the achievable SOC gain is estimated at 28–51 Tg C year^{-1}. Contrary to the reference scenario (20–43 Tg C year^{-1}), this simpler procedure does not consider the effects of regional differences in agro-ecological conditions nor soil quality on possible soil carbon sequestration rates within the different land use systems under consideration.

3.4. BEST MANAGEMENT PRACTICES

The identification of appropriate management practices to increase soil C reserves must be site specific. Available technological options will require evaluation and adaptation with reference to major soil type and land use system, and this preferably by agro-ecological region. At same the time, the recommended practices should be socially, politically, and economically acceptable.

Best management practices for cropland management will often encompass an adroit combination of the following options: (a) conservation tillage in combination with planting of cover crops, green manure and hedgerows; (b) organic residue and fallow management; (c) mulch farming, especially in dry areas; (d) water management, including *in situ* water conservation in the root-zone, irrigation, and

drainage to avoid potential risk of salinization and water-logging; (e) soil fertility management, including use of chemical fertilizers and organic wastes, liming
and acidity management; (f) introduction of agro-ecologically and physiologically
adapted crop/plant species, including agroforestry; (g) adapting crop rotations and
cropping/farming systems, with avoidance of bare fallow; and (h) stabilization of
slopes and terraces to reduce erosion risk. With respect to grassland management,
recommended practices include: (a) controlling of grazing to sustainable levels;
(b) fertilization; (c) introduction of improved pastures, including deep rooting
grasses and legumes; (d) fire management; as well as (e) soil degradation control. Further details may be found elsewhere (Batjes, 1999; Bruce et al., 1999; Lal,
1999; Sampson and Scholes, 2001).

Actual implementation of the above practices has met with varying success in
increasing soil nutrient and carbon stocks in Africa (Pierri, 1989; Breman and
Sosoko, 1998; FAO, 2001a). For example, the various options vary significantly with
respect to external inputs and may require a change in the way existing resources
are applied (Ringius, 2002), for instance with respect to the use of organic farm
residues, and they differ in terms of achievable rates of carbon sequestration and
their effect on crop yields. N-fertilizers, for example, are needed to increase crop
yields at a local level but their use may lead to increases in N_2O emissions, another
potent greenhouse gas, thereby reducing net C gains in the context of climate change
mitigation at a world level. Constraints and possible solutions, including the need
for farmer participatory approaches and integrated soil management, specific to
Africa's major agro-ecological zones have been presented elsewhere (Pieri, 1989;
Smaling et al., 1996; IFDC, 1997; FAO, 2001a,b). Overall, it is crucial to improve
the productivity and sustainability of existing agricultural lands to help reducing the
current rate of over-exploitation, new land clearing, and burning from which large
amounts of greenhouse gases from soil and biomass are released to the atmosphere
(Paustian et al., 1997; Scholes and Andreae, 2000).

4. Conclusions

Although only afforestation and reforestation projects will be eligible for carbon
crediting under the Clean Development Mechanism during the first 5-year commitment period, it would appear that carbon sequestration in 'agricultural soils and
other land-use change categories' could become eligible during subsequent commitment periods (UNFCCC, 2001). An estimated 0.5–1.1 Pg C can be sequestered
over the next 25 years in the soils of croplands and grasslands in Africa, provided
best management practices can be sustained, according to the reference scenario.
This could help mitigating an estimated 4–9% of annual African CO_2 emissions.
When discussing possibilities for carbon sequestration in relation to climate change
mitigation at a world level, however, the critical issue of soil organic matter maintenance in relation to improved agricultural productivity at a local level, so as to
ensure the livelihood of farmers and society at large, deserves careful attention.

A significant part of the agricultural community in Africa has little funds, operates outside of the market economy, and has no access to credit (Vlek, 1995). The socio-economic opportunities and challenges for Africa, in terms of soil carbon sequestration in relation to the Clean Development Mechanism proposed under Article 12 of the Kyoto Protocol, have been addressed elsewhere (Sokona et al., 1998; Ringius, 2002). Barbier (2000) and Koning et al. (2001) discussed the need for appropriate agricultural policies to facilitate widespread improvement of soil quality and soil productivity in Africa. Izac (1997) argued that the interests of resource-poor farmers, of individual African countries, and of the world community are often conflicting. A detailed discussion of these critical and complex issues is, however, beyond the scope of this paper.

This exploratory study puts boundaries on achievable levels of carbon sequestration in soils of croplands and grasslands in Africa. An important source of uncertainty is associated with the assumptions for likely increases in soil carbon sequestration upon the introduction of improved management or restoration practices (see introduction) and the available 0.5° by 0.5° resolution grid databases (Bouwman et al., 1999; Cramer and Fischer, 1997). In order to improve national assessment methodologies relating to land use options and UNFCCC requirements, it will be necessary to develop and implement generic tools that quantify the impact of land management and climate change scenarios on carbon sequestration in soils (Bazzaz and Sombroek, 1996; Paustian et al., 1997). Batjes (2001) discussed the usefulness of having an up-to-date, high resolution integrated Information System for Sustainable Management of Land Resources (LRIS) for Africa, linked to dynamic simulation models and socio-economic modules, that would allow regional and national assessments of areas considered suitable for implementing 'carbon sequestration projects'.

Reduction of atmospheric CO_2 concentrations by increased carbon sequestration in the soils of Africa appears particularly useful when addressed in combination with other pressing regional challenges, such as combating land degradation, improving soil quality and productivity, and preserving biodiversity, while simultaneously curtailing global greenhouse gas emissions associated with fossil fuel combustion and industry.

Acknowledgements

The invitation by Professor P.L.G. Vlek to attend the workshop on 'Tropical Agriculture in Transition – Opportunities for Mitigating Greenhouse Gas Emissions?' (Bonn, 7–9 November 2001), jointly organized by the Centre for Development Research (ZEF), Bonn, and the Fraunhofer Institute for Atmospheric Environmental Research (IFU), Garmisch-Partenkirchen, is gratefully acknowledged.

References

Alcamo, J., Kreileman, E., Krol Leemans, R., Bollen, J., van Minnen, J., Schaeffer, M., Toet, S. and de Vries, B.: 1998, 'Global modelling of environmental change: an overview of IMAGE 2.1', in J. Alcamo, R., Leemans, R. and E. Kreileman (eds.), *Global Change Scenarios of the 21st Century. Results from the IMAGE 2.1 Model*, Elsevier, Amsterdam, pp. 19–21.

Barbier, E.B.: 2000, 'The economic linkages between rural poverty and land degradation: some evidence from Africa', *Agriculture Ecosystem and Environment* **82**, 355–370.

Batjes, N.H.: 1996, 'Total carbon and nitrogen in the soils of the world', *European Journal of Soil Sciences* **47**, 151–163.

Batjes, N.H.: 1997, 'A world data set of derived soil properties by FAO-UNESCO soil unit for global modelling', *Soil Use and Management* **13**, 9–16.

Batjes, N.H.: 1999, *Management options for Reducing CO_2 Concentrations in the Atmosphere by Increasing Carbon Sequestration in the Soil*, NRP Report No. 410–200–031, Bilthoven, Dutch National Research Programme on Global Air Pollution and Climate Change, 114 pp.

Batjes, N.H.: 2001, 'Options for carbon sequestration in west African soils: an exploratory study with special focus on Senegal', *Land Degradation and Development* **12**, 131–142.

Batjes, N.H. and Bouwman, A.F.: 1989, 'JAMPLES a computerized land evaluation system for Jamaica', in J. Bouma and A.K. Bregt (eds.), *Land Qualities in Space and Time*, Wageningen, PUDOC, pp. 257–260.

Bazzaz, F. and Sombroek, W. (eds.): 1996, *Global Climate Change and Agricultural Production – Direct and Indirect Effects of Changing Hydrological, Pedological and Plant Physiological Processes*, Chichester, FAO and John Wiley & Sons, 358 pp.

Boden, T.A., Kaiser, D.P., Sepanski, R.J. and Stoss, F.W. (eds.): 1994, 'Trends '93: a compendium of data on global change', in G.M. Logsdon (ed.), Oak Ridge, Carbon Dioxide Information Analysis Center.

Bouwman, A.F., Derwent, R.G. and Dentener, F.J.: 1999, 'Towards reliable global bottom-up estimates of temporal and spatial patterns of emissions of trace gases and aerosols from land-use related and natural resources', in A.F. Bouwman (ed.), *Approaches to Scaling of Trace Gas Fluxes in Ecosystems*, Amsterdam, Elsevier, pp. 3–26.

Breman, H. and Sosoko, K.: 1998. *L' Intensification agricole au Sahel*. Paris, Editions Karthala.

Bruce, J.P., Frome, M., Haites, E., Janzen, H., Lal, R. and Paustian, K.: 1999, 'Carbon sequestration in soils', *Journal of Soil and Water Conservation* **54**, 382–389.

Cramer, W. and Fischer, A.: 1997, 'Data requirements for global terrestrial ecosystem modelling', in B. Walker, and W. Steffen (eds.), *Global Change and Terrestrial Ecosystems*, Cambridge, Cambridge University Press, pp. 529–565.

de Pauw, E., Nachtergaele, F.O., Antoine, J., Fischer, G. and Van Velthuizen, H.T.: 1996, 'A provisional world climatic resource inventory based on the length-of-growing-period concept', in N.H. Batjes, J.H. Kauffman and O.C. Spaargaren (eds.), *National Soil Reference Collections and Databases (NASREC)*, Wageningen, ISRIC, pp. 30–43.

FAO: 1993, *World Soil Resources: An Explanatory Note on the FAO World Soil Resources map at 1:25,000,000 Scale*, World Soil Resources Report 66 (rev. 1), Rome, Food and Agriculture Organization of the United Nations, 61 pp.

FAO: 1996, *Food Production and Environmental Impact*, Technical Background Document 11 (http://www.fao.org/wfs), Rome, Food and Agriculture Organization of the United Nations.

FAO: 2001a, *Soil Fertility Management in Support of Food Security in Sub-Saharan Africa*, Rome, Food and Agriculture Organization of the United Nations, 55 pp.

FAO: 2001b, *Soil Carbon Sequestration for Improved Land Management*, World Soil Resources Report 96, Rome, Food and Agriculture Organization of the United Nations, Rome, 58 pp.

IFDC: 1997, *Framework for National Soil Fertility Improvement Action Plans*, Lome, International Fertilizer Development Center, 10 pp.

IMAGE-team: 2001, *The IMAGE 2.2 Implementation of the SRES Scenarios. A Comprehensive Analysis of Emissions, Climate Change and Impacts in the 21st Century*, CD-ROM publication 481508018, Bilthoven, National Institute for Public Health and the Environment (RIVM).

Izac, A.M.N.: 1997. Developing policies for soil carbon management in tropical regions, *Geoderma* **79**, 261–276.

Jastrow, J.D. and Miller, R.M.: 1997, 'Soil aggregate stabilization and carbon sequestration: feedbacks through organomineral associations', in R. Lal, J.M. Kimble, R.F. Follet and B.A. Stewart (eds.), *Soil Processes and the Carbon Cycle*, Boca Raton, CRC Press, pp. 207–223.

Koning, N., Heerink, N. and Kauffman, S.: 2001, 'Food insecurity, soil degradation and agricultural markets in West Africa: why current policy approaches fail', *Oxford Development Stud.* **29**, 189–207.

Lal, R.: 1999, 'Global carbon pools and fluxes and the impact of agricultural intensification and judicious land use', in R. Dudal (ed). *Prevention of Land Degradation, Enhancement of Carbon Sequestration and Conservation of Biodiversity Through Land use Change and Sustainable land Management with a Focus on Latin America and the Caribbean*, Rome, Food and Agriculture Organization of the United Nations, pp. 79–94.

Lal, R., Kimble, J.M., Follet, R.F. and Cole, C.V. (eds.): 1998, 'The Potential of U.S. Cropland to Sequester Carbon and Mitigate the Greenhouse Effect', Chelsea, MI, Ann Arbor Press, pp. 128.

Oldeman, L.R.: 1994, 'The global extent of soil degradation', in D.J. Greenland and I. Szabolcs (eds.), *Soil Resilience and Sustainable land Use*, Wallingford, CAB Int., pp. 99–118.

Oldeman, L.R., Hakkeling, R.T.A. and Sombroek, W.G.: 1991, *World Map of the Status of Human-Induced Soil Degradation: An explanatory Note (rev. ed.)*, Wageningen, UNEP and ISRIC, 35 pp (with maps).

Paustian, K., Levine, E.R., Post, W.M. and Ryzhova, I.M.: 1997. The use of models to integrate information and understanding of soil C at the regional scale. *Geoderma* **79**, 227–260.

Paustian, K., Andrèn, O., Janzen, H.H., Lal, R., Smith, P., Tain, G., Tiessen, H., van Noordwijk, M. and Woomer, P.L.: 1998, 'Agricultural soils as a sink to mitigate CO_2 emissions', *Soil Use and Management* **13**, 230–244.

Pieri, C.: 1989, *Fertilité des terres de savanes – Bilan de trente ans de recherche et de développement agricoles au sud du Sahara*, Paris, Ministère de la Coopération et du Développement and CIRAD-IRAT, 444 pp.

Ringius, L., 2002. 'Soil carbon sequestration and the CDM: opportunities and challenges for Africa', *Climatic Change* **154**, 471–495.

Sampson, R.N. and Scholes, R.J.: 2000, 'Additional human-induced activities – Article 3.4', in R.T. Watson, I.R. Noble, B. Bolin, N.H. Ravindranath, D.J. Verardo and D.J. Dokken (eds.), *Land Use, Land-Use Change, and Forestry*, Cambridge, Published for the Intergovernmental Panel on Climate Change by Cambridge University Press, pp. 183–281.

Schlesinger, W.H.: 2000. 'Carbon sequestration in soils: some cautions amidst optimism', *Agriculture Ecosystems and Environment* **82**, 121–127.

Scholes, M. and Andreae, M.O.: 2000, 'Biogenic and pyrogenic emissions from Africa and their impact on the global atmosphere', *Ambio* **29**, 23–29.

Smaling, E.M.A., Fresco, L.O. and De Jager, A.: 1996, 'Classifying, monitoring and improving soil nutrient stocks and flows in African agriculture', *Ambio* **25**, 492–496.

Sokona, Y., Humphreys, S. and Thomas, J.-P.: 1998, 'What prosects for Africa?' in J. Goldemberg (ed.), *The Clean Development Mechanism: Issues and Options, New York*, United Nations Development Programme, pp. 109–118.

UNEP: 1992, *World Atlas of Desertification*, London, United Nations Environment Programme and Edward Arnold, 69 pp.

UNEP: 1997, 'Africa', in V. Vandeweerd, M. Cheatle, B. Henricksen, M. Schomaker, M. Seki, and K. Zahedi (eds.), *Global Environmental Outlook*, Nairobi, Oxford University Press and United Nations Environment Programme, pp. 25–41.

UNFCCC: 2001, *Report of the Conference of the Parties on its Seventh Session (Marrakesh, 29 October-10 November 2001)*, Report FCCC/CP/2001/13/add.1, United Nations Framework Convention on Climate Change [http://unfccc.int/resource/docs/cop7/13a01.pdf].

Vlek, P.L.G.: 1995, 'The soil and its artisans in sub-Saharan Africa', *Geoderma* **67**, 165–170.

WBBGU: 1998, 'Das Kyoto-Protkoll', in *Die Anrechnung biologischer Quellen und Senken im Kyoto-Protokoll: Fortschrift oder Rückschlag für den globalen Umweltschutz?*, Bremerhaven, Wissenschaftlicher Beirat der Bundesregierung Globale Umwelveränderungen, pp. 56–66.

MITIGATING GHG EMISSIONS IN THE HUMID TROPICS: CASE STUDIES FROM THE ALTERNATIVES TO SLASH-AND-BURN PROGRAM (ASB)

CHERYL PALM[1]*, TOM TOMICH[2], MEINE VAN NOORDWIJK[3], STEVE VOSTI[4], JAMES GOCKOWSKI[5], JULIO ALEGRE[6] and LOU VERCHOT[2]

[1] Tropical Soil Biology and Fertility Programme (TSBF), Nairobi, Kenya; [2] World Agroforestry Centre (ICRAF), Nairobi, Kenya; [3] ICRAF-Southeast Asia, Bogor, Indonesia; [4] University of California Davis, CA, USA; [5] International Institute of Tropical Agriculture (IITA), Yaounde, Cameroon; [6] ICRAF-Latin America, Lima, Peru
(*author for correspondence: P.O. Box 1000, Palisades, NY 10964 USA; e-mail: cpalm@iri.columbia.edu; fax: 1-545-680-4866; tel.: 1-845-680-4462)

(Accepted in Revised form 15 January 2003)

Abstract. Tropical forest conversion contributes as much as 25% of the net annual CO_2 emissions and up to 10% of the N_2O emissions to the atmosphere. The net effect on global warming potential (GWP) also depends on the net fluxes of greenhouse gases from land-use systems following deforestation. Efforts to mitigate these effects must take into account not only the greenhouse gas fluxes of alternative land-use systems but also the social and economic consequences that influence their widespread adoption. The global alternatives to slash-and-burn program (ASB) investigated the net greenhouse gas emissions and profitability of a range of land-use alternatives in the humid tropics. The analysis showed that many tree-based systems reduced net GWP compared to annual cropping and pasture systems. Some of these systems are also profitable in terms of returns to land and labor. The widespread adoption of these systems, however, can be limited by start-up costs, credit limitations, and number of years to positive cash flow, in addition to the higher labor requirements. Projects that offset carbon emissions through carbon sinks in land use in the tropics might be a means of overcoming these limitations. A synthesis of the findings from this program can provide guidelines for the selection and promotion of land-use practices that minimize net global warming effects of slash-and-burn.

Key words: carbon stocks, environmental and economic tradeoffs, net global warming potential, profitability, slash-and-burn.

1. Introduction

Tropical deforestation contributes up to 25% of the net annual CO_2 emissions (Watson et al., 2000) and 10% of the global N_2O emissions (Bouwman et al. 1995), primarily from the slashing and burning of the high-biomass vegetation and decomposition of the soil organic matter. The initial flux of CO_2, CH_4 and N_2O during and immediately following slash-and-burn is similar for most land-use systems. Less is known about carbon stocks and trace gas emissions during the time course of the various land systems established following deforestation (Houghton et al., 1993; Erickson et al., 2001; Erickson and Keller, 1997) compared to the immediate effects from the slash-and-burn process.

Environment, Development and Sustainability **6**: 145–162, 2004.
© 2004 *Kluwer Academic Publishers.*

Though studies over the past 10 years provide a substantial basis for making initial estimates of the trace gas fluxes and carbon stocks from these different land-use systems in the humid tropics (Mosier et al, this volume; Verchot et al., 1999; Davidson et al., 2000; Woomer et al., 2000; Erickson et al., 2001), most studies have focused on natural or old-growth forests, pastures in Latin America, paddy rice, and a few other cropping systems. There have been few studies from tree-based systems such as plantations, fallows from slash-and-burn agriculture, and agroforestry systems, yet tree-based systems often dominate the agricultural land-scape in the humid tropics (Wood et al., 2000). Efforts to mitigate the global climate change effects of tropical deforestation and land use must take into account not only the greenhouse gas fluxes of these land-use systems but also the social and eco-nomic consequences that influence the widespread adoption of the various land-use alternatives following slash-and-burn clearing.

In this paper we synthesize some of the major findings from a global project, the Alternatives to Slash-and-Burn Program (ASB), that can provide guidelines for the selection and promotion of land-use practices that minimize net global warming effects of slash-and-burn. The net global warming potential (GWP) of a range of land-use systems, including many tree-based systems, are estimated from information on carbon stocks and greenhouse gas emissions that were measured in this project. The GWP is based on the CO_2 fluxes resulting from changes in carbon in the vegetation and soils from deforestation and during the rotation time of the land-use systems and the N_2O and CH_4 fluxes from these systems. In addition, the profitability of the different land-use systems is provided to assess which land-use systems have the potential to generate income for the land users and thus have greater chances of adoption. The tradeoffs between global environmental services (e.g. reduced GWP) and private costs and benefits are discussed based on the findings for the different land-use systems.

2. The ASB Program

In 1991, ASB was initiated to address the agronomic, environmental, social, and political implications of slash-and-burn in the humid tropics (Brady, 1996). The overall goal of the program was to compare the environmental and social impacts of current land-use systems and to identify alternatives that were environmentally and agronomically or economically better. In addition, policies that would facilitate the adoption of these promising alternatives were considered.

Teams of national and international scientists were established in key locations, referred to as benchmark areas, around the humid tropical belt representing the range in biophysical and socioeconomic environments in which slash-and-burn is practiced. Standardized sets of parameters and measurements were established for assessing carbon stocks (Woomer and Palm, 1998), trace gas emissions (Palm et al., 2002), agronomic sustainability, profitability, and the institutional settings (Vosti et al., 2000) for the different land-use systems found in the benchmark areas.

The resulting 'ASB trade-off matrix' allows one to compare the environmental, production, and social cost and benefits of the different land-use systems (Tomich et al., 1998a; Vosti et al., 2000). The reader is also referred to Woomer et al. (2000), Palm et al. (1999, 2002), Hairiah et al. (2001), Gockowski et al. (2001), Tsuruta et al. (2000), Tomich et al. (2001, 2002), and Vosti et al. (2001) for more detailed descriptions and results of these studies.

2.1. DESCRIPTION OF THE ASB BENCHMARK SITES AND LAND-USE SYSTEMS

ASB benchmark areas were chosen to represent at the regional and global levels, large, active areas of deforestation caused by slash-and-burn practices (Palm et al., 1995). In this paper, we concentrate on the sites in Cameroon, the western Amazon, and Sumatra, Indonesia. These areas are in the humid tropical zone with annual rainfall generally greater than 2000 mm and dry seasons of less than six months. The sites are characterized by tropical rainforest vegetation and acid, infertile soils (primarily Oxisols and Ultisols). The site in Cameroon is defined over a population gradient with areas of lower population pressure representative of the equatorial Congo Basin rainforest of Congo, Gabon, Central African Republic, and Zaire while areas of higher pressure having greater affinity to the degraded humid forests of coastal West Africa. In Latin America, areas in the states of Rondonia and Acre in the western Amazon of Brazil represent areas of rapid deforestation as a result of colonization programs, while sites in Peru represent other areas of the western Amazon Basin with lower population density and poor infrastructure. Areas in Sumatra, Indonesia represent the equatorial rainforests of Indonesian and Malaysian archipelago where primary forests are being cleared by both indigenous practices and resettlement programs; in addition, degraded imperata grasslands invade the landscape as population densities increase. The contrasts in some of the site characteristics are presented in Table I.

A set of 'meta land-use categories' encompasses the different land-use systems that were described and studied at the ASB benchmark areas. These include forests, either undisturbed, managed, or logged; complex agroforestry systems that include a wide diversity of plant species; simple agroforestry systems and tree crop plantations that usually contain less than 10 plant species as major components of the system; crop-fallow agriculture rotations, including long-term fallows characteristic of shifting cultivation, short-term fallows and improved fallows; annual crops; and pastures and grasslands, including degraded grasslands and improved pastures.

3. Net GWP of ASB Land Uses

The effects of slash-and-burn and subsequent land use on net GWP are estimated from changes in carbon stocks over the time-course of the land-use change and fluxes of nitrous oxide (N_2O) and methane (CH_4).

TABLE I. Selected site characterization parameters for the ASB benchmark areas.

Characterization parameter	Western Amazon, Brazil/Peru	Southern Cameroon	Sumatran lowlands
Rainfall (mm yr^{-1})	1700–2400	1400–1900	2000–3000
Months dry season (less than 100 mm)	June–Sept	Apr–May; Oct–Feb	May–Sept
Dominant original vegetation	Tropical moist forest; semi-deciduous forest	Tropical moist forest; semi-deciduous forest	Tropical moist forest
Predominant soils (US Soil Taxonomy)	Paleud(ust)ults; Hapludustox;	Kandiudults	Hapludox; Kandiudox
Population (people km^{-2})	3–5	4–120	2–175
Farm size (ha household^{-1})	80–100	2–4*	5–10
Agriculture wages (US$ day^{-1})	6.25	1.21	1.67

* Excludes fallow land.

3.1. CARBON STOCKS AND TIME-AVERAGED CARBON

Carbon stocks were measured in the soils (0–20 cm) and aboveground vegetation in a total of 115 different locations in the benchmark areas according to standardized protocols (Woomer and Palm, 1998). The sites encompassed most of the meta land-use categories in each of the benchmark sites. Detailed results from those sites are reported in Kotto-Same et al. (1997), Fujisaka et al. (1998), Tomich et al. (1998a,b), and Woomer et al. (2000).

Data from those studies were used to describe the time course of carbon stocks over the rotation of the different land-use systems and to calculate the aboveground time-averaged C of each system as described and reported in Palm et al. (1999). This methodology allows the comparison of carbon stocks in systems that have tree-growth and harvesting rotations and is similar to the average storage method described in the IPCC Special Report on Land Use, Land-Use Change and Forestry (Watson et al., 2000).

The main findings from these carbon stock measurements indicate that that the C stocks in the vegetation of the primary forests averaged 300 t C ha^{-1}, that of logged or managed forests ranged from a high of 228 t C ha^{-1} in Cameroon to a low of 93 t C ha^{-1} in Indonesia (Table II). Time-averaged aboveground C for the different meta land uses ranged from 50 to 90 t C ha^{-1} in long-fallow shifting cultivation and complex agroforestry systems; 30–60 t C ha^{-1} in simple agroforestry systems, most tree plantations and medium-fallow rotations; and 3–12 t C ha^{-1} in short-fallow rotations, coffee plantations, annual crops, and pastures.

Soil carbon values (0–20 cm) compared to the 45 t C ha^{-1} found in the forest systems were 80–100% in agroforestry systems; 80% in pastures; 90–100% in long-fallow cycles; 65% in short-term fallows, and 50% or less in annual crops

TABLE II. Summary of the aboveground time-averaged C stock (mean and range) of the land-use systems sampled at the ASB sites, the range is given in parentheses (Palm et al., 1999).

Meta land-use systems	Country and specific land use	Time-averaged C of land-use system t C ha^{-1}
Undisturbed forest	Indonesia	306 (207–405)
	Peru	294
Managed/logged forests	Brazil/Peru	150 (123–185)
	Cameroon	228 (221–255)
	Indonesia	93.2 (51.9–134)
Shifting cultivation and crop-fallows	Cameroon	
	Shifting cultivation, 23-year fallow	77.0 (60.2–107)
	Bush fallow, 9.5 years	28.1 (22.1–38.1)
	Chromolaena fallow, 4 years	4.52 (2.68–6.38)
	Brazil/Peru	
	Short fallow, 5 years	6.86 (4.27–9.61)
	Improved fallow, 5 years	11.5 (9.50–13.4)
	23-year fallow	93 (80.5–101)
Complex/extensive agroforests		
Permanent	Cameroon	88.7 (57.2–120)
Rotational	Cacao	
	Indonesia	89.2 (49.4–129)
	Rubber	
	Cameroon	61 (40–83)
	Cacao	
	Indonesia	46.2 (28.9–75.2)
	Rubber	
Simple agroforests/ intensive treecrop	Brazil/Peru	
	Coffee monoculture	11.0 (8.73–12.5)
	Multistrata system	61.2 (47.5–74.7)
	Peach Palm, Oil Palm, rubber	47 (27–61)
	Cameroon	
	Oil Palm	36.4
	Indonesia	
	Pulp trees	37.2 (23.6–50.7)
Grasslands/crops	Brazil/Peru	
	Extensive pastures	2.85
	Intensive pastures	3.06
	Indonesia	
	Cassava/Imperata	<2

and degraded grasslands (Palm et al., 1999). These losses in soil C are similar to those reported by Detwiler (1986) in a review of soil C and land use in the tropics; though changes in pasture soils vary considerably ranging from 0% to 20% losses.

Combined losses of C from the vegetation and soils over the time course of the deforestation and subsequent land use can be as high as 320 t C ha^{-1} or a minimum of 110 t C ha^{-1} if an undisturbed forest is converted to annual cropping or permanent agroforests, respectively. Though much deforestation is now occurring from managed and previously logged forests that have already lost some C, rather than undisturbed forests, current losses would be somewhat less.

3.2. NITROUS OXIDE AND METHANE FLUXES

Estimates of N_2O and CH_4 fluxes require intensive, long-term sampling. This was not possible at most of the ASB sites. So, in order to obtain some estimate of the N_2O and CH_4 fluxes relative to net CO_2 fluxes from changes in land use, flux data from a long-term experiment in the western Amazon in Peru (Palm et al., 2002) were used for most of the meta land-use systems along with data from the literature for other land-use systems in the Amazon.

The land-use systems sampled in Yurimaguas, Peru included two annual cropping systems, one a high-input maize soybean rotation with 100 kg N ha^{-1} fertilizer applied to each maize crop, liming, and tillage and the other a low-input system with an upland rice–cowpea–legume cover crop rotation; four tree-based systems including a 13-year- and a 23-year-old secondary forest fallow, multistrata agroforestry system, and a peach palm tree plantation. Other than the 23-year-old secondary forest all other land-use systems were 13-years-old at the time of sampling. Gas fluxes were calculated from monthly samples taken over a 2-year period as described in Palm et al. (2002).

3.2.1. Nitrous oxide fluxes

The annual N_2O flux from 23-year-old secondary forest soil was 0.80 kg N ha^{-1} or 9.1 µg N m^{-2} h^{-1} (Palm et al., 2002), within the range of most other secondary forests reported for the Neotropics on acid, infertile soils (Davidson et al., 2000, 2001; Erickson et al., 2001). This value is about half that reported for primary forests on similar soils in the region (Davidson et al., 2001); therefore a value of 20 µg N m^{-2} h^{-1} was used as an estimate for primary forests, since none were evaluated in the Peru study (Table III). N_2O fluxes from the three tree-based systems in this study that had been established 13 years previously ranged from 6.4 to 10.2 µg N m^{-2} h^{-1}, again within the range for secondary forests in the region. Fluxes from the two cropping systems were higher than the tree-based systems, averaging 14.51 and 26.6 µg N m^{-2} h^{-1} for the low- and high-input cropping systems, respectively. The fluxes from the high-input cropping system were an order or two of magnitude lower than those measured for other fertilized systems in the humid and subhumid tropics (Davidson et al., 1996; Erickson and Keller, 1997; Veldkamp and Keller, 1997; Matson et al., 1998) and probably reflect the lower N application rates as well as the split application in this study (Palm et al., 2002).

TABLE III. The N_2O and CH_4 fluxes and changes in C stocks of slash-and-burn and different land-use systems in the Peruvian Amazon.

Land-use system	N_2O flux[a] ($\mu g\, N\, m^{-2}\, h^{-1}$)		CH_4 flux[a] ($\mu g\, C\, m^{-2}\, h^{-1}$)	Net C lost from vegetation[e] ($t\, C\, ha^{-1}\, 25\, yr^{-1}$)	Net C lost from soil[e] ($t\, C\, ha^{-1}\, 25\, yr^{-1}$)
	From burn	Land use			
Primary forest		20.0[b]	−40.0	0[f]	0[f]
25-year crop-fallow	1.82[d]	9.1	−30.0	135	5
Multistrata agroforest	1.82[d]	6.4	−24.2	165	9
Peach palm plantation	1.82[d]	10.2	−18.4	185	9
Low-input cropping	1.82[d]	14.5	−18.2	222	9
High-input cropping	1.82[d]	26.6	+15.2	222	22.5
Pasture	1.82[d]	7.0[c]	0	222	0

[a] From Palm et al. (2002) except for primary forest and pasture.
[b] From Davidson et al. (2001).
[c] From Verchot et al. (1999).
[d] From Erickson and Keller (1997); $4\,kg\, N\, ha^{-1}$ from slash-and-burn, distributed over 25 years.
[e] Includes effects of deforestation and assuming 225 and $45\,t\, C\, ha^{-1}$ in the initial forest vegetation and topsoil (0–20 cm) and time-averaged C of each land-use system for 25-year rotation.
[f] Assumes primary forest vegetation and soil are in equilibrium with respect to C stocks.

Nitrous oxide fluxes from pastures on acid, infertile soils of the Amazon were taken from the literature to provide a comparative flux from this important land use in the region. Annual N_2O fluxes from pastures range from 0.1 to about $2\,kg\, N\, ha^{-1}\, yr^{-1}$ (Verchot et al., 1999; Erickson and Keller, 1997). At intermediate value of $7.0\,\mu g\, N\, m^{-2}\, h^{-1}$ (or $0.6\,kg\, N\, ha^{-1}\, yr^{-1}$) was used for comparing pastures with the other land uses (Table III). Although a much higher flux, $5.7\, N\, ha^{-1}\, yr^{-1}$, was measured from a recently established pasture in the central Amazon (Luizao et al., 1989); it is now considered that this high value was probably transitory following deforestation. Similar patterns have been found following establishment of pastures in Costa Rica (Erickson and Keller, 1997).

3.2.2. Methane fluxes

The annual flux of CH_4 of $-2.6\,kg\, C\, ha^{-1}\, yr^{-1}$ or $-30.0\,\mu g\, C\, m^{-2}\, h^{-1}$ from the secondary forest (Table III; Palm et al., 2002) falls within the range reported for primary forests in the Amazon (Steudler et al., 1996; Verchot et al., 2000). The CH_4 consumption rates of the tree-based systems and the low-input cropping systems ranged from 54% to 86% that of the secondary forest. In contrast, there was a net CH_4 production of $15.2\,\mu g\, C\, m^{-2}\, h^{-1}$ in the high-input cropping system. Both the higher flux of N_2O and production of CH_4 in the high-input cropping system were associated with higher soil bulk density and %wfps due to the deterioration of the soil structure from the long-term tillage operations in this system.

Others have reported that conversion of humid tropical forest to cropping systems decreased the CH_4 consumption rates by 75% (Keller et al., 1990) but did not result in a shift to net annual CH_4 production. Conversion of humid forests to pastures often results in net CH_4 production rather than consumption (Keller and Reiners, 1994; Steudler et al., 1996) though Verchot et al. (2000) found production CH_4 in the rainy season but annual net CH_4 consumption. Given this range that included slight CH_4 production and CH_4 consumption, a value of 0 was used for CH_4 flux from pasture soils in this comparative analysis.

3.3. NET GWP OF SLASH-AND-BURN SYSTEMS: AN EXAMPLE FROM PERU

In order to make an overall comparison of the net GWP from the emissions of CO_2 from the biomass burning as a result of deforestation and the subsequent changes in carbon stocks in the soil and vegetation and the fluxes of CO_2, N_2O, and CH_4 from the different land-use systems, information on time-averaged carbon stocks from the different land-use systems and N_2O, and CH_4 fluxes from the Peru site was combined (Table III). This was the only site with extensive C stock and gas flux measurements required to make these estimates. So, while these estimates of GWP are not meant to be representative of the entire humid tropics, the values are intended to give an indication of the relative effects of CO_2, N_2O, and CH_4, as well as the relative effects of the different phases of deforestation, and the different types of land use on GWP. Information in the previous sections does show that the C stock changes and gas fluxes from the different land uses in Peru generally fall within the ranges reported for other parts of the Amazon and the resulting estimates of GWP should be considered to be within the ranges expected for the region.

The net GWP of the different land-use systems in Peru were calculated using the following assumptions and conditions:

(1) Carbon stocks of the aboveground vegetation and topsoil (0–20 cm) of the forest systems that were slashed and burned were 225 and 45 t $C\,ha^{-1}$, respectively. Losses of CO_2 from biomass burning were assumed to be 100% of the aboveground C of the forest and were averaged over a 25-year time frame, or 9 t $C\,ha^{-1}\,yr^{-1}$.

(2) N_2O flux from biomass burning with deforestation was assumed to be about 4 kg $N\,ha^{-1}$ (Erickson and Keller, 1997), distributed over a 25-year time span would equal 0.16 kg $N\,ha^{-1}\,yr^{-1}$ (1.82 µg $N\,m^{-2}\,h^{-1}$).

(3) Net losses of C in the above ground vegetation and topsoil (0–20 cm) over the 25 years were then determined as the initial C stocks of the forest minus the time-averaged value for each system that were calculated and standardized for the 25-year time frame for all systems. Only one value is given for the two fallow systems since their time-averaged C would be the same for the 25-year rotation. It is important to note that changes in C stocks in belowground vegetation are not considered here and so the values given in Table III are underestimates.

(4) N_2O and CH_4 fluxes from the land-use systems in Peru were assumed to be averages for the 25-year time span. The measurements from Peru were flux estimates for land-use systems 13 years following slash-and-burn and do not include the time course since deforestation from which to calculate time-average fluxes. Other studies from chronosequences indicate that N_2O fluxes from pastures in the Amazon decline as much as 50% over 20 years (Erickson and Keller, 1997; Verchot et al., 1999) while fluxes from secondary forests increase by as much as 100% over the first 20 years. These trends suggest that the fluxes from the 13-year-old systems in Peru, being at the midpoint of the 25-year time frame, would provide a reasonable time-averaged flux for 25-year period rotations. Fluxes for the 'old-growth' forest and pastures were taken from the literature as noted in Sections 3.2.1 and 3.2.2.

(5) Net GWP was calculated by converting changes in C stocks and fluxes of N_2O and CH_4 to mol CO_2 m^{-2} yr^{-1}, using net radiative forcing values of 1, 310, and 21 for CO_2, N_2O, and CH_4, respectively (Watson et al., 2000).

The most notable result from the analysis of the net GWP indicates that the CO_2 released from the vegetation as a result of biomass burning from deforestation (75 mol C m^{-2} yr^{-1}; dashed line in Figure 1) far outweigh the subsequent emissions of CO_2, N_2O, and CH_4 emissions from the soils of the different land-use systems. A similar conclusion was reached for the Sumatra benchmark sites by Tomich et al. (1998b). The GWP due to net CO_2 emission from the decomposition of soil organic matter following deforestation, 0–8 mol C m^{-2} yr^{-1}, was as high or higher than that of the combined N_2O and CH_4 despite the higher net radiative forcing values for the latter two gases (Figure 1) and the GWP from CH_4 production in the high-input cropping system 0.41 mol C m^{-2} yr^{-1} or consumption in the other systems, -0.28 to -0.46 mol C m^{-2} yr^{-1}, were negligible in comparison to the GWP from CO_2.

Once the slash-and-burn process has occurred then the sum of the radiative forcing effects of CO_2 and N_2O+CH_4 are highest in the high-input cropping system, due to larger losses of soil organic matter, a higher N_2O flux as a result of fertilization, and a loss of the soil CH_4 sink and in fact a net CH_4 source from the soil in this system. Considering only the portion of GWP due to the combined $N_2O + CH_4$ fluxes, it is interesting to note that estimates for the primary forest (3.27 mol C m^{-2} yr^{-1}) are second only to the high-input cropping system (5.57 mol C m^{-2} yr^{-1}). This is due to the high N_2O flux from the forest soil and despite the fact that this system has the highest CH_4 sink. In contrast to most perceptions, the GWP due to $N_2O + CH_4$ for the other systems are lower than that of the primary forest system. Even the low-input cropping system was 30% less than that of the forest, while the remaining tree-based systems had values 60% less than the primary forest.

Most of the tree-based systems at the ASB benchmark sites are not intensively fertilized so the radiative effects of N_2O are not large. Many intensive tree crop systems in the humid tropics, such as coffee, peach palm, and banana are fertilized at rates often exceeding 300 kg N ha^{-1} yr^{-1} (Szott and Kass, 1993; Veldkamp and

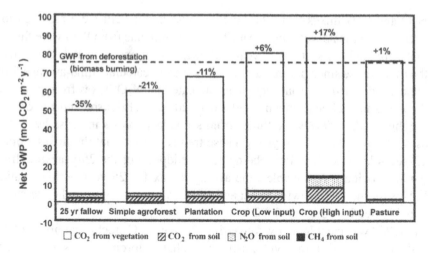

Figure 1. Sources of the net GWP over a 25-year period for the different land-use systems in the Peruvian Amazon. The dashed line represents the GWP resulting from deforestation and biomass burning. Details of the calculations are shown in Section 3.3.

Keller, 1997). There is insufficient information to determine if gaseous N losses will increase significantly from these tree-based systems with fertilization but emissions should depend largely on the rates and timing of fertilizer application relative to plant demand and age of the system.

In comparing the net GWP of the different land-use systems with the GWP from the deforestation and biomass burning, the tree-based systems all reduce this initial flux by 11–35% (Figure 1); this decrease being attributed primarily to the sequestration of C in the vegetation. In contrast, the net GWP of the two cropping systems increased the initial GWP from burning by 6% to almost 20% due to losses of soil C and, in the case of the high input cropping system, higher N_2O losses and net CH_4 production. The net GWP of the pasture system essentially remained the same as that resulting from biomass burning.

Efforts to mitigate this dominating effect of the release of CO_2 from the slash-and-burn process should obviously focus on reducing rates of deforestation or establishing tree-based land-use systems that sequester more C in the vegetation and soil compared to annual cropping systems and pasture. Efforts to mitigate the high GWP from the high-input cropping system, from the combined CO_2, N_2O, and CH_4 fluxes from the soil, would be to synchronize better the timing of N fertilizer applications with those of plant demand and to improve the soil structure and increase soil organic matter content through conservation tillage operations.

4. Social and Economic Factors of Slash-and-Burn Systems

It is one thing to indicate which land-use systems have the highest GWP and to make recommendations as to how these effects can be reduced by managing the systems differently; it is another thing to expect land users to adopt these systems or practices because of the global environmental effects. Many of the recommendations for mitigating the greenhouse gas effects of land use in the tropics, such as reducing deforestation, establishing and maintaining tree-based systems, and increasing soil organic matter, all include additional inputs of labor and capital. Whether future profits from these different land-use systems will eventually offset these additional costs will help determine which, if any, of the systems land-users will adopt.

The categories included in the ASB social and economic analysis are profitability, including returns to land and labor, and establishment costs; household food security, including nutritional value, food entitlement, and risks (Vosti et al., 2000). Of these categories, financial profitability, labor requirements, and food security were considered essential for the assessment. Systems that generate inadequate levels of financial profits, require more labor input than household can provide or hire or that competes with other traditional systems, or compromise the ability to produce or buy sufficient food will not be attractive to farmers. In addition to profitability and household food security, the institutional capacity such as potential bottlenecks and equity issues may affect the adoptability of the different land-management systems.

4.1. PROFITABILITY OF ASB LAND-USE SYSTEMS

The ASB methods for profitability applied the 'policy analysis matrix' approach of Monke and Pearson (1989) for assessing economic outcomes and policy distortion. For the purpose of comparing profitability the private, or farmer, prices were used. Private prices are those actually paid by farmers and include distortions such as taxes, subsidies, and non-tariff barriers. Private prices, as opposed to social prices that estimate prices in absence of such distortions, are most relevant for considering the likelihood of adoption.

Values for assessing profitability at each site were obtained by extensive field studies to determine inputs (including labor) and output levels for the various land-use system components and prices were drawn from the fieldwork and secondary sources. Financial profitability includes establishment costs and costs and benefits of the production activities of each land-use system. These costs/benefits are then discounted to provide summary measures (e.g. net present value, NPV, that was used in the ASB analyses) that can be used to compare LUS across and especially within benchmark sites. For ASB purposes of comparing across land-use systems and benchmark sites the NPV of costs and returns were calculated over a 20- to 25-year time span for each system, using discount rates of 10–15%. Summary measures of financial profitability were expressed as returns to land and to labor, reported in 1996US$. Returns to land represent the implicit rental rate on land dedicated to a specific LUS and returns to labor represent the break-even daily wage for family labor input to a system. The daily wage rates were US $6.00, 1.21, and 1.70 for Brazil, Cameroon, and Indonesia, respectively (Table I). Those land-use systems with returns to labor above that of the wage rate will be more readily adopted.

Details of the economic findings and analysis of trade-offs between environmental and economic indicators from Brazil, Cameroon, and Indonesia can be found in Vosti et al. (2001); Gockowski et al. (2001), and Tomich et al. (1998b, 2001, 2002), and are summarized in Table IV. There is a broad range in the profitability, calculated at private prices, across and within the meta land-use categories.

In general, food crop systems were not profitable. The short-fallow crop rotations, characteristic of areas with higher demographic pressure, were not profitable in Indonesia. Such systems have basically disappeared from the landscape in Brazil and are becoming scarce in Indonesia, perhaps a reflection of the negative returns to land and returns to labor below that of the wage rate. This land-use system is still practiced to some degree, perhaps as a means of assuring food security. In contrast, the longer fallow systems characteristic of the Cameroon site were profitable with high returns to land and returns to labor similar to the wage rate; such systems are a prevalent feature of the landscape there. Annual cropping systems were only found in Indonesia and were profitable only if fertilizers were not used but then the agronomic sustainability of the systems was compromised.

Agroforestry and plantation systems were found to some extent at all sites and in most cases were profitable and provided the highest returns to land and labor. But even within the different tree-based systems some systems were not profitable. As examples, oil palm plantations in Indonesia were highly profitable compared to net losses in monoculture rubber plantations. The rubber agroforests, in contrast, were potentially the most profitable (Tomich et al., 2001; Williams et al., 2001). The more diversified tree-based systems in Cameroon, such as cocoa-fruit agroforests compared to cocoa plantations, also provided higher returns than the more intensive systems due to the earlier harvests and diverse crops for revenue (Gockowski et al., 2001). Pasture systems of the western Amazon or Brazil were also profitable, with returns to land being two orders of magnitude higher in improved as compared to traditional pastures and returns to labor were tripled (Vosti et al., 2001).

4.2. ECONOMIC AND ENVIRONMENTAL TRADE-OFFS OF ASB LAND-USE SYSTEMS

Many of the land-use systems in the humid tropics are profitable and therefore provide an incentive to deforest, which in turn leads to increased greenhouse gas emissions. If deforestation is to occur then from a global environmental perspective it would be most desirable to establish those systems with the lowest net GWP; for the forest margins of the humid tropics these would be the tree-based systems that sequester the most carbon as indicated in Section 2.3. For lands already deforested, converting to land-use systems with higher C stocks should be advocated.

Trade-off matrices were used to identify those ASB land-use systems that may provide both reduced greenhouse gas emissions, through increased C stocks, and potential profitability. The tree-based systems with relatively high C sequestration values (Table II) and higher profitability (Table IV) included oil palm plantations and cocoa agroforests, particularly those interplanted with valuable fruit trees for

TABLE IV. Profitability in terms of returns to land and labor and labor requirements for the different meta land-use systems for the ASB benchmark sites in Indonesia, Brazil, and Cameroon (from Tomich et al., 2001; Vosti et al., 2001; ASB, 2000).

Meta land use	Returns to land (farmer prices, $ ha⁻¹)			Returns to labor (farmer prices, $ person⁻¹ day⁻¹)			Time-averaged labor (day ha⁻¹ yr⁻¹)		
	Sumatra, Indonesia[a]	Acre, Brazil[b]	Southern Cameroon[c]	Sumatra, Indonesia[d]	Acre, Brazil[b]	Southern Cameroon[c]	Sumatra, Indonesia[d]	Acre, Brazil[b]	Southern Cameroon[c]
Forest									
Managed	3 to 7	416	NA	5	1	NA	0.3	1	NA
Logged	−54 to −335	NA	NA	−7 to 1	20	NA	31	1.2	NA
Agroforests									
Complex	1 to 918	NA	424 to 1409	2 to 3	NA	1.6 to 2.4	111 to 150	NA	43 to 97
Simple	−70 to 115	870 to 1955	736 to 1471	1.5 to 2.5	9 to 13	1.8 to 2.4	108 to 133	27 to 59	72
Crops-fallow									
Short fallow	−90 to 32	−17	283 to 623	1.2	6	1.7 to 1.8	15 to 25	23	44 to 115
Annual crops	−30 to 227	NA	NA	1.8	NA	NA	98 to 104	NA	NA
Pasture	NA	2 to 710	NA	NA	7 to 22	NA	NA	12	NA

[a] From Tomich et al. (1998). [b] From Vosti et al. (2001). [c] From ASB (2000). [d] From Tomich et al. (2001).

Cameroon (Gockowski et al., 2001); oil palm plantations and rubber agroforests in Sumatra (Tomich et al., 2001); and coffee and rubber or timber agroforestry systems in Brazil (Vosti et al., 2001).

Many of the land-use systems with low C stocks, such as short-term fallows and annual cropping systems or degraded grasslands, have either net losses or low-potential profitability so conversion to some of the profitable tree-based systems make both environmental and economic sense. An analysis by Tomich et al. (1997) shows that agro-reforestation of the degraded *Imperata* grasslands in Indonesia to *Acacia mangium* plantations or rubber agroforests were indeed profitable. These tree-based systems have time-averaged C stocks of $60 \, t \, C \, ha^{-1}$ compared to 5 for the grasslands (Table II). The low C stock pasture systems of Latin America, in contrast to the degraded grasslands of Southeast Asia, are more profitable and require less labor than most of the higher C stock tree-based systems. Finding incentives for mitigating the greenhouse gas effects of these profitable pasture systems will be more difficult.

4.3. LABOR AND CAPITAL CONSTRAINTS TO THE MITIGATION OF GREENHOUSE GASES IN ASB SYSTEMS

Despite the profitability and positive environmental aspects of many of the tree-based systems at the ASB sites, there are economic and tenure issues that may prevent large-scale adoption of these compared to other land uses. An example from Sumatra, Indonesia shows the annual labor requirements over the 25-year rotation of agroforests was 125 days $ha^{-1} \, yr^{-1}$ while those of logging or crop-fallows are 30 and 20 days, respectively (Table IV; Tomich et al., 2001). Evidence from Southeast Asia demonstrates that local people will invest in establishing tree-based systems if they have secure claims over the products, access to markets and natural risks, such as fire, are not too high (Tomich et al., 2002).

The simple agroforestry systems in Brazil had labor requirements 2–5 times higher than pasture systems and this combined with lower returns to labor than improved pasture systems would suggest less likelihood of adoption (Table IV; Vosti et al., 2001). As pointed out by Gockowski et al. (2001) in areas of low population density that are characteristic of most slash-and-burn areas, land-use systems that are land saving will only be adopted if they also increase the returns to labor. When these systems are evaluated at private costs then labor is generally the scarce resource.

Because agroforestry and other tree-based systems typically involve a time lag in production, time plays a factor in farmers' evaluation of monetary returns. A simple measure of this constraint to adoption is the number of years to positive cash flow. In Indonesia, the number of years to positive cash flow was 2 and 10 for annual crops and agroforestry systems, respectively (Tomich et al., 2001). In Brazil, although an intensive coffee agroforestry system provides high returns to labor, adoption may be limited by start-up costs, credit limitations and number of years to positive cash flow, in addition to the higher labor requirements (Vosti et al., 2001). In Cameroon, farmers have adapted to this time lag for production with perennial crops by inter-planting food crops, especially plantains and tannia *(Xanthosoma sagittifolium)* during the initial establishment phase. This provides revenues while at the same time providing the shade necessary for the dry season survival of the tree crop seedlings.

5. Carbon Offset Projects as Mechanism for Mitigating Greenhouse Gas Effects of Slash-and-Burn Systems

Efforts to mitigate the greenhouse gas effects of slash-and-burn systems, either through reduction in rates of deforestation or establishment of tree-based sys-tems, may require incentives to the land-users. The process of deforestation, in general, is profitable, so land-users would need some sort of compensation and the adoption of tree-based systems may require government support in terms of land or tree tenure, access to markets, or other institutional support. Opportunities to provide the necessary compensation or institutional support to reduce rates of deforestation or to establish land-use systems with higher C stocks may eventually develop from the proposed clean development mechanism (CDM) in Article 12 of the 'Kyoto Protocol'. The Kyoto Protocol raises the possibility of offsetting C emissions with C 'sinks' in land use, land-use change, and forestry (LULUCF). Establishment of tree-based systems on cropland and grasslands has been identified as the largest potential sink of C globally though land-use change (Watson et al., 2000).

While the CDM may offer gross financial benefits, there has been little analysis of the opportunity costs of foregone resource exploitation and development oppor-tunities. Simply stated, would payments for C sinks at $25 per Mg C, one estimate of the trading price for C, be sufficient for land-users in the tropical forest margins of the humid tropics to reduce deforestation? Tomich et al. (2002) estimated that farmgate (or forestgate) payment per Mg of C needed to offset incentives for forest clearing in Indonesia would be $0.10 per Mg for community-based forest manage-ment, under $4 per Mg for large-scale oil palm plantations, and as high as $10 per Mg for rubber agroforests. This suggests that a world price of $25 per Mg of C could shift incentives from forest conversion to conservation, if these payments reach the people making the decisions and agreements are enforceable. In Cameroon the net present value of carbon sequestered when converting short-fallow cropping systems to cocoa agroforests was shown to depend on the marginal value of Mg C, the social discount rate, the production cycle of the agroforest, and the rate of C sequestration over time (J. Gockowski and S. Dury, personal communication). At $20 per Mg of C, the net present values for sequestered C ranged from $550 to $740 per ha. Carbon credits for smallholder-based agroforestry systems offer a potentially powerful pol-icy option for building assets among the rural poor if institutional mechanisms can be devised.

It is not clear how such institutions for transferring C credits would work and little is known about the actual transaction costs when smallholder communities are involved in C trade. Included in these costs would be baseline and monitoring measurements in addition to those costs normally associated with project development and implementation. Transaction costs are important: if they are too high compared to the global price of C stocks, smallholders' incentives will be inadequate to induce a change in land use.

Multidisciplinary data generated through the ASB has provided a means for assessing the tradeoffs between global environmental and private economic aspects of land-use systems in the humid tropics. The analysis showed that many tree-based systems had moderate levels of carbon storage and overall reduced net GWP compared to annual cropping and pasture systems, and thus provide some global benefit. A subset of these systems would also be attractive to small scale land users in the tropics because they are profitable in terms of returns to land and labor. The widespread adoption of these systems, however, can be limited by start-up costs, credit limitations, and number of years to positive cash flow, in addition to the higher labor requirements. Projects that offset carbon emissions through carbon sinks in land use in the tropics might be a means of overcoming these limitations. Studies are needed across a range of circumstances in the humid tropics to improve estimates of the direct opportunity costs of shifting to or conserving land uses that can store more C as well as an assessment of the transactions costs for implementing such projects.

Acknowledgements

Work reported here is part of the global ASB. ASB has been financially supported by the Global Environment Facility (GEF) with UNDP sponsorship and by DANIDA. Additional support for the Indonesian benchmark site and greenhouse gas measurements in Peru were provided by the Australian Centre for International Agricultural Research (ACIAR) and for the Brazilian benchmark site by the Inter-American Development Bank, EMBRAPA, and Japan. In Cameroon additional funding was provided from the European Commission.

References

ASB: 2000, *Summary report and synthesis of phase II in Cameroon*, Nairobi, Kenya, ICRAF, 72 pp.
Brady, N.C.: 1996, 'Alternatives to slash-and-burn: a global perspective', *Agriculture, Ecosystems, and Environment* **58**, 3–11.
Bouwman, A.F., van der Hoek, K.W. and Olivier, J.G.J.: 1995, 'Uncertainties in the global source distribution of nitrous oxide', *Journal of Geophysical Research* **100**, 2785–2800.
Davidson, E.A., Matson, P.A. and Brooks, P.D.: 1996, 'Nitrous oxide emission controls and inorganic nitrogen dynamics in fertilized tropical agricultural soils', *Soil Science Society of America Journal* **100**, 1145–1152.
Davidson, E.A., Keller, M., Erickson, H.E., Verchot, L.V. and Veldkamp, E.: 2000, 'Testing a conceptual model of soil emissions of nitrous and nitric oxide', *BioScience* **50**, 667–680.

Davidson, E.A., Bustamente, M.M.C. and de Siqueira Pinto, A.: 2001, 'Emissions of nitrous oxide from soils of native and exotic ecosystems of the Amazon and Cerrado regions of Brazil', in Optimizing Nitrogen Management in Food and Energy Production and Environmental Protection: Proceedings of the 2nd International Nitrogen Conference on Science and Policy, *The Scientific World* **1**(S2), 312–319.

Detwiler, R.P.: 1986, 'Land use change and the global carbon cycle: the role of tropical soil', *Biogeochemistry* **2**, 67–93.

Erickson, H.E. and Keller, M.: 1997, 'Tropical land use change and soil emissions of nitrogen oxides', *Soil Use and Management* **13**, 278–287.

Erickson, H.E., Keller, M. and Davidson, E.A.: 2001, 'Nitrogen oxide fluxes and nitrogen cycling during postagricultural succession and forest fertilization in the humid tropics', *Ecosystems* **4**, 67–84.

Fujisaka, S., Castilla, C., Escobar, Rodrigues, G.V., Veneklass, E.J., Thomas, R. and Fisher, M.: 1998, 'The effects of forest conversion on annual crops and pastures: Estimates of carbon emissions and plant species loss in a Brazilian Amazon colony', *Agriculture, Ecosystems, and Environment* **69**, 17–26.

Gockowski, J., Nkamleu, G.B. and Wendt, J.: 2001, 'Implications of Resource-use Intensification for the Environment and Sustainable Technology Systems in the Central African Rainforest' in *Tradeoffs or Synergies? Agricultural Intensification, Economic Development and the Environment*, D.R. Lee and C.B. Barrett (eds), CAB International, Wallingford, UK, pp. 197–217.

Hairiah, K., Sitompul, S.M., van Noordwijk, M. and Palm, C.A.: 2001, 'Methods for sampling carbon stocks above and below ground', in M. Van Noordwijk, S.E. Williams and B. Verbist (eds.), *Towards integrated natural resource management in forest margins of the humid tropics: local action and global concerns.* ASB-Lecture Notes 1–12, ASB_LN 4B, International Centre for Research in Agroforestry (ICRAF), Bogor, Indonesia, http://www.icraf.cgiar.org/sea/Training/Materials/ASB-TM/ASB-ICRAFSEA-LN.htm

Houghton, R.A., Unruh, J.D. and Lefebvre, P.A.: 1993, 'Current land cover in the tropics and its potential for sequestering carbon', *Global Biogeochemical Cycles* **7**, 305–320.

Keller, M. and Reiners, W.A.: 1994, 'Soil-atmosphere exchange of nitrous oxide, nitric oxide, and methane under secondary succession of pasture to forest in the Atlantic lowlands of Costa Rica', *Global Biogeochemical Cycles* **8**, 399–409.

Keller, M., Mitre, M.E. and Stallard, R.F.: 1990, 'Consumption of atmospheric methane in soils of Central Panama: Effects of agricultural development', *Global Biogeochemical Cycles* **4**, 21–27.

Kotto-Same, J., Woomer, P.L., Moukam, A. and Zapfak, L.: 1997, 'Carbon dynamics in slash and burn agriculture and land use alternatives of the humid forest zone in Cameroon', *Agriculture, Ecosystems, and Environment* **65**, 245–256.

Luizao, F., Matson, P.A., Livingston, G., Luizao, R., and Vitousek, P.: 1989, 'Nitrous oxide flux following tropical land clearing', *Global Biogeochemical Cycles* **3**, 281–285.

Matson, P.A., Naylor, R. and Ortiz-Monasterio, I.: 1998, 'Integration of environmental, agronomic, and economic aspects of fertilizer management', *Science* **280**, 12–115.

Mosier, A., Wassmann, R., Verchot, L., King, J. and Palm, C.: this volume, 'Methane and nitrogen oxide fluxes in tropical agricultural soils: Sources, sinks and mechanisms'.

Monke, E. and Pearson, E.: 1991, 'Introduction', in S. Pearson, W. Falcon, P. Heytens, E. Monke and R. Naylor (eds.), *Rice in Indonesia*, Cornell University Press, Ithaca, New York, USA.

Palm, C.A., Izac, A.-M. and Vosti, S.: 1995, '*Alternatives to slash-and-burn: Procedural guidelines for characterization*', Nairobi, Kenya, ASB, ICRAF.

Palm, C.A., Woomer, P.L., Alegre, J., Arevalo, L., Castilla, C., Cordeiro, D.G., Feigl, B., Hairiah, K., Kotto-Same, J., Mendes, A., Moukam, A., Murdiyarso, D., Njomgang, R., Parton, W.J., Ricse, A., Rodrigues, V., Sitompul, S.M. and van Noordwijk, M.: 1999, 'Carbon sequestration and trace gas emissions in slash and burn and alternative land uses in the humid tropics', Nairobi, Kenya, *ASB Climate Change Working Group Final Report, Phase II*, ASB Coordination Office, ICRAF.

Palm, C.A., Alegre, J.C., Arevalo, L., Mutuo, P., Mosier, A. and Coe, R.: 2002, 'Nitrous Oxide and Methane Fluxes in Six Different Land Use Systems in the Peruvian Amazon'. *Global Biogeochemical Cycles*, **16**:1073, doi:10.1029/2001GB001855.

Steudler, P.A., Melillo, J.M., Feigl, B.J., Neill, C., Piccolo, M.C. and Cerri, C.C.: 1996, 'Consequences of forest-to-pasture conversion on CH_4 fluxes in the Brazilian Amazon Basin', *Journal of Geophysical Research* **101**(D13), 18547–18554.

Szott, L.T. and Kass, D.C. 1993, 'Fertilizers in agroforestry systems', *Agroforestry Systems* **23**, 153–176.

Tsuruta, H., Ishizuka, S., Ueda, S. and Murdiyarso, D.: 2000, Seasonal and spatial variations of CO_2, CH_4, and N_2O fluxes from the surface soils in different forms of land-use/cover in Jambi, Sumatra, in D. Murdiyarso and H. Tsuruta (eds.), *The Impacts of Land-use/cover Change on Greenhouse Gas Emissions in Tropical Asia*, Global Change Impacts Centre for Southeast Asia and National Institute of Agro-Environmental Sciences, Bogor, Indonesia.

Tomich, T.P., Kuusipalo, J., Menz, K. and Byron, N.: 1997, 'Imperata economics and policy', *Agroforestry Systems* **36**, 233–261.

Tomich, T.P., van Noordwijk, M., Vosti, S.A. and Witcover, J.: 1998a, 'Agricultural Development with Rainforest Conservation Methods for Seeking *Best Bet* Alternatives to Slash-and-Burn, with Applications to Brazil and Indonesia', *Agricultural Economics* **19**, 159–174.

Tomich, T.P., van Noordwijk, M., Budidarsono, S., Gillison, A., Kusumanto, T., Murdiyarso, D., Stolle, F. and Fagi, A.M.: 1998b, '*Alternatives to Slash-and-Burn in Indonesia. Summary report and synthesis of Phase II*', Nairobi, Kenya, ASB, ICRAF.

Tomich, T.P., van Noordwijk, M., Budidarsono, S., Gillison, A., Kusumanto, T., Murdiyarso, D., Stolle, F. and Fagi, A.M.: 2001, 'Agricultural intensification, deforestation, and the environment: Assessing tradeoffs in Sumatra, Indonesia', in *Tradeoffs or Synergies? Agricultural Intensification, Economic Development and the Environment*, D.R. Lee and C.B. Barrett (eds.), CAB International, Wallingford, UK, pp. 221–244.

Tomich, T.P., de Foresta, H., Dennis, R., Ketterings, Q., Murdiyarso, D., Palm, C.A., Stolle, F., Suyanto, S. and van Noordwijk, M.: 2002, 'Carbon offsets for conservation and development in Indonesia?', *American Journal of Alternative Agriculture* **17**, 125–137.

Veldkamp, E. and Keller, M. 1997, 'Nitrogen oxide emissions from a banana plantation in the humid tropics', *Journal of Geophysical Research* **102**, 15889–15898.

Verchot, L.V., Davidson, E.A., Cattanio, J.H., Ackerman, I.L., Erickson, H.E. and Keller, M.: 1999, 'Land use change and biogeochemical controls of nitrogen oxide emissions from soils in eastern Amazon', *Global Biogeochemical Cycles* **13**, 31–46.

Verchot, L.V., Davidson, E.A., Cattanio, J.H. and Ackerman, I.L.: 2000, 'Land-use change and biogeochemical controls on methane fluxes in soils of eastern Amazon', *Ecosystems* **3**, 41–56.

Vosti, S.A., Witcover, J., Gockowski, J., Tomich, T.P., Line Carpentier, C., Faminow, M., Oliveira, S. and Diaw, C.: 2000, *Working Group on Economic and Social Indicators: Report on Methods for the ASB Best Bet Matrix*, Nairobi, Kenya, ASB Coordination Office, ICRAF, 41 pp.

Vosti, S.A., Witcover, J., Line Carpentier, C., Oliveira, S.J. and dos Santos, J.C.: 2001, 'Intensifying Small-Scale Agriculture in the Western Brazilian Amazon: Issues, Implications and Implementation', in D.R. Lee and C.B. Barrett (eds.), *Tradeoffs or Synergies? Agricultural Intensification, Economic Development and the Environment*, Wallingford, UK, CAB International, pp. 245–266.

Watson, R.T., Noble, I.R., Bolin, B., Ravindranath, N.H., Verardo, D.J. and Doken, D.J. (eds.): 2000, *Land Use, Land-Use Change and Forestry. Intergovernmental Panel on Climate Change*, Cambridge, UK, Cambridge University Press.

Williams, S.E., van Noordwijk, M., Penot, E., Healey, J.R., Sinclair, F.L. and Wibawa, G.: 2001, 'On-farm evaluation of the establishment of clonal rubber in multistrata agroforests in Jambi, Indonesia', *Agroforestry Systems* **53**, 227–237.

Wood, S., Sebastian, K. and Scherr, S.J.: 2000, '*Pilot Analysis of Global Ecosystems: Agroecosystems*', IFPRI and WRI, Washington, DC, USA.

Woomer, P.L. and Palm, C.A.: 1998, 'An approach to estimating system carbon stocks in tropical forests and associated land uses', *Commonwealth Forestry Review* **77**, 181–190.

Woomer, P.L., Palm, C.A., Alegre, J., Castilla, C., Cordeiro, D.G., Hairiah, K., Kotto-Same, J., Moukam, A., Ricse, A., Rodrigues, V. and van Noordwijk, M.: 2000, 'Slash-and-Burn Effects on Carbon Stocks in the Humid Tropics', in R. Lal, J.M. Kimble and B.A. Stewart (eds.), *Global Climate Change and Tropical Ecosystems*, Advances in Soil Science, Boca Raton, FL, USA, CRC Press, Inc. pp. 99–115.

AN AMAZON PERSPECTIVE ON THE FOREST–CLIMATE CONNECTION: OPPORTUNITY FOR CLIMATE MITIGATION, CONSERVATION AND DEVELOPMENT?

GEORGIA CARVALHO[1,2]*, PAULO MOUTINHO[2],
DANIEL NEPSTAD[1,2], LUCIANO MATTOS[2] and MÁRCIO SANTILLI[2]
[1]*Woods Hole Research Center, P.O. Box 296, Woods Hole, MA, USA;*
[2]*Instituto de Pesquisa Ambiental da Amazônia, Belém, Pará, Brazil*
*(*author for correspondence, e-mail: gcarvalho@whrc.org;*
fax: 508-540-9700; tel.: 508-540-9900, ext. 144)

(Accepted in Revised form 15 January 2003)

Abstract. Amazonia contains more carbon (C) than a decade of global, human-induced CO_2 emissions (60–80 billion tons). This C is gradually being released to the atmosphere through deforestation. Projected increases in Amazon deforestation associated with investments in road paving and other types of infrastructure may increase these C emissions. An increase of 25–40% in Amazon deforestation due to projected road paving could counterbalance nearly half of the reductions in C emissions that would be achieved if the Kyoto Protocol were implemented. Forecasted emission increases could be curtailed if development strategies aimed at controlling frontier expansion and creating economic alternatives were implemented. Given ancillary benefits and relative low costs, reducing deforestation in Amazonia and other tropical areas could be an attractive option for climate mitigation. Projects that help contain deforestation and reduce frontier expansion can play an important role in climate change mitigation but currently are not allowed as an abatement strategy under the climate regime. Creating incentives for forest conservation and decreased deforestation can be a unique opportunity for both forest conservation and climate mitigation.

Key words: Amazonia, climate change, deforestation, greenhouse gas emissions, land use change.

1. The forest–climate connection: land use change in Amazonia and climate change

The atmospheric temperature is increasing at a 0.2°C rate per decade as a result of greenhouse gas emissions, such as carbon dioxide (CO_2) and methane (CH_4) (IPCC, 2001). Net carbon (C) accumulation in the atmosphere amounts to c. 3 billion tons per year and there are no signs that this situation will be reversed in the near future. Amazon forests play an important role in this scenario of global warming. The trees in Amazon forests contain 60–80 billion tons of C, an amount equivalent to more than a decade of global human-induced emissions. Typical Amazon forests contain, on average, around 350 tons of biomass per hectare, which corresponds to approximately 175 T of C per hectare (Houghton et al., 2001). When deforestation and forest fires disturb these forests, a vast amount of CO_2 is liberated into the atmosphere. Deforestation in Amazonia alone, releases 200–300 million tons of C (2–4% of world emissions) annually (Fearnside, 1997; Houghton et al., 2000). This amount corresponds to about 2/3 of Brazil's emissions and is more than twice as much as that emitted nationwide through the burning of fossil fuels (95.1 million tons in 2000[1]). Furthermore, Amazon emissions associated with land use may increase two-fold during severe drought years if the C liberated by logging and extensive forest fires were accounted for (Nepstad et al., 1999).

Brazil's energy sector is relatively clean with hydropower supplying more than 52% of national energy needs and representing 87% of installed electricity capacity in 1999,[2] and 10% of its automobile fleet is fueled by alcohol.[3] However, a national energy shortage is driving thermoelectric plant construction and other energy investments that may provoke growth in C emissions of 6.6 million tons per year (Tolmasquim et al., 2001). So Brazil's principal contribution towards greenhouse gas emissions reduction will have to be through controlling deforestation in Amazonia. On the other hand, a significant increase in deforestation and fires in Amazonia could undo most of the anticipated gains from the implementation of the Kyoto Protocol if the current policy plans for Amazonia are fully implemented (Figure 1), as described below.

Brazil's historical contribution to the global warming problem is small compared to that of industrialized countries which based their development process on activities that emitted great quantities of greenhouse gases over the last two centuries. Emissions reduction efforts should, thus, be the primary responsibility of developed countries. That does not mean that developing countries should be exempt from any action, given that their economic growth and land use activities could lead them to the same emission level as developed countries in as little as a 30 years (Becker, 2001).

Brazil can, in the near future, look for ways of increasing energy efficiency while at the same time addressing land use based emissions, particularly by reducing deforestation in Amazonia. Amazon deforestation reduction can be accomplished without hampering the region's essential development process, since the major form of deforestation is forest conversion to cattle pastures of low productivity (Arima and Uhl, 1997). On the contrary, Amazonia presents tremendous development potential based on standing forests, and agricultural intensification on the half million square

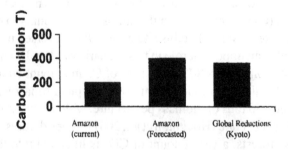

Figure 1. Annual net carbon emissions (million t C) due to current and forecasted deforestation in Amazonia as opposed to anticipated reductions in the global scale (implementation of the Kyoto protocol excluding US participation); forecasted future emissions assume a higher deforestation rate associated with proposed infrastructure projects and drought-induced forest fire.

kilometers of land that have already been deforested. In the following sections we argue that a development trajectory of expanded economic prosperity and reduced deforestation in Amazonia is attainable. But planned investments in transportation infra-structure could lead to the opposite trend.

2. Undoing Kyoto: forecasted emissions for Amazonia

Land use changes in Brazilian Amazonia have historically been stimulated by public policy interventions, especially investments in road infra-structure and the estab-lishment of fiscal incentives for those investing in the region. The current process of increased occupation and environmental degradation in the region is to a large degree the result of development policies pursued since the mid 1960s that encour-aged frontier expansion and the uncontrolled exploitation of the region's resources. At the core of Brazil's Amazon development strategy were infra-structure devel-opment projects, such as roads providing access to frontier regions, and large hydroelectric reservoirs built to supply energy to other regions of the country (Mahar, 1988; Fearnside and Barbosa, 1996).

Roads are the main vectors of deforestation in Amazonia (Alves, 2001; Nepstad et al., 2001). Since 1995, the Brazilian government renewed its focus on infra-structure as the basis for its Amazon development policy in an updated version of policies that led to frontier expansion in the past (Carvalho et al., 2002). The Cardoso administration has outlined government plans that emphasize the expan-sion of the economic infra-structure in the country, especially development and modernization of the transportation corridors to allow the country to decrease trans-portation costs and help its exports become more competitive in the global economy (Becker, 1999). This package, named Avança Brasil, if completely implemented, will add over 6000 km of paved highways to the region's paved road network, including the BR-163 Cuiabá-Santarém, BR-319 Porto Velho-Manaus, and BR-174 Manaus-Boa Vista highways which cut through largely undisturbed forest areas. These road-paving projects could lead to increased deforestation, forest impover-ishment through logging, and higher incidence of forest fire (Figure 2) (Carvalho et al., 2001; Nepstad et al., 2001). In addition, there are social and land conflicts that result from frontier expansion (Schmink and Wood, 1992).

Looking at historical patterns of deforestation associated with construction and pavement of roads in Amazonia we have found that between 28% and 55% of the forests along paved roads were cut within 15–25 years of paving, while a maximum of 7% of the forests along unpaved highways were cut during this same period (Nepstad et al., 2001) (Figure 3). Using this historical relationship we have estimated the area that will be deforested after the paving of the roads proposed by the Avança Brasil program. Our analysis suggests that between 120 000 and 270 000 km^2 of forests could be converted in the next 25–35 years if the government implements the Avança Brasil plans (Nepstad et al., 2001). Added to the current deforestation rates of approximately 17 000 km^2 yr^{-1} (INPE, 2000).

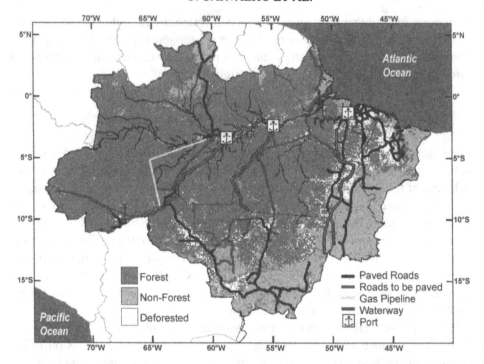

Figure 2. Map of Brazilian Amazonia depicting deforested area and road infrastructure (investments as described in the *Avança Brasil* program).

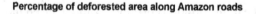

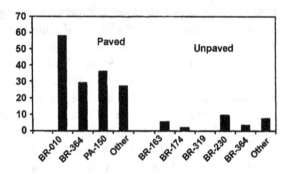

Figure 3. Deforested area along Amazon roads (50 km belt on either side) as of 1992.

The increased deforestation rate would lead to an upsurge in CO_2 emissions in the order of 2–5 billion tons of C in the same 25–35 year period, or an additional 85–140 million T/C/year[4] (Figure 1). Nevertheless, the forecasted emission increase could be curtailed if these projects were implemented in conjunction with development strategies aimed at controlling deforestation and frontier expansion and creating sustainable economic alternatives for the population of Amazonia.

3. Opportunities to modify the land use in Amazonia:
reduced deforestation and climate mitigation

The forecasted deforestation increase would be of 5000–8000 km^2 yr^{-1}, which at 175 T/C ha^{-1} in Amazon forests would result in the increase of 85–140 million T/C yr^{-1} from Amazon deforestation, in addition to the current 200 million T/C yr^{-1} estimated by Houghton et al. (2000, 2001). The forecasted emissions do not take into account the possible regrowth along paved roads (approximately 5.5 T C ha^{-1} yr^{-1}) given the difficulty in estimating how much regrowth there would be along roads.

The Kyoto Protocol allows for a variety of ways in which emissions can be abated, including the clean development mechanism (CDM) (IPCC, 2000).[5] The CDM will permit that emission reduction projects implemented in developing countries sell certificates of emissions reductions to parties with emission reduction targets within the Kyoto Protocol (Annex B countries), thus lowering implementation costs for developed countries. There are various project options being considered under the CDM, both in the energy sector and the forestry sector, such as combined energy production or fuel switching, industrial applications and land-use change, including tree plantations and forest regeneration. In addition, The Kyoto Protocol explicitly encourages parties to "protect and enhance sinks and reserves, promote sustainable forest management practices, afforestation and reforestation" under Article 2.

Climate regime negotiations and related discussions of forest sector based projects have focused on the modalities that will likely be allowed under the CDM, namely C sequestration through afforestation and reforestation (IPCC, 2000). Although the text that was agreed upon in Bonn, in July 2001, excludes avoided deforestation and conservation as modalities under the CDM, at least for the first commitment period (2008–2012), it is important to examine some of the potential positive effects of creating incentives for the development of these activities under the climate regime. A climate regime that addresses the connection between land use in the tropics, would be much more capable of mitigating climate change in the long run, since these emissions represent between 20% and 25% of global emissions (IPCC, 2000).

The scale of the capacity for climate change mitigation of afforestation and reforestation pales in comparison to the potential benefits that might be achieved through initiatives to reduce deforestation (Moura Costa and Wilson, 2000). For instance, one of Brazil's largest C sequestration projects, the "Projeto PLANTAR", in the state of Minas Gerais will reduce C emissions by approximately 3.2 million tons over a 21 year period, through the production of pig iron using charcoal made from *Eucalyptus* plantations rather than coal, at the relatively high investment cost of US$ 1102 ha^{-1} (the cost of the Ton of C would be approximately US$ 9).[6] At the end of two decades, the project will have avoided C emissions equivalent to only 1.5% of the emissions that come from annual deforestation in the Amazon. In contrast, mature tropical forests store approximately 175 T C ha^{-1} (Houghton et al., 2001). Taking into consideration the additional environmental benefits of maintaining mature tropical forests (biodiversity, hydrological cycle, soil conservation), and the significance of land use based emissions in Brazil, controlling deforestation can be a better option for climate mitigation, especially if it becomes recognized (and compensated) as an abatement activity under the climate change regime in the future.

If activities that create an incentive to reduce deforestation and conserve forests were considered under the emerging climate regime, several initiatives proposed or being developed at relatively small scales in Amazonia could become much more economically attractive and prevalent in the region. Among these initiatives one can include sustainable forest management techniques, sustainable extraction of non-timber forest products, low impact logging practices, agroforestry systems, forest fire control, and forest conservation. Since these activities tend to decrease forest conversion rates they can lead to a net decrease in emissions from land use change in the region at a much larger scale than projects that rely on C sequestration through afforestation or reforestation. Other scientists and environmental organizations agree that forest conservation projects in the Amazon and around the world can be one of the best options for emissions mitigation (WRI, 1998a,b).

Projects that encourage conservation and reduce deforestation are in many ways similar to fossil fuel emissions reduction projects in that they reduce emissions (rather than removing CO_2 from the atmosphere, as is the case with planted forests). In addition, these activities have other environmental and social benefits, such as decreasing migration of young rural population to cities, protecting biodiversity and conserving watershed and soils (Frumhoff et al., 1998; WRI, 1998a,b)

There are serious technical difficulties and risks associated with land use change and forestry projects, including establishing whether the project has a real mitigating effect (additionality), defining baseline and system boundaries (leakage) to determine the net contribution of the project, and monitoring projects for duration and permanence of mitigating effects. However, these problems and risks plague both conservation/reduced deforestation projects as well as sequestration projects which are currently allowed under the climate regime (Brown et al., 1998; Chomitz, 2000; Moura Costa et al., 2000). On the other hand, the positive effects derived from conservation or reduced deforestation projects are stronger than those from reforestation and afforestation projects according to some analyses (Bass et al., 2000; Hardner et al., 2000; Moura Costa et al., 2000; Seroa da Motta et al., 2000).

One way to decrease deforestation in the Amazon is through increased governance. Increasing governance in the region has the potential to change the development trajectory in the region, leading to greater control over the natural resource base before it has been depleted by initial development boom. Two policy initiatives in Brazilian Amazonia could lead to increased governance while decreasing deforestation rates, encouraging conservation and alternative development strategies, and reducing emissions from land use change. These initiatives are complementary, since one focuses on improving monitoring of deforestation, while the other attempts to create market based incentives to decrease forest conversion.

The first initiative reviewed was Mato Grosso's statewide effort to license and improve monitoring of land use activities and preliminary results show that it is already decreasing the deforestation rate in that state. The second initiative a policy proposal aimed at establishing a credit program for family agricultural producers that creates incentives for the adoption of sustainable agricultural practices, called the PROAMBIENTE, can also potentially contribute to rural development, increase employment and create alternative sources of income for forest dwellers. Both initiatives could be expanded if there were incentives available for this type of project under the climate regime.

Mato Grosso's attempt at implementing a licensing and monitoring system to control deforestation, is already having measurable positive results in reducing deforestation and if expanded throughout Amazonia it could reduce greenhouse gas emissions by millions or, perhaps, tens of millions of tons C annually. Begun in 1999, the experience in Mato Grosso has been successful at decreasing deforestation (35% lower for the period 2000–2001 in comparison to the previous two years). The reduction appears to be the result of the successful implementation of the forest code[7] through monitoring efforts by the state environmental agency, which is in charge of the program (Fearnside, 2002). Previous to the implementation of the licensing and monitoring program, the interannual variation of the deforestation rate in Mato Grosso was similar to that in the remainder of Legal Amazonia, increasing slightly over the period between 1994 and 1999. But since the implementation of the program Mato Grosso's deforestation rate has decreased while the remainder of the region's deforestation rates increased (INPE, 2001; Fearnside, 2002). The sharp decrease (of 35%) verified since 1999 is likely explained by the monitoring program, which focused on the most troubled areas of the state, especially in transition forests (where the deforestation rate decreased by 43.7%) (Fearnside, 2002).

According to Fearnside (2002) a preliminary calculation of the reduction of greenhouse gas emissions from the Mato Grosso licensing and monitoring program is approximately 43 million T C yr^{-1} for the period 2000–2001. Although not all of the avoided emissions can be attributed to Mato Grosso's licensing and monitoring program, there is evidence that a significant portion of the reduction can be explained by its implementation. The overall costs of the program of licensing and monitoring in Mato Grosso are very low compared to the results. The program costs have been approximately US$ 3 million yr^{-1} (6 million Reais yr^{-1}) since 1999. The resources have been provided partly by the World Bank (between 0.6 and 1 million yr^{-1}), partly by the Rain Forest Trust Fund (of the G-8) (US$ 5 million yr^{-1}) and partly by the government of the state of Mato Grosso (which has contributed personnel, salaries and infra-structure).

The second policy initiative, still in a proposal stage, is The PROAMBIENTE. A proposed credit program devised by the family producers association (FETAG – Federações dos Trabalhadores na Agricultura) of the Amazon states, in technical partnership with Instituto de Pesquisa Ambiental da Amazônia (IPAM, the Amazon Institute of Environmental Research) and the Federação dos Órgãos para Assistência Social e Educacional (FASE). The program, still in the initial planning phases, proposes to create credit lines that compensate farm families for their investments in their land that protect or restore environmental services.

The rationale behind the PROAMBIENTE is to help producers make the transition from the traditional slash and burn agricultural practices that currently prevail in the Amazon frontier toward more diversified and sustainable agricultural and extractive practices, thus slowing down forest conversion and emissions. Unlike existing agricultural credit programs, the PROAMBIENTE would create an incentive for more sustainable economic activities by compensating, directly or indirectly, family based producers for good agricultural practices and associated environmental services such as forest conservation and management, reduction of forest fire and fragmentation, maintenance of stream and river margins, soil conservation, recuperation of degraded areas and biodiversity conservation.

The PROAMBIENTE proposes to reallocate the resources of an existing rural credit line, the FNO,[8] so as to cover the costs of agricultural production and related technical project, while allowing for the creation of a new fund that would pay of the debt incurred by farm families as they demonstrate progress in protecting or restoring environmental services. The proposal is innovative both in its origin (it is being proposed by producers themselves), and because it would be one in the first instances of a market based economic instrument (credit) being used to modify the behavior of family based producers to help contain deforestation.

The PROAMBIENTE is still at the proposal stage, but it is estimated that it has the potential to create 7500 jobs in the first 10 pilot areas, increasing family income and helping family farmers transition to perennial agroforestry production systems. Its contribution to climate change mitigation would come primarily from the estimated 430 000 tons of avoided C emissions (estimated based on the average deforestation rate of 1 ha yr^{-1} for family based farms in Amazonia × 175 T/C ha^{-1}, assuming the participation of 250 properties in each of the 10 pilot areas). Beside the contribution from avoided deforestation, there is the potential 25 000 tons of net C uptake in agroforestry systems and secondary forests (2 T/C ha^{-1} yr^{-1} × 10 ha per property × 250 properties in each of 10 pilot areas) (Mattos and Nepstad, 2002).

Other environmental benefits from the agroforestry based production encouraged by the PROAMBIENTE, would be lower incidence of fire, lower sedimentation of streams and/or rivers, protection of the integrity of regional rainfall patterns, increase in biodiversity, and soil recuperation and conservation.

The PROAMBIENTE proposes that monitoring of its environmental component be carried out through independent audits. Avoided deforestation and C sequestration would be assessed by measuring biomass using satellite imagery after a baseline were established (based on a composite image over several years). In addition, each participant would undergo an environmental certification process and production practices could be proxy indicators of environmental services related to water, biodiversity, soils and flammability.[9] The estimated costs of the PROAMBIENTE are US$ 640/property/year (or US$ 6.4 ha^{-1} yr^{-1}) and the total cost for the 10 pilot areas (250 properties each) over the 15 year period is estimated at US$ 27 million.

It is important to highlight that although the PROAMBIENTE is an innovative proposal with good potential, it is still in a developmental stage. The viability of the PROAMBIENTE will ultimately depend on a variety of factors. Among these factors are a solid technical project that takes into account the economics of Amazon agriculture, the support of local governments through improvements in infra-structure (e.g. local road conditions and market accessibility) in each pilot area, development of niche markets for these environmentally responsible products, and improvement of basic rural infra-structure (especially health and education) to help retain families in rural areas.

These and other important initiatives underway in the Amazon demonstrate the potential for frontier governance that could sharply reduce the rate of deforestation and forest impoverishment through logging and fire while fostering enduring economic prosperity, and we have described previously (Carvalho et al., 2002; Nepstad et al., 2002). For instance, there is progress in prevention and control of accidental fires, through the Brazilian Fire Control Program for Amazonia (PROARCO), that limits burning during the peak of the dry season. The program implementation in 2000 corresponded with a two fold reduction in the number of fires detected by satellite in 1999 and 2000 throughout most of the more densely populated eastern and southern Amazonia and this reduction cannot be solely explained by changes in precipitation levels. Strengthening and expansion of protected area systems, prohibition of deforestation on areas with low agricultural potential, and local economic planning processes are some of the other elements to a scenario of frontier governance.

Policy initiatives such as Mato Grosso's licensing and monitoring program and the PROAMBIENTE, could present investors with an attractive opportunity to earn emission reduction credits at rates that are competitive with C sequestration projects, and that have significant ancillary social (rural development) and environmental benefits (forest and biodiversity conservation).

4. Conclusion

This article highlights the connection between tropical forests and climate change. In fact, emissions originating from tropical deforestation rates forecasted for the next 3 decades could undo much of the emissions reductions achieved under the Kyoto Protocol. Although the climate regime as it currently stands does not have any mechanisms that create an incentive to enhance conservation and reduce deforestation, tropical forests remain an important part of the climate equation. Conservation, development, and climate abatement goals can and should be complementary. Projects that reduce deforestation by licensing and monitoring and encourage conservation of forests are key to healthy ecosystems and a healthy climate. Combining these policies with strategies to improve the livelihoods of local populations, be it through the use of improved agricultural techniques, encouragement of more productive agroforestry systems, forest fire control or sustainable forest management techniques provide a unique opportunity to foster development while mitigating climate change.

Projects that help contain deforestation and reduce frontier expansion can play an important role in climate change mitigation but currently are not allowed as an abatement strategy under the climate regime. A broader climate regime that creates incentives for forest conservation and decreased deforestation can present a unique opportunity to further both climate mitigation and conservation goals. From an Amazon perspective, if the climate regime continues to ignore the connection between forests and climate and the growing contribution of deforestation and land use change to climate change, it will miss the opportunity to effectively address climate change.

Acknowledgements

The authors would like to thank the Ford and Avina Foundations and USAID for their support of research related to this paper. The map was prepared by Paul Lefebvre and Michael Ernst. We thank Yabanex Batista, Reiner Wassmann and two anonymous reviewers for their comments on previous versions of the manuscript.

Notes

[1] See http://www.eia.doe.gov/emeu/iea/tableh1.html.

[2] See http://www.fe.doe.gov/international/brazover.html. However, the lakes formed by hydroelectric power plants can produce great quantities of CH_4, which has a more powerful greenhouse gas effect than CO_2 (Fearnside, 1995).

[3] According to Denatran there are 3 million vehicles fueled by ethanol out of Brazil's national fleet of circa 30 million. See http://www.denatran.gov.br for statistics on Brazilian fleet.

[4] The forecasted deforestation increase would be of 5000–8000 km^2 yr^{-1}, which at 175 T/C ha^{-1} in Amazon forests would result in the increase of 85–140 million T/C yr^{-1} from Amazon deforestation, in addition to the current 200 million T/C yr^{-1} estimated by Houghton et al. (2000, 2001). The forecasted emissions do not take into account the possible regrowth along paved roads (approximately 5.5 T C ha^{-1}yr^{-1}) given the difficulty in estimating how much regrowth there would be along roads.

[5] For more details on the rules for the flexibility mechanisms of the Kyoto Protocol www.unfccc.int.

[6] See http://www.plantar.com.br.

[7] The Brazilian forest code requires that at least 80% of rural properties be protected as forest reserves; it also requires licenses for and deforestation, logging and burning MP Version 2080-631.

[8] The 1988 Constitution established that 3% of funds collected from income tax by the federal government would go to Constitutional Funds to help diminish regional development gaps. The FNO is the Constitutional Fund for the North (Amazonia) and receives 0.6% of the total funds. In the 2001 fiscal year the total amount of resources were 562 million Brazilian Reais (BRL), or approximately 208 million USD, for fiscal year 2002–2003 the forecasted resource amount to approximately BRL 400 million/148 million USD. Of this total amount, approximately 77 million BRL (28.5 million USD) are allocated to the FNO Especial credit line, funds earmarked to finance small and family based producers (see http://www.basa.com.br).

[9] To ensure uniformity, certification could follow the standards and the certifiers would have to be regulated as suggested by Moura Costa et al. (2000).

References

Alves, D.: 2002, 'An analysis of the geographical patterns of deforestation in Brazilian Amazonia the 1991–1996 period', in C. Wood and R. Porro (eds.), *Patterns and Processes of Land Use and Forest Change in the Amazon*, Gainesville, University of Florida Press.

Arima, E. and Uhl, C.: 1997, 'Ranching Brazilian Amazon in the national context: economics, policy and practices', *Society and Natural Resources* **10**, 433–451.

Bass, S. et al.: 2000, *Rural Livelihoods and Carbon Management*, London, IIED.

Becker, B.: 1999, *Cenários de curto prazo para o desenvolvimento da Amazônia*, Cadernos do NAPIAm, n.6.

Becker, B.: 2001, 'Amazonian frontiers at the beginning of the 21st Century', in D.J. Hogan and M.T. Tolmasquim (eds.), *Human Dimensions of Global Environmental Change: Brazilian Perspectives*, Rio de Janeiro, Academia Brasileira de Ciências, pp. 299–324.

Brown, P., Kete, N. and Livernash, R.: 1998, 'Forest and land use projects', in J. Goldemberg (ed.), *Issues and Options – the Clean Development Mechanism*, United Nations Development Program, New York, pp. 163–173.

Carvalho, G., Barros, A.C., Moutinho, P. and Nepstad, D.: 2001, 'Sensitive development could protect Amazonia instead of destroying it', *Nature* **409**, 131.

Carvalho, G.O., Nepstad, D., McGrath, D. Del CarmenVera Diaz, M., Suntilli, M. and Barros, A.C.: 2002, 'Frontier expansion in the Amazon: balancing development and sustainability', *Environment* **44**(3), 34–42.

Chomitz, K.: 2000, *Evaluating Carbon Offsets from Forestry and Energy Projects: How do They Compare?* Washington, D.C., The World Bank Development Research Group, Infrastructure and Environment, Policy Research Working Paper.

Fearnside, P.: 1995, 'Hydroelectric dams in the Brazilian Amazon as sources of greenhouse gases', *Environmental Conservation* **22**, 7–19.

Fearnside, P.: 1997, 'Greenhouse gases from deforestation in Brazilian Amazonia: net committed emissions', *Climatic Change* **35**, 321–360.

Fearnside, P.: 2002, 'Controle de desmatamento no Mato Grosso: Um novo modelo para reduzir a velocidade da perda da Floresta Amazonica', *paper presented at the Seminar Applications of Remote Sensing and Geographic Information Systems in the Brazilian Amazon* April 02–03, Brasília-DF, Brazil.

Fearnside, P. and Barbosa, R.: 1996, 'The Cotingo Dam as a Test of Brazil's System for Evaluating Proposed Developments in Amazonia', *Environmental Management* **20**, 631–648.

Frumhoff, P., Goetze, D. and Hardner, J.: 1998, *Linking Solutions to Climate Change and Biodiversity Loss Through the Kyoto Protocol's Clean Development Mechanism,* Union of Concerned Scientists.

Hardner, J., Frumhoff, P. and Goetze, D.: 2000, 'Prospects for mitigating carbon, conserving biodiversity, and promoting socioeconomic development objectives through the clean development mechanism', *Mitigation and Adaptation Strategies for Global Change* **5**(1), 61–80.

Houghton, R.A., Skole, D., Nobre, C.A., Hackler, J., Lawrence, K. and Chomentowski, W.: 2000, 'Annual fluxes of carbon from deforestation and regrowth in the Brazilian Amazon', *Nature* **403**, 301–304.

Houghton, R., Lawrence, K., Hackler, J. and Brown, S.: 2001, 'The spatial distribution of forest biomass in the Brazilian Amazon: a comparison of estimates', *Global Change Biology* **7**, 731–746.

INPE: 2000, *Monitoring of the Brazilian Amazon Forest by Satellite 1999–2000*, S.J. dos Campos, Brasil, www.inpe.br/Informacoes_Eventos/amz1999_2000/Prodes/index.htm.

IPCC: 2000, *IPCC Special Report, Land Use, Land Use Change and Forestry*, http://www.grida.no/climate/ipcc/land_use/index.htm.

IPCC: 2001, *Climate Change 2001: Synthesis Report*, Cambridge, Cambridge University Press, http://www.ipcc.ch/.

Mahar, D.:1988, *Government Policies and Deforestation in Brazil's Amazon Region*, Washington, DC, World Bank.

Mattos, L. and Nepstad, D.: 2002, *An Agricultural and Environmental Credit Line for Amazon Farmers*, London, IV Katoomba Group Meeting.

Moura Costa, P. and Wilson, C.: 2000, 'An equivalence factor between CO_2 avoided emissions and sequestration – description and applications in forestry', *Mitigation and Adaptation Strategies for Global Change* **5**(1), 51–60.

Moura Costa, P., Stuart, M., Pinard, M. and Phillips, G.: 2000, 'Elements of a certification system for forestry-based carbon offset projects', *Mitigation and Adaptation Strategies for Global Change* **5**(1), 39–50.

Nepstad, D.C., Verissimo, A., Alencar, A., Nobre, C.A., Lima, E., Lefebvre, P., Schlesinger, P., Potter, C., Moutinho, P., Mendonza, E., Cochrane, M. and Brooks, V.: 1999, 'Large-scale impoverishment of Amazonian forests by logging and fire', *Nature* **398**, 505–508.

Nepstad, D., Carvalho, G., Barros, A.C., Alencar, A., Capobianco, J.P., Bishop, J. Moutinho, P., Lefebvre, P., Lopes Silva, Jr. U. and Prins E.: 2001, 'Road paving, fire regime and the future of Amazon forests', *Forest Ecology and Management* **154**, 395–407.

Nepstad, D., McGrath, D., Alencar, A., Barros, A.C., Carvalho, G., Santilli, M. and del C. Vera Diaz, M.: 2002, 'Frontier governance in Amazonia', *Science* **295**, 629–631.

Schmink, M. and Wood, C.: 1992, *Contested Frontiers in Amazônia,* New York, Columbia University Press.

Seroa da Motta, R., Ferraz C. and Young, C.: 2000, 'Brazil: CDM opportunities and benefits', in *Financing Sustainable Development with the Clean Development Mechanism,* Washington, DC, World Resources Institute, pp. 18–31.

Tolmasquim, M., Cohen, C. and Szklo, A.: 2001, *Development of Energy Consumption and CO_2 Emissions in the Brazilian Industrial Sector According to the Integrated Energy Planning Model (iepm),* Rio de Janeiro, COPPE/UFRJ.

WRI: 1998a, *The Clean Development Mechanism and the Role of Forests and Land-Use Change in Developing Countries,* Washington, DC, World Resources Institute.

WRI: 1998b, *Biodiversity and Climate: Key Issues and Opportunities Emerging from the Kyoto Protocol,* Washington, DC, World Resources Institute.

SOIL RESPIRATION AND CARBON STORAGE OF AN ACRISOL UNDER FOREST AND DIFFERENT CULTIVATIONS IN RIO DE JANEIRO STATE, BRAZIL

JOHN E.L. MADDOCK[1], MARIA B.P. DOS SANTOS[1],
SONIA R.N. ALVES DE SÁ[2] and PEDRO L.O. DE A. MACHADO[3]*

[1]*Departamento de Geoquímica, Instituto de Química, Universidade Federal Fluminense, Niterói,
RJ, Brazil;* [2]*Departamento de Físico-Química, Instituto de Química, Universidade Federal Fluminense,
Niterói, RJ, Brazil;* [3]*Embrapa Solos, Rio de Janeiro, RJ, Brazil*
(*author for correspondence, e-mail: pedro@cnps.embrapa.br; fax: +55 21 2274 5291;
tel.: +55 21 2274 4999)

(Accepted in Revised form 15 January 2003)

Abstract. Soil respiration rates of a clay-loam textured Acrisol under different uses (Atlantic forest, manioc, horticulture and pasture) from Rio de Janeiro State were measured. The relationship between carbon dioxide (CO_2) emissions and soil physico-chemical properties were investigated. Rates of CO_2 emission of two sites (Atlantic forest and horticulture) were also evaluated in different seasons in 1997 and 1998. In the forest site, monthly means of measured respiration rates showed good correlation with soil temperature in the range 19.6–24.1°C ($r^2 = 0.89$). In the horticulture site, no change was observed with soil moisture alone, in the range 3.0–13.2 wt%. In the horticulture soil, even when the surface soil was very dry, respiration rates increased in the hot, wetter summer but remained higher than the mean flux from forest soil. The CO_2 emission flux of the Acrisol under different use showed good correlation with soil temperature ($r^2 = 0.72$) and moisture ($r^2 = 0.61$).

Key words: horticulture, manioc, pasture, soil temperature, soil moisture, soil chemical properties.

1. Introduction

In climax ecosystems soil respiration is taken to be approximately balanced by photosynthetic uptake of carbon dioxide (CO_2) to produce biomass. During crop growth on cultivated land, CO_2 uptake may exceed soil respiration but it is assumed that carbon in harvested and waste vegetation returns to the atmosphere as CO_2 within an annual cycle. Soils emit CO_2 which is generated by the respiration of plant roots and organisms, such as bacteria, fungi, worms and insects, which live in it. This CO_2 is produced as a result of metabolic oxidation of the soil organic carbon and, in normal, aerobic soils, the rate of emission is thus a measure of the carbon mineralization rate. It is expressed as a flux (e.g. g CO_2 m^{-2} d^{-1}) and this emission, on a global scale, is comparable to other major CO_2 fluxes (Buyanovsky and Wagner, 1995). The global increase in atmospheric CO_2, the main greenhouse gas, comes mainly from fossil fuels (6.5 Gt C yr^{-1}), together with about 1.6 Gt C yr^{-1} from deforestation (Smith, 1999). Major emission of CO_2 is caused by above ground biomass burning during deforestation in tropical regions but, more slowly, changes occur in carbon stocks stored in soils when land use change occurs, as for example in deforestation in order to create pasture or crop plantations. This can be assessed by transient changes in CO_2 emission rates. Soil emissions from Amazonian deforestation represent a quantity of carbon approximately 20% as large as Brazil's annual emission from fossil fuels (Fearnside and Barbosa, 1998). Little is known however about the CO_2 emission from soils under the Atlantic Forest. Today, the Atlantic Forest is restricted to $98\,800\,km^2$ of remnants, or 7.6% of its original extension (Morellato and Haddad, 2000). In order to quantify the transient effects, and simply to define mineralization rates characteristic of various land use practices, the steady state CO_2 emission, or soil respiration rates of these, are also of interest. We present respiration rates and organic carbon contents of an Acrisol for some lowland sub-tropical environments in the State of Rio de Janeiro, Brazil.

2. Materials and methods

2.1. SITE DESCRIPTION

Short term (20 min) respiration rates were determined on various occasions in the following environments: (i) *Mata Atlântica* lowland dense evergreen rainforest (IBGE, 1993); (ii) adjacent manioc plantation on land cleared of forest 40 years ago; (iii) long term, unmanaged cattle pasture; and (iv) 'organic' horticulture site. The manioc plantation has received only occasional, sparse fertilizer application in recent years, and none during the period when respiration was measured. During the period when respiration rates were measured in the horticulture soils, a fast growing legume plant (pigeonpea, *Cajanus cajan*, L.) was cultivated as green manure for soil conditioning and nitrogen addition. The soil type was a well drained clay-loam textured Acrisol (Argissolo Vermelho-Amarelo, Brazilian Soil Classification) with $300\,g$ clay kg^{-1}. According to Menezes et al. (1980) the area is characterized by Precambrian rocks and Quaternary sediments and the rock bodies consist predominantly of gneiss and magmatic material. Sites were on or neighbouring the Atlantic coastal plain ($43°40'W$ and $22°44'S$) at altitudes of less than $200\,m.s.l.$, within $50\,km$ of the city of Rio de Janeiro. The climate is Aw (Köppen, 1936), humid semitropical with a yearly average temperature of $23°C$ and $1500\,mm$ of average rainfall. The coolest months, May–August are drier than the hottest months, December–March, but there is no rain in free dry season. Mean midday air temperature under canopy and monthly average rainfall for May–August, were $18.5°C$ and $65\,mm$, respectively, and for December–March, $23.5°C$ and $260\,mm$, respectively.

2.2. MEASUREMENTS

CO_2 emission was measured using static chambers and values calculated as described by Maddock and Santos (1997). Cylindrical PVC chambers (diameter $= 30$ cm, height $= 14$ cm), fitted with copper capillary vents (i.d. 0.6 mm $\times$ 20 cm length), and syringe sampling septa were employed. The bevelled rims of these were inserted in the soil to depths of 2–20 mm, in order to obtain air tight seals to the soil. Ambient air samples and 30-ml samples of the air from the chambers, withdrawn at intervals of 0, 5, 10, 15 and 20 min after sealing to the soil, were taken with 50-mL polypropylene syringes and analyzed for CO_2, by gas chromatography, using a Porapack Q column and TCD detector in a Shimadzu Model 6AM GC. Fluxes were calculated from the data of CO_2 concentrations in the chamber and sampling times.

Approximately, 50 cm outside the chamber walls soil samples were collected at 0–10 cm depth, and at the time of each flux measurement. The methods described by Embrapa (1979) were adopted to determine the following soil properties: soil moisture by gravimetric method, and soil texture by hydrometer method. Soil temperature was measured using a geothermometer (Taylor and Jackson, 1986). Averages of soil temperature, moisture and flux from at least five points were therefore used as data for each measurement occasion, in this work. Air-dried samples passed through a 2-mm sieve were employed for the chemical analysis. Organic carbon was determined by wet combustion method and total nitrogen content by Kjeldahl digestion, and pH in H_2O (1 : 2.5) according to Embrapa (1979) also. Additionally, soil bulk density was measured by the core method using 5 cm diameter $\times$ 5 cm deep metal rings and the total carbon analysed for the whole soil was used to calculate the organic carbon storage (C storage) on a unit area basis to 0–10 cm.

3. Results and discussion

3.1. TEMPORAL VARIATIONS

Considerable variations were observed between fluxes measured simultaneously at different, randomly selected, points within one site.

TABLE I. Monthly mean CO_2 emission, soil temperatures and humidities of forest soils.

Month	Season	Soil temperature (°C)	Soil moisture (wt%)	CO_2 emission $\pm \sigma$ (10^{-3} mol m^{-2} h^{-1})
May/97		20.0	30.9	6.6 ± 3.8
June	D	20.3	—	5.0 ± 2.0
July	R	19.6	15.1	4.5 ± 2.2
August	Y	19.6	16.0	4.6 ± 2.0
September		21.0	15.9	4.8 ± 2.2
October		21.4	23.7	4.0 ± 3.3
November	H	24.1	25.4	14.8 ± 4.8
December	U	24.0	24.2	11.0 ± 5.3
January/98	M	24.1	28.0	12.9 ± 4.1
February	I	23.2	22.5	12.6 ± 3.5
April	D	23.6	22.2	12.2 ± 5.5

Respiration rates of forest soils were measured over one year to study the effects of time variable factors, namely soil temperature and humidity. In the forest site, monthly means of measured respiration rates showed good quadratic correlation with soil temperature in the range 19.6–24.1°C ($r^2 = 0.89$) (Table I and Figure 1) while, although in general there was a change with soil humidity, correlation was not strong. In the horticulture site no change was found with soil moisture alone,

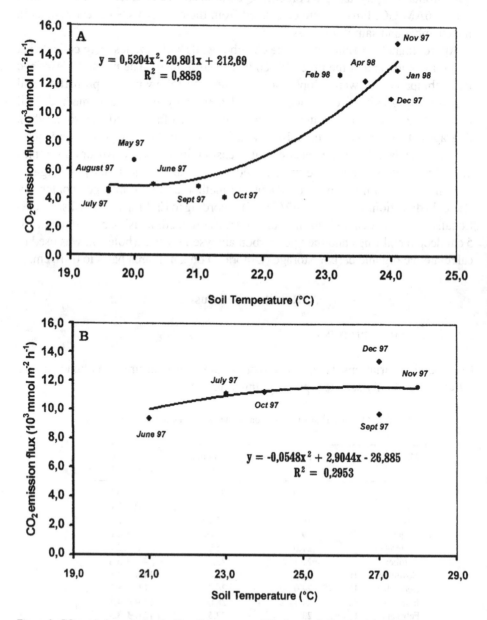

Figure 1. CO_2 emission correlated with soil temperature (A) of an Acrisol under Atlantic Forest and under horticulture (B) at different terms in 1997 and 1998.

in the range 3.0–13.2 wt%. However, the soil temperature range observed was small. Higher humidities were often accompanied by lower soil temperatures and vice versa, but it was found that respiration rates did increase significantly on those occasions when both humidity and temperature increased. This is given in Table I. La Scala et al. (2000) found no linear correlation between CO_2 emissions of a Ferralsol (bare soil), in Brazil at the same latitude as the present study, and soil temperature and moisture. On the other hand, in their study, measurements were conducted on various days, all in November 1998.

3.2. VARIATIONS AMONG RESPIRATION RATES TYPICAL OF EACH SOIL ENVIRONMENT

Means of respiration rates (CO_2 flux) and physico-chemical properties determined for each land use type are presented in Table II. The differences in means of respiration rates between the soils under different land use can be attributed partly to differences in mean soil temperatures during the periods of observation at each site. On the other hand, the forest soil responded very little to increases of temperature without humidity, some of the differences must be due to innately different soil properties. This is especially evident for the pasture and horticulture sites where

TABLE II. Soil CO_2 emission, soil temperature, soil moisture and chemical properties of an Acrisol under different use; standard deviations were given where applicable.

Parameters (units)	Sites			
	Forest	Pasture	Manioc	Horticulture
CO_2 flux (10^{-3} mmol m^{-2} h^{-1})	7.9 ± 4.9	12.1 ± 8.9	5.4 ± 1.7	10.5 ± 3.1
Soil temperature (°C)	20.7 ± 2.3	26.0 ± 0.5	24.8 ± 1.7	26.0 ± 2.3
Soil moisture (wt%)	21.5 ± 6.0	15.4 ± 1.0	20.1 ± 5.8	11.8 ± 5.3
Organic carbon (g kg^{-1})	20.0	12.0	15.0	21.0
Carbon storage* (mg ha^{-1})	16.94	8.42	9.00	11.60
Total nitrogen (g kg^{-1})	2.10	1.20	0.73	0.91
C/N	9	10	23	23
Soil pH (water)	3.9	5.4	5.4	6.0

TABLE III. Seasonal variation of respiration rates of an Acrisol under horticulture in 1997.

Month	Season	Respiration rate (10^{-3} mol m^{-2} h^{-1})	Soil temperature (°C)	Soil moisture (wt%)
May	D	10.50		5.28
June	R	9.34	21	3.00
July	Y	11.10	23	3.00
August		—	25	—
September		9.66	27	9.58
October	HU	11.20	24	13.18
November	M	11.59	28	7.84
December	ID	13.40	27	3.00

high fluxes of CO_2 were emitted from dry soils. Table III shows that for horticulture soil, respiration rates increased in the hot, wetter summer but remained higher than the mean flux from forest soil (Table II) even when surface soil was very dry. CO_2 emission flux of the Acrisol under different use also showed good correlation

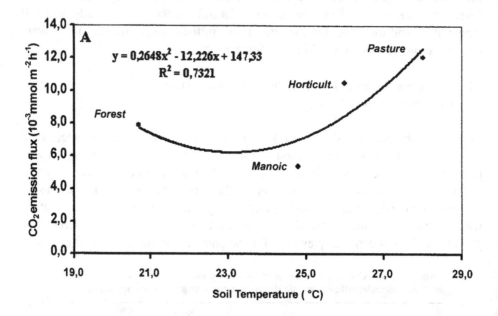

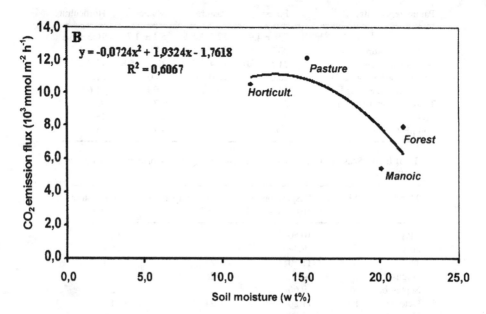

Figure 2. CO_2 emission correlated with soil temperature (A) and soil moisture (B) of an Acrisol under different use.

with soil temperature ($r^2 = 0.72$; quadratic regression; Figure 2) and moisture ($r^2 = 0.61$; quadratic regression; Figure 2). Rates of CO_2 emission vary as a function of soil temperature and soil water content due to the dependence of the density of micro-organisms on these soil physico-chemical conditions (Schlesinger, 1995).

The evaluation of the individual contribution of soil chemical properties to the CO_2 emission of the Acrisol under different use showed that there exists no correlation in any obvious manner with the different soil properties measured: pH,

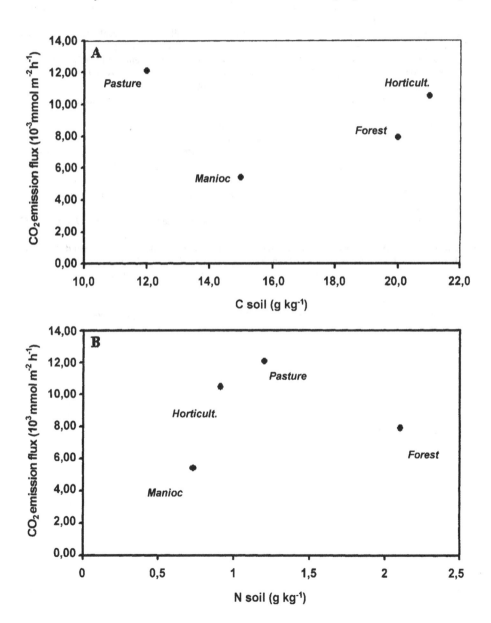

Figure 3. CO_2 emission correlated with total carbon (A) and nitrogen (B) of an Acrisol under different use.

total carbon and nitrogen, C/N and carbon storage. This is shown, by examples, in Figure 3. Soil respiration was found to be linearly correlated with total soil carbon in a study of a Ferralsol from São Paulo State (La Scala et al., 2000).

References

EMBRAPA – Centro Nacional de Pesquisa de Solos: 1979, *Manual de métodos de análise de solo*, Rio de Janeiro, Embrapa-CNPS, 212 pp.

Maddock, J.E.L. and dos Santos, M.B.P.: 1997, 'Measurements of small fluxes of Greenhouse gases to from the earth's surface, using static chambers', *An Acad Bras Ciências* **68**(Suppl. 1), 95–99.

Buyanovsky, G.A. and Wagner, G.H.: 1995, 'Soil respiration and carbon dynamics in parallel native and cultivated ecosystems', in R. Lal, J. Kimble, E. Levine and B.A. Stewart (eds.), *Soils and Global Change*, Advances in Soil Science, Boca Raton, CRC Press Inc., pp. 209–217.

Fearnside, P.M. and Barbosa, P.I.: 1998, 'Soil carbon changes from conversion of forest to pasture in Brazilian Amazonia', *Forest Ecological Management* **108**, 147–166.

IBGE: 1993, '*Mapa de vegetação do Brasil 1 : 5,000.000*', Rio de Janeiro, IBGE, 1 map.

Köppen, W.: 1936, '*Grundriss der Klimakunde*', Berlin, Guyter Verlag, 388 pp.

La Scala Jr., N., Marques Jr., J., Pereira, G.T. and Corá, J.E.: 2000, 'Carbon dioxide emission related to chemical properties of a tropical bare soil', *Soil Biology Biochemistry* **32**, 1469–1473.

Menezes, S. de O., Vieira, A.C., Souza, J.M. and Toledo, J.B.: 1980, '*Nota sobre a geologia da Folha Universidade Rural, RJ*', relatório, Imprensa Universitária, Itaguaí, Brazil, Universidade Federal Rural do Rio de Janeiro.

Morellato, L.P.C. and Haddad, C.F.B.: 2000, 'Introduction: The Brazilian Atlantic Forest', *Biotropica* **32**, 786–792.

Schlesinger, W.H.: 1995, 'An overview of the carbon cycle', in R. Lal, J. Kimble, E. Levine and B.A. Stewart (eds.), *Soils and Global Change*, Advances in Soil Science, Boca Raton, CRC Press Inc., pp. 209–217.

Smith, K.: 1999, 'After the Kyoto Protocol: Can soil scientists make a useful contribution', *Soil Use Management* **15**, 71–75.

Taylor, S.A. and Jackson, R.D.: 1986, 'Temperature', in A. Klute (ed.), *Methods of Soil Analysis*, Madison, American Society of Agronomy Inc., Soil Science Society of America Inc., pp. 927–940.

THE CLEAN DEVELOPMENT MECHANISM: MAKING IT OPERATIONAL

MATTHEW MENDIS* and KEITH OPENSHAW

Energy and Environmental Management Division (EEM) of the International Resources Group (IRG), 1211 Connecticut Avenue, NW, Suite 700, Washington DC 20036, USA (*author for correspondence, e-mail: Mmendis@IRGltd.com; fax: (202) 289 7601; tel.: (202) 289 0100)

(Accepted in Revised form 15 January 2003)

Abstract. The Clean Development Mechanism (CDM) is a facility for trading 'certified emission reductions (CERs) between developing and developed countries, thus saving non-renewable carbon emissions by promoting renewable energy, energy efficiency and/or carbon sequestration projects in LDC's. The purpose of the CDM is to help these latter countries meet their obligations under the Kyoto Protocol while at the same time promoting "sustainable" development in the former countries, thereby reducing the build-up of greenhouse gases (GHG). This paper examines the progress in achieving a workable CDM in time for the first commitment period (2008–2012), and the kinds of initiatives that can be pursued in the agricultural, land-use and forestry sector in tropical countries.

The critical element for the success of the CDM is the participation of a broad cross-section of buyers (ultimately from developed countries) and sellers (from developing countries) of CERs. Trading is the final step, which starts with project formulation, through successful implementation and then certification. This paper lays out a market-based framework for promoting CDM transactions between private sector project developers and traders and public sector policy makers, with regulators, governed by CDM rules, overseeing the smooth running of the CDM. However, as there are numerous players; it is proposed that trial CDM projects be demonstrated with the support of National/International bodies to iron out the problems and come up with practical solutions so that carbon trading can become a reality.

Most rules for the CDM were clarified at the 7th Conference of Parties (CP) in October/November of 2001 in Marrakesh. An executive Board (EB) was appointed and this EB is in charge of proposing workable ground rules to promote the CDM. These have to be submitted for approval by the 8th CP in late 2002. Three broad kinds of projects qualify for the CDM, these are: renewable energy projects that will be alternatives to fossil fuel projects; sequestration projects that offset GHG emissions; and energy efficient projects that will decrease the emissions of GHG. It is possible to have a combination of these initiatives. A fourth type covering GHG reduction is omitted. As elaborated in the text, in order to qualify for the CDM, the proposed projects may have to have additional costs when compared to the alternative(s). Two time frames have been agreed for CDM projects to qualify for the first commitment period, namely 7 years (with an option of two 7-year renewals) and 10 years. Also, for land-use, land-use change and forestry projects only afforestation/reforestation initiatives are recognized as being permissible for the first commitment period. These rules seem rather shortsighted, as forestry and/or renewable energy projects usually require more than 21 years to be fully effective. Also, the major cause of deforestation is clearing land for agriculture, not harvesting wood. Therefore, improving agricultural productivity may be the best way to reduce deforestation and its subsequent release of GHGs. These conditionalities should be re-examined when rules for the second commitment period are decided. However, various agricultural and land-use projects are discussed under the existing guidelines which could qualify as CDM projects in the first period.

Key words: agriculture, carbon sequestration, certified emission reductions, energy efficiency, forestry, Kyoto Protocol, renewable energy, tropical countries.

1. Introduction

The Kyoto Protocol allows for the trading of greenhouse gas (GHG) offsets and for Joint Implementation (JI) among developed countries and Eastern European countries with economies in transition, collectively referred to as Annex 1 Parties. The Protocol also makes provisions for the trading of GHG offsets from non-Annex 1 Parties (principally all developing countries) to Annex 1 Parties within the context of the Clean Development Mechanism (CDM), a mechanism whereby sustainable development is pursued using renewable energy resources, reductions from non-fossil fuel sources or initially through the decreased use of fossil fuels or switching to less GHG emitting fossil fuels (Ellerman et al., 1998). This paper focuses on this trading of GHG offsets between developing and developed countries through the CDM. It also discusses possible agriculture and land-use initiatives in tropical countries to mitigate GHG emissions.

The critical factor for the successful operation of CDM transactions is an active international market for certified emission reduction units (CERs)[1] as a result of interventions to reduce or offset GHG emissions. An international market has to facilitate partnerships between several bodies, namely project developers, investors, independent auditors, national authorities in host and recipient countries and the international agencies that are responsible for implementing the Kyoto Protocol. Because of the differing interests of the various players, there is a danger that CDM transactions may get bogged down in bureaucracy, resulting in investors or other parties losing interest. Thus, a successful international market framework for CDM transactions must be driven by a number of fundamental principles. The framework must:

(a) Result in *agreed sustainable development that meets national objectives* for the host country Kyoto Protocol) and not just emission reduction (ER) through CERs for the recipient country.
(b) Help *maximize the generation or supply* of cost-effective CERs.
(c) Provide *reliable information and secure access* for the buyers of CERs.
(d) Provide *legal recourse* for both buyers and sellers of CERs.
(e) Meet the needs of *a wide spectrum* of potentially diverse project types and proponents.
(f) Provide *a real incentive* for a broad base of investors to invest in CDM projects and not just attract a limited band of "green" investors.
(g) Result in CDM projects that are *additional to defined baselines.*[2]

Significant time has been spent to study, debate and develop a framework for the CDM (Matsuo, 1998). Most of the discussions and efforts have focused on the political dimensions, equity issues, baseline definitions and regulatory aspects of the CDM while downplaying the framework for investments and market transactions that are necessary for the CDM to be effective, but new literature more fully discusses these topics.[3] Now it is important to move from protocol development to creating a mechanism for CDM project formulation and implementation. In this context, there are areas that would benefit from further elaboration. These include:

- Market forces and mechanisms that must be employed if volume and efficiency are to be achieved in the production and sale of CERs.[4]
- Investor's attitudes, behavior and requirements, which must be clearly reflected in the emerging regulatory governance of the CDM.
- Clear and transparent rules and guidelines for the definition of "baselines" and "additionality" associated with CDM projects.
- The responsibility for and integration of CDM monitoring, verification and certification processes within the overall investment and implementation project cycle.
- Legal recourse for failed or delayed transactions undertaken within the CDM framework.
- Revocation of CDM activities and its ramifications for host countries.

This paper provides a guideline for the building blocks of a practical international market framework for investing in CDM projects, taking into account the Marrakesh rulebook. Furthermore, it gives examples in the agriculture and land-use sectors in tropical countries for such CDM projects. It builds on and summarizes earlier works of Hassing and Mendis (1998, 1999), that present structures and mechanisms for operating and managing CDM projects. The paper provides guidance to host-country governments who wish to participate in CDM activities as well as investors seeking to understand the CDM business.

2. CDM governance

Clear rules, regulations and guidelines for the CDM help stimulate the identification and development of projects that are eligible for the CDM.[5] National and international rules and regulations that meet the requirements of the Kyoto Protocol provide the governance for the contribution of the CDM instrument to sustainable development, project financing and global GHG mitigation. All non-Annex 1 (developing) countries, that ratify the Kyoto Protocol are eligible to participate in the CDM. However, a host country's willingness to participate in the CDM must be clearly established by the country and is voluntary in accordance with Article 12.5 (a). Additionally, the eventual acceptance of a project as a CDM "certified project activity" is dependent on the international and national rules that govern the CDM. Thus, identification and development of projects for the CDM should not be undertaken independently of the governance of the CDM, but in association with it.

2.1. INTERNATIONAL CDM EXECUTIVE BOARD

Article 12 of the Kyoto Protocol indicates that the responsibility for the design of the international CDM regulatory framework falls under the authority of the Conference of the Parties serving as the meeting of the Parties (CP/MP) to the Kyoto Protocol. It clearly indicates that the CDM is subject to the authority and guidance of the CP/MP and will be supervised by an Executive Board (EB) of the CDM. At the CP/6 (part-two) meeting held in Bonn in July 2001, it was agreed to facilitate a prompt start of the CDM and to call for nominations for membership to the EB, prior to the seventh session held in Marrakech, Morocco from 26 Oct. to 9 Nov. 2001. At this CP/7 session, 10 members (and 10 alternates) of the EB were elected, one from each of the five UN regional groups, two from parties included in Annex 1, two from non-Annex 1 parties and one representative of the small island developing states. The EU has secured two seats on the Board from France and Germany and Annex 1 parties have a blocking minority on the board.[6]

2.2. INTERNATIONAL CDM EB RESPONSIBILITIES

Article 12 endows the EB of the CDM with a number of responsibilities. These include:

- Specific eligibility rules.
- Enforcing the rules for both project eligibility and project baselines.
- Overseeing *operational entities* for the independent verification and certification of CDM project activities.
- Ensuring that CDM project ERs or enhanced carbon stores *are additional to* any that would occur in the absence of the certified project activity.
- Ensuring that "certified" or validated CDM projects have real, measurable and long-term benefits related to the mitigation of climate change.
- Establishing committees, panels or working groups to assist in the performance of its functions.
- Assisting in arranging funding for certified project activities as necessary.
- Ensuring that a share of the proceeds from certified project activities (maximum 2%) are used for countries that are particularly vulnerable to the adverse effects of climate change to meet the costs of adaptation. An additional tax rate to cover CDM administrative expenses remains to be fixed.

The Marrakesh Accord envisages that the CDM EB will use the services of the UNFCCC Secretariat to help it administer and implement the objectives, rules and guidelines of the CDM. This operational body must serve to implement the international regulatory framework for the CDM and interact with national entities undertaking CDM activities. This is covered by the registry rules under the Marrakesh Accord.

The CP/6 [part-two] meeting also recognized the need to simplify modalities and procedures for small-scale projects. It stated that the EB (elected at the seventh session) should develop, recommend and present simplified modalities and procedures to the CP at the eighth session (in late 2002). These small-scale CDM projects cover the following three groups of initiatives.

- Renewable energy projects with a maximum output capacity equivalent of up to 15 MW (or an appropriate equivalent).
- Energy efficiency improvement project activities which reduce energy consumption, on the supply and/or demand side by up to the equivalent of 15 GW h per year.
- Other project activities that both reduce anthropogenic emissions by sources (e.g. afforestation/reforestation projects) and directly emit less than 15 000 tonnes of carbon dioxide equivalent (4090 tC) annually.

2.3. NATIONAL CDM BOARD

The Marrakesh Accord states that the host countries shall designate a 'National Authority' for the CDM. A national CDM regulatory framework need not be consistent across recipient or host countries: what is important is that it is not contradictory to the international CDM regulatory framework. Therefore, while the process for recognition and approval of proposed CDM activities is likely to vary from country to country, the resulting eligibility of projects for the CDM should have a consistent base across countries. In principle any project within the rules is eligible; countries can however, define project types that they do not want to be done. The responsibilities of such a National CDM Board could include:

- Defining CDM project types that: (a) meet national priorities; (b) contribute to sustainable development; and (c) result in real, measurable and long-term benefits related to mitigating climate change.
- Facilitating investments in approved national CDM project activities.
- Establishing rules and guidelines for monitoring CDM project activities to ensure data availability needed for independent verification of the resulting ERs that are supplementary to the international rules.

2.4. OPERATIONAL ELEMENTS OF A CER MARKET FRAMEWORK

In addition to the international and national regulatory framework outlined above, a number of operational elements should emerge to support the development of a CER market. These include:

- One or more institutions/specialist/consultants that provide technical inputs for the identification, formulation and development of CDM projects and project baselines.

- One or more institutions/banks/development agencies that provide or secure financial resources for developing and financing CDM projects.
- Approved agents/agencies (*operational entities*) that are capable of providing validation, verification and certification services for CDM projects.
- Markets and information sites where potential sellers and buyers can obtain the price and other relevant information relating to the supply of and demand for CERs.
- Brokers that bring potential buyers and sellers together to assist in the buying and selling of CERs and the recording of binding transactions.

Many of these elements have evolved in anticipation of CDM and market for CERs. Some of these elements are active in Activities Implemented Jointly (AIJ) transactions while others have operated in other related emissions trading markets. The rules governing the operation of these elements will derive from the rules established by the CDM EB, the National CDM authorities and the Annex 1 Supervisory Committee etc.

2.5. LEGAL ENVIRONMENT

A necessary but not sufficient pre-condition for attracting investments in CDM projects is a legal environment that fosters general investment. Certainly, an environment in which rules and regulations are not in place to protect and foster investments is discouraging for CDM initiatives. Additionally, environments in which internal and/or external investors have little or no legal recourse will also inhibit CDM project investments. Thus, favorable legal and regulatory investment environments should attract CDM initiatives and vice versa.

3. Project identification and formulation

The first and most important step in the CDM project cycle is the identification and formulation of potential CDM projects. A false diagnosis could lead to the expenditure of considerable time, effort and resources, resulting in an ineligible or unacceptable project for the CDM. The lessons learned from the AIJ Pilot Program regarding project eligibility and the definition of baselines is reviewed in order to provide clear and early guidance for determining the eligibility of proposed CDM projects. This should help conserve and channel scarce resources to project activities that ultimately will be certified by the CDM.[7] Figure 1 illustrates the key actors/stakeholders and relevant issues associated with the project identification and formulation step.

3.1. PROJECT PROPONENTS

There is a broad spectrum of project proponents and parties involved in the project identification and formulation stages. Potential key players in this first step of the

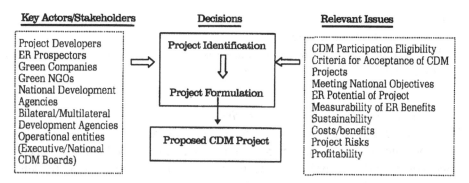

Figure 1. Project identification and formulation.

CDM project cycle are listed in Figure 1. They include project developers who may or not be investors, private and public financiers, government agencies, monitoring organizations and CDM personnel. The relevant issues to be considered before a CDM project is accepted concern of its eligibility, compliance with national objectives and sustainability and the costs and benefits, risks and profitability. Private companies or individuals will not invest unless they can see tangible benefits.

The on-going pilot AIJ initiative has resulted in the emergence of players in all the above categories (except CDM personnel). However, to date, many potential key players have been reluctant to invest a great deal of time and resources as the rules of the AIJ pilot and the CDM are not clearly defined or well established. Most have ventured into the AIJ arena with support from Annex 1 development or environmental agencies using public funds. Until recently, the mobilization and use of private funds was very limited, but it has picked up due to the activities of the Prototype Carbon Fund, the Dutch CDM tender program and several funds operated by private banks.

For the market in CDM projects to flourish, the rules of the CDM should be clear, transparent and workable. This is a key role for the CDM EB and/or National CDM boards. Then key players, recognizing the potential value of this market, will independently invest and undertake the front-end work of project identification and formulation needed to successfully develop CDM projects. By fostering the involvement of independent players, the CDM can avoid the need for expending public funds for project development, while relying on the marketplace to mobilize and apply funds to develop projects for the CDM.

3.2. PROJECT ELIGIBILITY CRITERIA

Article 12 of the Kyoto Protocol stipulates four principal eligibility criteria for CDM projects; these have been elaborated at Marrakesh. They are:

- Non-annex 1 Parties "will benefit from project activities resulting CERs".
- Projects must assist non-annex 1 Parties "in achieving *sustainable development* and contributing to the ultimate objective of the Convention".

- Projects must result in "real, measurable and long-term benefits related to the mitigation of climate change".
- Projects must result in "reductions in emissions that are additional to any that would occur in the absence of the certified project activity".

3.2.1. National benefits from project activities

The first two criteria are clearly established to assist developing countries in achieving economic, social, environmental and "sustainable" development objectives while reducing GHG emissions. Under the Marrakesh rules, the host country defines what constitutes 'sustainable development'. Thus there is a risk that some governments knowingly or unknowingly accept projects with negative effects. Therefore, it is essential to have transparent procedures where all interested parties can provide inputs. This criterion is particularly important for some agricultural/land-use projects that may produce GHG ER or sink enhancement, but which may not have any additional benefits for the host country. Examples of this may be a project to reduce methane emissions from rice cultivation that does not decrease the cost of rice production or increase rice yields, or a project that increases the stock of carbon in a natural forest without enhancing benefits for the local people or the nation. Such projects could be considered under other funding initiatives such as the *Global Environmental Facility* (GEF). However, even in the above two examples, some local benefits may ensue. For example, reduction of methane emissions from rice cultivation may come about by applying biogas slurry to rice fields rather than raw animal manure, lowering the water level in rice paddies and other management techniques, using quick maturing rice strains etc. Such interventions could have local and national benefits. Again, increasing the carbon stock in a forest may enhance its biodiversity and amenity value. *But, eligible land-use projects to be credited to the first commitment period are confined to afforestation/reforestation projects.*

The determination of whether projects help host countries achieve sustainable development must lie with the host country. There is no operational or "objective" method to determine if a project contributes to a country's sustainable development. Attempts are underway to find indicators for sustainable development and some examples in the agricultural/land-use fields are cited below in Table I, but general acceptance of any resulting indicators will have political ramifications and will need to be validated through the political process. Therefore, host countries must decide for themselves if a proposed CDM project is likely to contribute to sustainable development. However, in the absence of a common standard of a project's contribution to sustainable development, it is possible that where competition between projects occur, sustainable development could be neglected in favor of maximizing climate mitigation benefits. Hassing and Mendis (1998) propose limiting eligibility for CDM participation to project activities with clearly proven sustainable development impacts; however, this was not accepted in the negotiations.

In the long run, truly sustainable development can be achieved by using or recycling renewable resources and recycling non-renewable ones. This is especially so

TABLE I. Possible fields for CDM projects related to agriculture/land-use.

Agriculture	Growing energy crops for direct and indirect energy use (see below).
	Increasing agricultural productivity through the use of organic fertilizers, nitrogen fixing trees and improved microclimate. *Increased productivity could release marginal agricultural land to grow perennial crops.*[a]
	Conservation agriculture (zero cultivation, crop rotations, soil carbon conservation).
	Sustaining agricultural productivity by substituting organic nitrogen fertilizers for inorganic nitrogen fertilizers made by energy intensive use of fossil fuels.
	Using biomass and other (renewable) energy forms for irrigation pumping to increase agricultural productivity.
	Decreasing emissions of methane through using biogas slurry on rice paddies rather than dung.
	Use of residues for energy, rather than burning *in situ*.
Land use/ agro-forestry/ forestry	Promoting agro-forestry as one way to stabilize shifting cultivation, thereby allowing much of the land in the shifting cultivation cycle to revert back to forest or to be planted with trees.
	Growing wood crops for direct and indirect energy use (see below) and/or construction materials to substitute for fossil fuels or energy-intensive construction materials (steel, concrete).
	Planting of trees to improve microclimate, provide animal feed and/or wood products.
Energy	Use of renewable biomass as an energy source, both specifically grown or from residues for:
	(a) direct use (household and non-household use);
	(b) to convert into more convenient gaseous (biogas, producer gas), liquid (methanol, ethanol, plant oils) and solid fuels (densified biomass, charcoal);
	(c) an energy feedstock to produce electricity.
	Improving the efficiency of biomass energy producing systems and end-use devices. These initiatives could help replace fossil fuels or keep the user from substituting fossil fuels for biomass fuels.

[a]Increasing agricultural production also reduces the pressure on forest land, thus saving the conversion of some areas to arable agriculture. However, this is considered to be management of existing land, trees or forests and for non-annex 1 countries is not counted as a CDM initiative although the emission of carbon is prevented by not clearing forests. This is further discussed in the text.

for energy. Fossil fuels are finite and their use is the principal cause of atmospheric carbon accumulation. Many of the proposed projects, including CDM ones, are interim measures such as switching from one relatively large carbon emitting fossil fuel to another that emits less carbon per unit of energy (e.g. from coal to natural gas) and/or improving the efficiency of the end-use device, storing CO_2 from fossil fuel burning, reducing venting of natural gas and using gaseous petroleum products from wells or refineries for energy purposes in place of flaring them. These measures will only slow down CO_2 accumulation not prevent it. Thus, steps are also necessary to promote renewable energy, particularly biomass as a legitimate and available fuel source.

Today, the largest source of renewable energy, accounting for about 15% of the World's energy demand, comes from biomass. In many developing countries it is the most important household fuel and is a significant industrial fuel in many rural areas. There is considerable scope to expand biomass use either by tapping under-used biomass resources or increasing the supply. At the same time, end-use efficiency measures could be taken to mitigate demand. These kinds of initiatives should be uppermost in project proposals if sustainable development is to be realized.

3.2.2. Measurability and long-term benefits related to climate change
The criterion of "real, measurable and long-term benefits related to the mitigation of climate change" requires that the ERs associated with projects for the CDM must posses some specific characteristics.

- The ERs must be based on *real* reductions of GHG emissions and/or carbon sequestration that are directly associated with the CDM project activity.
- The benefits of the project will not be countered by an offsetting measure, e.g. the clearing of a forest area to provide land for farmers in compensation for land provided by them for a tree planting project. Thus, CDM activities must give net benefits to help mitigate climate change.
- The ERs that are produced must be *measurable* or quantifiable using reliable measuring, sampling or mass balance techniques. In essence, this requires that projects that produce ERs that are not directly measurable should not be eligible for validation as CDM projects.
- The ERs must be *long term*.[8] Risks associated with the permanence of GHG reductions are directly related to whether reductions are reversible at a future point in time.[9]

3.2.3. Additionality of reductions
GHG emissions from CDM projects must be lower than those that would have occurred in the absence of the CDM activity. To estimate accurately the additionality of reductions of a CDM project, it is important to have an accurate portrayal of the baseline[10] or "business as usual". However, if the CDM project is undertaken, then the baseline is counter-factual and an accurate measurement of the emissions from the "business as usual" or baseline cannot be achieved. To date, the most practical experience with the estimation of emissions from "business as usual" interventions is in the context of AIJ pilot phase and the GEF. Current assessments seem to indicate that a menu of options for estimating the baseline is likely to be more appropriate than a single approach. Considerable work has been done on defining and measuring baselines and additionality, especially in the context of the OECD (Bode et al., 2000; Bosi, 2000; Ellis, 1999; Puhl, 1998).

An important financial consideration emerges in the process of defining baselines that must also be considered within the context of the "additionality" CDM requirement. Specifically, a CDM project should also have "investment additionality" in comparison to the baseline option. If a proposed CDM project is financially more attractive than the intervention that would occur in the baseline, then the argument can be made that the CDM project belongs in the baseline and should replace the assumed baseline initiative. The issue of investment additionality is the subject of considerable debate. However, without these criteria, there is no basis to determine if baselines are an accurate reflection of expected profit maximizing behavior. The possibility may arise to define baselines that maximize the projects eligibility for CDM and thereby result in ERs that are not additional to what would have occurred in the absence of the CDM. However, the Marrakesh rules do not explicitly ask for a check on investment additionality. Also, proposed/actual CDM activities may not necessarily operate with profit maximization as an overriding criterion.

4. Baseline definitions

The definition of the baseline, against which the ERs of a proposed CDM project are assessed, is a very important step in the CDM project cycle. A project baseline defines a level of expected emissions/carbon stores that is used to assess the mitigation performance of an alternative project. It is the basis from which the ERs for a CDM project activity must be measured. The quantity of ERs that a potential CDM project activity can generate provides the basis for attracting the additional investments that may be needed to support the CDM project activity. Therefore, the development of a baseline for a CDM project lies at the heart of the validation process.

To date most of the experience with project baselines for estimating ERs has been gained in the context of the AIJ pilot phase and the GEF (Ellis, 1999; Michaelowa, 1999). This experience has shown that current approaches for baseline setting provide little guidance for the CDM. Key issues such as defining the expected baseline, setting the period for which baselines should be valid and defining system boundaries for the baseline activity should be clearly addressed. Agreement has been reached for the time frame. The crediting period for CDM projects is either:

– A maximum of seven years which may be renewed at most two times (that is up to 21 years), provided that for each renewal, a designated operational entity determines and informs the EB, that the original project baseline is still valid or has been updated to take account of new data where applicable; or
– A maximum of ten years with no option of renewal.

Additionally, the issue of macro-economic policies and regulations that inhibit the adoption of CDM type activities should also be carefully evaluated in order to minimize the potential for rewarding bad policies with CDM projects. Figure 2 identifies the key actors/stakeholders and the relevant issues associated with the process of baseline definition for CDM projects.

The Marrakesh Accord defines some issues on baselines, inter alia that they shall be project-specific. There is still work to be done on the baseline methodologies. Such methodology will have to be considered by the EB and if accepted included in the rulebook.

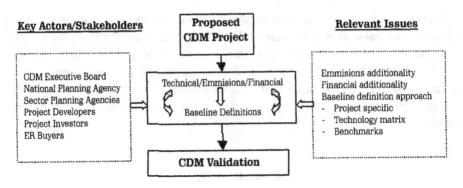

Figure 2. Baseline definition.

4.1. RESPONSIBILITY

The principal responsibility for defining the baseline associated with a specific project will lie with the project developer/investor. However, the underlying assumptions and data that support a baseline definition must be derived from national or international authorities. For example, the sector growth rates, performance of baseline technologies, cost of baseline technologies and emission /sequestration rates of baseline technologies will need to be derived from national data and ultimately validated at the national level. To the extent possible, the national authority for the CDM may establish the baseline parameters that are ultimately needed to help define project specific baselines. However, the guidelines for establishing these baseline parameters must derive from and be approved by the CDM EB in order to ensure a degree of consistency across host countries. This is especially so for small-scale CDM projects already outlined in Section 2.2 above. The CDM EB has to develop and recommend to the CP at its 8th session simplified modalities and procedures for the following small-scale project activities:

- Renewable energy projects with the equivalent output capacity of 15 MW.
- Energy efficiency projects which save up to the equivalent of 15 GW h per year.
- Projects that reduce anthropogenic emissions and directly emit $<15\,\mathrm{kt\,a^{-1}}$ of CO_2 equivalent.

As part of the baseline study, an environmental examination has to be undertaken for each project, including trans-boundary impacts, if host country laws demand it. If such an examination reveals that the impacts could be significant, *then a full environmental impact assessment (EIA) is required.* Such an EIA will solicit the opinions and concerns of local stakeholders and responsible government agencies and draw up an environmental action plan to mitigate the possible negative effects.

4.2. APPROACHES

There are several approaches that can be used to define a baseline, but for small-scale projects, they have to be simple, but effective. These vary in the degree of aggregation and level of accuracy. At present there is an ongoing debate regarding the scope of baseline definition methodologies, the accuracy of various methodologies, the pros and cons of each methodology and the methodologies that are best suited to the needs of the CDM. What is important is that if the market for ERs from the CDM is to grow, it will require that baseline definitions are consistent, transparent and relatively easy (i.e., not costly) to apply. Three methodologies for defining baselines that meet the above criteria are presently gaining recognition in the literature. They are: (1) project specific; (2) technology matrix; and (3) benchmarking. However, the Marrakesh Accord states that baselines should be project specific. Thus only this methodology is discussed, especially with reference to agricultural/land-use projects.

4.2.1. Project specific

The project specific baseline approach has been established and used extensively by the GEF. It requires that for each proposed CDM project activity, a specific baseline project must be defined which provides the equivalent "normal" economic benefits, as does the proposed CDM project activity. The GHG emissions/storage of the baseline project are then estimated and compared against that of the proposed CDM project activity. For agricultural and land-use projects under CDM, this is the most appropriate approach.

For example, if a tree planting project is proposed to provide feedstock for an electric power plant, then the baseline study would assess the initial organic carbon store in wood and soil on the proposed area to be planted. It would then measure the stock and annual off-take of wood when the plantation reaches equilibrium, plus the average amount of organic carbon in the soil beneath the different age classes in the plantation. The difference between the amount of organic carbon on-site at the outset of the project and the store in the wood and soil at equilibrium gives the amount of sequestered organic C due to the project's intervention. In addition, there is the annual production of sustainable biomass energy. This will have to be measured in terms of the amount of fossil fuel it displaces and not the amount of carbon contained in the wood (50% by dry weight). Knowing the efficiency for the wood-fired boiler, the amount of feedstock required to replace carbon in a fossil fuel boiler can be determined. The carbon that this wood replaces can be calculated each year over the lifetime of the system, taking into consideration the products of incomplete combustion under each system.

One question that has to be addressed is the fuel and the technology that the wood-fired boiler is to replace. This could be done either by assuming it would replace the most common technology and fuel or that it would replace the average of all the fossil fuel technologies for power generation. This should be decided by the CDM board from the outset of the proposed initiative.

One concern is that the store of carbon in the wood may be eliminated at any time if a more profitable use for the land (or wood) can be found, or if the power company finds a cheaper source of wood or other biomass. If such a case were to happen, then there could be a deduction for the store of carbon in the wood and soil. Alternatively, if the wood were used say as a boiler fuel in the process of clearing the land, it would not be credited as a carbon offset. However, the maximum lifetime of a CDM project is 21 years. Therefore, if the plantation is still standing after this period of time, then no deductions can be made, even if a land-use change occurs later.

For small-scale projects, all that needs to be measured is the annual supply of wood fuel to particular end-uses, together with the amount of fossil fuel carbon it replaces. This neglects the store of carbon in the plantation, but this can be compensated by paying a premium for the traded carbon or not deducting a percentage of the 'carbon tax' for CDM expenses.

As stated above, where biomass sequestration and/or biomass energy projects are proposed for specific areas, it is important to undertake a baseline survey to measure the existing store of organic carbon in the biomass and soil. Without such measurements, it is impossible to judge whether there is a change in organic matter as a result of the project. An increase in sequestration should be measured against the cost of alternative initiatives. It is possible that growing an energy crop could lead to a decrease in the carbon store. This should be compared to the net benefits (if any) from using renewable energy as a substitute for fossil fuels. In some instances, the time period for a project may be critical.

4.3. VALIDITY PERIOD

The period for which a baseline is valid should be equivalent to the period for which the underlying baseline project is in fact replaced by the CDM project. This is either: (a) 10 years or (b) from 7 years to 14 or to a maximum of 21 years. However, this is not an operational definition as the underlying baseline project is counter-factual. The actual baseline conditions may be dynamic and should be updated periodically. It is quite conceivable that a technology used in a CDM project could become the baseline technology shortly after the initiation (and possibly because) of the CDM project. For example, new clones may be introduced in a tree planting project increasing average yield by 25% and a new wood gasification method used in a power plant improving conversion efficiency from 25% to 33%. Also, for tree planting projects there should be monitoring of the plantation area throughout the growth cycle.

The magnitude of the GHG reduction/sequestration credited to a CDM project varies depending on the validity period as well as the definition of the "expected" emissions/sequestration. Clearly, there are significant implications on the outcome of a CDM project relating to the validity period and the expected emissions/sequestrations. A 21-year tree planting project should have about 3 times the ERs than a 7-year project. The CDM EB has specified the validity period options. It should now establish criteria and guidelines for determining the expected GHG savings/sequestrations. These criteria and guidelines must take into account the need to encourage the market for CDM projects while limiting the potential for rents on resulting ERs.

5. Validation of CDM projects[11]

Figure 3 illustrates the key actors/stakeholders and relevant issues that are associated with the validation of a CDM project. Projects that are proposed for the CDM have to be certified or validated as eligible projects before they are able to secure all their financing (financial closure). The validation procedure is outlined in Figure 3.

This process is necessary in order to capitalize on the financial value of the project's ERs. Validation as a CDM project activity is tantamount to receiving a license to produce and sell ERs. Without the CDM validation, a project developer would not be able to negotiate an ER purchase agreement (ERPA) with potential buyers. This is similar to a power production license, which must precede the negotiation of a power purchase agreement (PPA).

5.1. CONDITIONS

The conditions for CDM validation are, to a large extent, dictated by the Marrakesh Accord, Article 12 of the Kyoto Protocol and by the national objectives of host countries.

5.2. RESPONSIBILITY

The responsibility for CDM validation lies with the Operational Entity, i.e. an independent certifier; the final decision rests with the CDM EB. The project developer has to submit a project design document (PDD), including approval by all involved

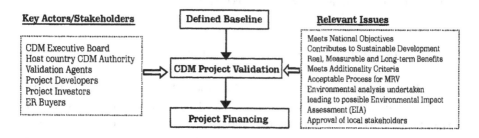

Figure 3. Validation of CDM projects.

government agencies. The PDD must provide the documentary evidence to demonstrate that the proposed project meets the validation criteria as specified in the Marrakesh Accord rules.

A critical element of the CDM validation process is the assessment of the assumed baseline and the estimated production of ERs by the proposed project. The process must validate the estimated ER production of the proposed CDM project against the assumed baseline conditions. The validation, once issued, would serve as a license to produce and sell ERs that can be fully certified. As discussed in Section 4, the validity period for the baseline defines the period for which the ERs from a CDM project are considered to be additional. Thus, the period of validation for a CDM project will be tied to the validity period for the associated baseline.

6. Financing CDM projects

Reaching financial closure for CDM projects is very similar to reaching financial closure for a project in general with one exception. The potential financial value of the resulting ERs from the CDM project must also be secured. The "additionality" requirement of CDM projects will translate to additional costs for CDM projects in comparison to the baseline options. Therefore, recovering these additional costs through the sale of the project's ERs is critical for financial closure of CDM projects. Structuring and having an (ERPA) Purchase agreement will be necessary to reach financial closure for CDM projects. This section explores a number of financial instruments and opportunities for securing an ERPA. If a sound ERPA cannot be secured for a CDM project, it is unlikely to reach financial closure.[12]

6.1. SECURING AN ERPA

A number of common project elements need to be addressed in advance of securing an ERPA. These project elements do not have to be designed from scratch. Useful analogies can be found in the context of energy performance contracting as well as PPAs.

6.2. ESTABLISHING THE MONITORING REPORTING AND
VERIFICATION (MRV) PROCESS

At the heart of any financial instrument based on the sale of ERs are the rules that establish how the ERs are measured and monitored have been set out in the Marrakesh Accord; they are to be refined by the EB. This is because these rules have a fundamental impact on the capacity of a CDM project to generate ERs. There are two basic kinds of ER units, namely those that are sequestered in some form of storage and those that reduce the emissions of GHG.

Sequestered ERs in biomass. For Agriculture and Land-use CDM projects, atmospheric carbon is stored mainly in woody biomass and in the soil, although grasses have a store in their roots. For annual crops, carbon is stored over a short time period before it returns to the atmosphere. By convention, this is not considered as a store, but the part of it not respired by plants and animals could be used for energy purposes before it is recycled. Examples of this are the use of bagasse for boiler fuel in a sugar factory and the use of dung for cooking in households. Principally, atmospheric carbon is stored in woody biomass, both above and below ground. There is about 50% of carbon in dry wood. Above-ground wood, especially tree stems, can be measured with a reasonable degree of accuracy and multiplying factors can be added to account for the branches, twigs, roots, shrubs and small diameter wood. These multiplying factors and the measuring techniques will have to be agreed to by the CDM EB on the advice from experts in the mensuration field. Vine et al. (2001) have produced a paper on monitoring, evaluating, verification and certification of forestry projects, which discusses the monitoring challenges etc. and Openshaw (2000b) describes a forestry baseline survey in Benin.

Sequestered ERs in soils. For the same type of soil and climatic conditions, there is a greater amount of organic carbon stored in forest soils than in grassland soils and there is more organic carbon stored in grassland soils than in arable agricultural soils. A tropical high forest can have about twice as much organic carbon in its soil ($200\,t\,ha^{-1}$ C down to 5 m) than in similar arable soils ($100\,t\,ha^{-1}$ C down to 5 m, Bouwman, 1990). Organic soil carbon comes mainly from the attrition of roots and rootlets, but some comes from the decomposition of surface matter and the death of soil fauna. The build-up of organic carbon is slow but steady when an area is converted from annual to perennial crops. Thus, in order to judge the incremental organic carbon built up in soils, soil samples should be taken throughout the lifetime of the CDM project and in the different age classes. This may be time consuming and bodies such as the GEF could commission studies by research organizations to generate standard tables to apply to various soil types under different climatic conditions. What is certain is that if the incremental accumulation of organic carbon in soils under woody biomass is neglected, then up to half the atmospheric carbon sequestration in plantations could be missed.

Reduced GHG emissions. These ERs are not physical products that can be stored, counted and transferred (although avoiding deforestation can be counted as a store of organic C). A renewable energy project built to replace or instead of a fossil fuel project is substituting the carbon in the fossil fuel, not the energy content in the renewable entity. Therefore, estimates have to be made of the kind and amount of fossil fuel the project will replace. This depends on the assumed efficiencies of the two systems under existing technologies and the kind of fuel that is to be replaced. Thus, an agreement on the MRV process needs to be made between two contracting parties, one of them the performance contractor/project owner or seller and the other the investor/recipient or buyer. There is need for independent auditors to monitor and verify the reported ERs because, one or both contracting parties may have an incentive to set up a process that would deliver over-stated amounts of ERs at no additional costs to either party. However, this would work to the detriment of the CDM and the objectives of the UNFCCC. This is why independent certification of the ERs must, by necessity, check against these potential abuses.

The MRV process for projects requires a detailed ex-ante description of the applicable process that is consistent with CDM requirements and will require approval by the national authority responsible for the CDM. The proposed MRV plan should be part of the information provided and approved during the validation process. The energy efficiency industry has created similar protocols for its own purposes, and work to adopt these examples for the purposes of ER performance monitoring is underway.[13]

6.3. FINANCIAL MODELS FOR CAPITALIZATION

A number of common financial models can be applied to capitalize the potential ER revenue stream from CDM projects. ER benefits could be capitalized by equity inputs. Where debt or third party equity financing is sought in return for the sale of a project's ERs, CDM project developers, prior to financial closure, must negotiate an ERPA with potential buyers. A 'futures market' for carbon may develop, where annual carbon credits from accredited CDM projects are traded at a discount and used to finance the additional cost of the project.

6.4. ACHIEVING FINANCIAL CLOSURE

Financial closure is achieved when all contractual arrangements related to project financing are finalized. These include the project area for renewable energy production/sequestration, fuel supply, establishment/construction, operation and maintenance, performance monitoring and product sales. During this process, performance risks, responsibilities and liabilities are allocated and key licenses and contracts for the import of equipment, plant construction and plant operation are secured. From the perspective of a project's ERs, it is important that the ERPA cover all relevant aspects that relate to its financial value. These include:

- Protocols for measuring, reporting and verifying resulting ERs, both stored and produced.
- Quantity, price and delivery dates of ERs.
- Responsibilities and liabilities in case of non-performance with regard to ERs.
- Required approvals from the host government and the CDM as applicable.
- The implications of a status change in the project's CDM validation and baseline reference.
- Procedures to resolve impacts of future changes (in the regulatory environment or baseline definition) on a project's capacity to generate ERs.

7. Project implementation and operation

Upon achieving financial closure, a project moves quickly to the project implementation and operation phase. The effort and time required during this phase of the project cycle is highly dependent on the specifics of each project. For CDM projects, an important element is ensuring that monitoring and reporting procedures are implemented according to any agreed or required protocols. This is particularly important if the resulting project ERs are to be verified and certified at a later date. Failure to do so could lead to the voiding of potential ERs. Thus, the principal objective during the project implementation and operation phase is to ensure the production of ERs and to establish a clear and practical audit trail for the monitoring and verification of the ERs. Possible CDM Projects related to agriculture and land-use are now outlined. It must be stressed that these projects are only valid if they are not in the baseline scenario of country programs because, for one reason or another, it is considered that they are not viable and/or practical without some form of 'seed' money.

7.1. POSSIBLE CDM AGRICULTURAL PROJECTS

A principal cause of deforestation in the tropics is clearing land for agricultural expansion. Therefore, increasing unit agricultural productivity could improve the living standards of the farmers, while at the same time reducing deforestation. There are a number of technical and physical constraints, which may inhibit productivity increases. These include lack of land tenure/ownership, poor infrastructure, inadequate extension services, weak markets, low farm-gate prices, insufficient access to and knowledge about improved seeds, fertilizers, integrated pest management etc. and unreliable water availability. Some or all of these impediments have to be addressed if agricultural productivity is to be increased. At the same time, non-farm opportunities must be expanded to offer employment to the increasing rural population.

However, during the first commitment period, afforestation/reforestation projects are the only eligible Land-use/Land-use Change/Forestry (LULUCF) initiatives that are allowed under the CDM, (Article 12: CP/6 part 2 page 9, paragraph 8), although cropland management, forest management, grazing land management and re-vegetation are eligible activities under LULUCF for Annex 1 (developed/transition) countries (CP/6 part 2 page 11, paragraph 4). This is a rather strange decision as most if not all net-deforestation (and forest degradation) occurs in developing countries as a result of population pressure and the need to grow export cash crops. Deforestation in developing countries accounts for about 15% of net atmospheric carbon additions and if sustainable development and GHG mitigation are to be tackled seriously, then confronting the root causes of deforestation should be of prime importance.

There are potential CDM projects that could assist in increasing agricultural productivity. These include: (a) ensuring a reliable supply of renewable energy for water pumping in irrigation schemes; (b) preventing venting of methane from manure by first extracting it in biogas digesters and then adding the slurry to fields: the methane could be used for energy purposes, replacing fossil fuels; (c) planting nitrogen fixing woody plants in agro-forestry formations; (d) improving the microclimate by planting shelterbelts. These latter two initiatives depend on whether such kinds of tree planting activities are considered to be afforestation/reforestation.

Through irrigation, if two or three annual crops can be grown on areas that previously only grew one crop, and/or if unit productivity increased significantly, then there should be more residues available for animal feed, mulching and energy purposes. Similarly by improving non-irrigated productivity, more potential renewable energy and/or feed are available. One consequence could be an increase in farm animal numbers and hence more organic fertilizers etc. are accessible. At the same time the demand for new farmland from clearing forests may diminish significantly.

On the other hand in some areas/countries, there may be too much application of mineral fertilizers, while concurrently, animal manure may be dumped into rivers. This could cause excessive eutrophication (undue nutrient content in water bodies) and methane venting from the dumping of manures in anaerobic conditions. Eutrophication of water bodies leads to a diminution of the light penetration layer (and hence the growth of light dependent organisms), to widespread anoxia (absence of oxygen which encourages methane production amongst other things) and the shortening of the food chain. By improving manure management through the extraction of methane before it is added to the crop, especially rice, methane venting could be diminished, renewable energy made available and eutrophication decreased. This would improve the food chain in water bodies and increase the sustainable off-take of fish etc.

The CDM requires the verification of the reduction of GHG and/or the use of renewable energy in place of fossil fuels. It is straightforward to measure captured methane from digesters and the increase in grain, animal and residue yields. Estimates could be made of vented methane from water bodies and paddy fields and the amount attributed to manure and perhaps eutrophication. However, in order to qualify for CDM transactions a group of farmers etc. may have to combine for consideration under the CDM small-scale initiatives.

But until cropland projects are allowed under LULUCF for non-annex 1 countries, and until the prevention of deforestation through improved agricultural productivity can be counted as a legitimate ER activity, such important GHG reduction initiatives will have to be considered under other options such as the GEF or bilateral/multilateral donor funding.

7.2. POSSIBLE CDM LAND-USE, AGRO-FORESTRY AND FORESTRY PROJECTS

At present, CDM initiatives do not include improving the management of existing forests, but are confined to afforestation/reforestation projects. Whether this encompasses the planting and management of trees in various formations such as on-farm, or in shelterbelts, or whether it is just confined to woodlots or plantations is still not clear. If the broad definition is used to include the planting of any kind of trees in groups then all these initiatives can be pursued. As the rules for CDM sink projects will only be decided in 2003, there is time to discuss the effects and repercussions of different definitions; however, it is unlikely that the definition of a forest area will change much from the one made at Marrakesh under LULUCF, namely a minimum of 0.05 ha. Trees can be grown to assist both arable and pastoral agriculture, improve the microclimate (reduce wind velocity, decrease transpiration, moderate temperature and mitigate dust/sand storms) and to provide goods and services including biomass energy, timber and non-wood products. In agro-forestry formations, trees planted in fields could occupy 5% of the farm land, therefore, it is possible, but may be not probable that such tree plantings could be included.

Woody biomass is a store of carbon, both above and below ground. And as stated above, more organic carbon accumulates in the soil beneath trees than on similar land under arable agriculture or in grasslands. Of course, there is an equilibrium point when no more carbon is stored. That is when new carbon fixation is cancelled out by attrition of trees. This carbon will eventually return to the atmosphere if and when the tree population as a whole is liquidated. However, if trees are cut at or near the point of maximum growth, and then the area is replanted or regenerated *ad infinitum*, the overall production and store of carbon will be greater than if a forest is allowed to grow to maturity and then left. This is because annual growth approaches zero at maturity.

Depending on the end-use and species of trees, they can be on rotations from one year or less to 300 years or more. Farm trees tend to be on short rotations, whereas trees in national parks or biodiversity areas are usually left until they die naturally. Also, their spatial arrangement can vary from open to closed formations or be planted as individual trees. On very short rotations there is little storage of carbon, especially above ground, but average annual production of woody biomass may be relatively high: this will increase and peak at the point of maximum mean annual increment. After this peak the average annual production gradually declines. Also, other things being equal, production is function of water availability. Thus, with an average annual rainfall of 500 mm, net primary production (NPP) of above-ground biomass in a natural forest is of the order of 3–4 tonnes, whereas with a rainfall of 2000 mm., NPP could be 13–14 tonnes (Western, 1981).

In agro-forestry formations, the production of nitrogen-rich mulch or animal fodder may be the prime concern. This is to assist agricultural productivity or to provide animal feed. In addition, some stick wood will be produced. If for example trees are grown in alleys with maize on a one to two year rotation, crop yields could be doubled. In Malawi, with a rainfall of about 1000 mm, the annual maize production without agro-forestry interventions was about $1 t^{-1} ha^{-1}$ and $1 t^{-1}$ of residues. With trees in agro-forestry formations, maize production was $2 t^{-1} ha^{-1}$, crop residues $2 t^{-1} ha^{-1}$ and stick wood $1 t^{-1} ha^{-1}$. Of course it took two to three years to reach this state and extra labor inputs are required. The total additional production and store of organic carbon in biomass and the soil is about $2 t^{-1} C ha^{-1}$ (Magembe, pers. comm.). This 1 ha agro-forestry initiative may save about 1 ha of woodland from being converted to agriculture. The store of organic carbon in this 'saved woodland' may be about $50\,t^{-1}\,ha^{-1}$ in the wood and an additional 30 and $50\,t^{-1}$ C in the soil. Thus, this agro-forestry initiative could save up to $100\,t^{-1}\,C\,ha^{-1}$ and at the same time produce extra residues and stick wood. But at present the CDM will only allow direct sequestration and production, not ERs in woodlands saved from deforestation. Therefore, such projects may not be attractive with only a relatively small amount of ERs to trade per hectare.

If trees are grown for fruit, leaves for silk worms, or tapped for resin etc., the main source of wood may be from prunings and there will be a relatively large store of organic carbon in the wood and forest soil. For a mango plantation, the annual production of (dry) fruit after 20 years may be about $3–4\,t^{-1}\,ha^{-1}$ and that of wood from prunings, about $1\,t^{-1}\,ha^{-1}$. The store of above- and below-ground wood in this 20-year-old orchard may be about $200\,t^{-1}\,ha^{-1}$ $(100\,t^{-1}\,C^{-1})$ and there would be about an additional accumulation of 20 t organic carbon in the soil. This latter would continue to accumulate until it levels off at about an additional $80–100\,t^{-1}\,ha^{-1}$ C (Openshaw, 2000a). Such projects are profitable once they have reached maturity, but there is a waiting period of 5–10 years before break-even point is achieved. This is why funds from carbon trading could help in the initial period when no income is forthcoming.

Trees grown for wood products or for energy can be on rotations ranging from 3 to 80 years or more. The store of wood in live trees will be in direct proportion to age, and annual production will depend on rainfall, the species and its growth habit. If the average rainfall is about 1000 mm, a eucalyptus plantation on an 8-year rotation should have an annual production of above-ground wood of about $10\,t^{-1}\,ha^{-1}$ $(5\,t^{-1}\,C^{-1})$ and the store of organic carbon in above- and below-ground wood should be about $60\,t^{-1}\,ha^{-1}$ $(30\,t^{-1}\,C^{-1})$ once the planting cycle is complete and there is an equal representation of all ages. In addition, there will be a gradual accumulation of organic carbon in the soil. This may be at the rate of about $1\,t^{-1}\,ha^{-1}$ per year if the planting took place on abandoned farm land: this accumulation may proceed for 20–30 years (AED, 1997). Not all the harvested wood may be used for energy. Wood used for poles and furniture etc. will continue to store carbon, but ultimately it will be burnt or disintegrate into compounds of C, H and O.

In these three examples, farmers may be unable to proceed with such initiatives because the initial waiting period is too long before income is generated or there is uncertainty about the new technology or markets. It is possible that these initiatives could qualify under the CDM and that CER units could be traded. But realistically, it would have to be an aggregate of farmers etc. who would combine as a unit, or a group of investors. A minimum of at least 500 ha may be the smallest unit that is practical.

7.3. POSSIBLE CDM RENEWABLE ENERGY PROJECTS

The most likely candidates for the CDM in the agricultural/land-use/forestry sector are renewable energy projects as substitutes for fossil fuels. These projects are similar to the previous ones under Sections 7.1 and 7.2. However, their main function is the production of renewable energy or at least this is a principal purpose. A bonus for woody biomass energy projects is that there is an element of carbon storage. Such projects could grow biomass directly for energy purposes, grow cash crops with the residues being used for energy, or grow biomass to convert into more energy-intensive products such as ethanol or charcoal.

Examples of these three kinds of projects are respectively: growing wood for agricultural processing such as tea drying or as a feedstock for electrical generation; sugar cane production, with the bagasse being used for heat and power production; and charcoal for steel manufacture as a coke/coal substitute or ethanol production from sugar or maize as a substitute for motor fuel or methyl tertiary butyl ether (MTBE, an octane enhancer). A fourth kind of project is one that uses residues to convert biomass into more convenient energy forms while providing useful products. These include making biogas from biomass waste, with the slurry being an excellent fertilizer and growing plant oils for motive power while producing by-products of biomass and fertilizers. A variation of the above renewable energy projects is upgrading biomass energy units to make them more efficient or substitute more efficient units thereby increasing their substitution effect. These may be classified as energy efficiency projects rather than biomass energy projects.

But in all these examples, the cost of producing the heat, power or fuel has to be more expensive than the alternative fossil fuel interventions, otherwise they are considered as non-additional cost projects and do not qualify for the CDM. Also, the additional cost of an individual project has to be not more than the net income from trading CERs, otherwise there is no financial incentive to promote it. For example, the production of plant oil for motive power may be more than the net income from CERs and there may be more profitable uses for the oil such as soap making.[14]

7.4. THE TIME FRAME FOR RENEWABLE BIOMASS ENERGY PROJECTS

There are two time options for CDM projects, one for 10 years with no renewals and the other for 7 years with two possible renewals. For many biomass projects, even 21 years may be too short a time period to complete the optimal cycle. If a power plant is based on plantation grown wood, it may take 5–10 years before the plantation is ready to supply the optimum capacity. Also, the economic lifetime of a power plant should be at least 20 years. Therefore, such a project, including the growing of the feedstock and the building of the power plant could have a minimum lifetime of 25 years. Even if the power plant is based on bagasse, it may take at least 3 years to complete the plant, thus the minimum lifetime is 23 years. Only for improving the energy efficiency of existing biomass energy projects of 7 or 10 years may be appropriate. The existing time frame for CDM projects may make afforestation/reforestation initiatives marginal. Even growing crops for ethanol production may be marginal considering the cost of an ethanol plant and its anticipated 25–30 year lifetime. Thus, the CDM EB should re-examine this conditionality. However, if forward markets exist, this may be one way round this dilemma.

8. ER monitoring, verification and certification[15]

The ERs of a validated CDM project must be carefully monitored, verified and certified prior to being eligible for transfer to an Annex 1 country (that is listed in Annex B of the Kyoto Protocol). The requirements for monitoring, verification and certification are interrelated while the entities responsible for monitoring, verification and certification are distinct. The key actors/stakeholders and the relevant issues associated with these last three steps in the CDM project cycle are highlighted in Figure 4. A brief discussion of the important processes and issues related to CDM project monitoring, verification and certification is presented below.

8.1. ER MONITORING

The CDM project owner/operator has the primary responsibility for the monitoring of ERs. The owner/operator is the initial seller of the ERs and, therefore, must have the principal responsibility for putting in place the required procedures and measures for monitoring the project's resulting ERs. The owner/operator may contract or may have agreed to contract a third party to carry out the ER monitoring functions. This may be true especially for measuring biomass and soil carbon and perhaps for small-scale projects. However, because of concerns for other proprietary information, most project owner/operators may not be willing to allow ER buyers or other external parties to monitor daily project operations. Additionally, this may be impractical and costly and best done by the owner/operator as long as the data produced can

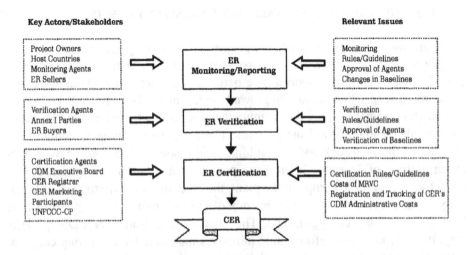

Figure 4. Monitoring, verification and certification of ERs.

be audited and verified by the ER buyer/agent and the national and international CDM authorities.

The monitoring of CDM project ERs is analogous to the monitoring and accounting of other product streams associated with the project. However, as ERs are not easily measured physically, it is important that corresponding measurable physical outputs of the project be identified that can serve as surrogate indicators for ER production. For example, measuring woody biomass to estimate the sequestered carbon in wood. Again, a biomass power project that produces electricity to displace electricity generated from oil can be monitored for its ERs by monitoring the actual electricity production of the biomass project and using the baseline definition of the GHG emissions associated with the generation of electricity from oil. However, depending on its complexity, biomass CDM projects may require the development of approved monitoring protocols in advance of project implementation and ER certification.

The end result of the ER monitoring process has to be an ER monitoring report (EMR) that is subject to auditing and verification by CDM authorities. The EMR would serve much like the accounting books of a firm that are subject to auditing and verification by a public accountant or potential corporate buyers. The EMR would provide the documentary basis for ER verification and ultimate certification.

8.2. ER VERIFICATION

The ERs claimed by the owner/operators of CDM projects must be subject to verification by independent operators licensed by the EB. This is particularly important because there is no exchange of a physical commodity that transpires that would allow the CDM authority or buyer to verify unilaterally the production of valid ERs. Given the nature of ERs, the CDM EB (or an authorized representative) and ER buyers would have to inspect the operating records of the CDM project. The auditing of the EMR can facilitate this process.

The process of ER verification is very similar to the process of independent product inspection and testing prior to payment by a buyer. The buyer must be assured of receiving what was contracted for prior to payment. Similarly, the independent verifier must also be assured that the ERs have been produced according to its guidelines and conditions as agreed to in the initial validation of the CDM project. Additionally, as the resulting ERs will ultimately be certified and transferred (exported), the national authority needs to record the transaction and may want to extract administrative and royalty fees. This is similar to administrative and royalty fees (resource payments) that are charged for minerals, wood and other commodities.

Independent certified ER verification agents could undertake the actual verification process. The CDM EB, in accordance with guidelines provided by the CP/MP, would accredit these agents. The ER verification agents would work with the project's EMR to carry out their audit and verification duties. It is anticipated that this would be an annual process. However, the frequency could be more or less depending on the nature of the project and the conditions agreed to in the ERPAs.

8.3. ER CERTIFICATION

Upon completing the verification process, the resulting ERs are ready for certification by the operational entity entitled by the CDM EB. Certification will ensure that only ERs that meet the criteria of the CDM are ultimately certified and become CERs. If all of the earlier steps for CDM project validation and ER monitoring and verification have been accomplished successfully and in accordance with approved guidelines and regulations, then the certification of the resulting ERs is procedural. However, under the process agreed in Marrakesh, the issuance of the CERs will be done by the EB, which levies an adaptation tax of 2% in kind. This tax will not be charged on projects in the least developed countries and small island states. The process of ER certification will also provide the data necessary to track CERs as they are transferred and traded. A mature market for CERs will result in multiple transfers and trades and will need to operate much like a commodity market. In fact, CERs could be traded in existing commodity markets as is currently being done with a limited number of carbon offset certificates from Costa Rica.

Certification signifies that a reported ER represents a "real and measurable" ER according to approved protocols and that the information used to calculate the ER is a true representation of the project's performance. Upon obtaining certification, the ER is registered with the national and international CDM authorities and receives CER status. Considering the approval and acceptance process in the recipient country, the CDM registry links all CERs to their originating project (including country and time-stamp). Additionally, the CDM registry would need to keep track of the transfer and ownership of all CERs to ensure that CERs are not double counted by Annex 1 countries in meeting their compliance targets. If it is found that counties are in non-compliance, then the certificate will not be issued (Werksman, 1999).

Because most of the 'Land-use' projects will be of long duration, (up to 21 years) and will provide an annual stream of ERs, it is envisaged that a 'futures market' will develop. Of course, these ERs cannot be transferred until the year they are actually available.

9. Conclusions

The steps in the CDM project development cycle are straightforward and logical. However, the smooth operation of the CDM still requires the resolution of a number of critical and politically sensitive issues. The process for defining the baseline against which a CDM project will be assessed must be clear and transparent. The rules and guidelines for defining baselines and determining the additionality of

CDM projects must be consistent. The process by which the ERs of CDM projects are monitored, verified and certified must be accepted by all Parties and must be manageable and financially affordable. A vibrant and active market for ERs from CDM projects will only evolve when the key issues associated with the CDM project development cycle are resolved and accepted by all Parties. However, the best approach for resolving many of these crucial issues is to build on the results of the pilot program of AIJ and initiate a demonstration CDM program in a number of developing countries.

Sustainable development has been highlighted as an important issue in this and many other publications. And indeed, non-annex 1 countries that wish to undertake CDM projects have to verify that such projects are sustainable. This is why renewable energy projects should be at the forefront of the CDM initiative. However, there are a number of concerns related to the promotion of biomass energy projects and/or the sequestration of carbon in woody biomass. At present, only afforestation/reforestation projects qualify in the LULUCF category. Yet the largest cause of deforestation is clearing land for agricultural expansion, because agricultural productivity has not kept pace with population increase. Such deforestation accounts for about 15% of the additional accumulation of atmospheric carbon. Therefore, slowing down deforestation by promoting agricultural productivity is essential. But agricultural projects are not included in the CDM, unless they are associated with producing energy. This seems to be a significant omission.

Several examples of potential projects have been given relating to agriculture and land-use. These are representative of many projects that could be proposed for the CDM. Most of these projects require a considerable duration before they are mature. Yet the present CDM rules only allow for a maximum time period of 21 years. This is too short in many instances and may deter biomass CDM projects and promote those with less 'sustainable development' credentials. This time rule should be re-examined.

The CDM has the potential to help reduce global GHG emissions while assisting developing countries achieve sustainable development. A market framework for CDM transactions will permit the participation of a broad cross-section of critical players while attracting the financial resources that are necessary for the success of the CDM. However, the CDM EB was only appointed at CP 7 in Nov. 2001 and all the rules and regulations will not be finalized before at least late 2002. At present, the process seems cumbersome, and for all but the simplest of projects not amenable to investors. As stated above, the CDM EB should initiate a number of demonstration projects in developing countries to jump-start the market for ERMs. These demonstration projects could be undertaken on a bilateral basis with the specific intention of engaging interested developing country Parties in a collaborative process of testing and resolving a workable process for the CDM. The process of implementing demonstration CDM initiatives will force the identification and resolution of many of the key issues surrounding the CDM. It can provide the UNFCCC's Subsidiary Body for Scientific and Technical Advice the critical inputs that it will need to help define a workable process for the CDM.

Notes

[1] The term CER unit CER is derived from Article 12 of the Kyoto Protocol. It refers to a greenhouse gas emission reduction (ER) that has received the certification of the CDM and can therefore be "used to assist in achieving compliance in the first commitment period" (2008–2012) by the Parties listed in Annex B of the Protocol.

[2] Note: Some of these principles are mutually exclusive [b & g]. Thus trade-offs have to be made clear.

[3] There is some new literature on investments and market transactions, see Kaplan and Mein (2000); WBCSD (2001); PWC (2000); Sandor (2000); WBSSP (1997).

[4] E. Haites (1998) and others estimate the potential annual market for CDM at between US$ 5.3 and 25.9 billion. The range of estimates reflects different price and quantity assumptions.

[5] An established regulatory framework distinguishes the CDM from the *ad hoc* character of the Activities Implemented Jointly (AIJ) pilot phase. In the AIJ pilot phase, project identification and implementation occurred in the absence of a uniform and formal regulatory infrastructure. Guidelines for AIJ projects were established in bilateral discussions and varied considerably from one Party to another.

[6] It should be noted that Annex 1 countries have a parallel Supervisory Committee with the same composition of ten members as the EB. These two committees may have some cross-membership, but even if not, they should work closely together.

[7] To help conserve resources, many AIJ project developers resorted to forwarding projects that have already undergone considerable preparation and repackaged them for AIJ consideration. Most likely, this process will be repeated at the outset of the CDM to bring projects forward quickly for consideration.

[8] The question of long term is relative, for all actions are reversible. The substitution of renewable energy for fossil fuels, or for one fossil fuel such as coal with another such as natural gas may be reversed under a different government or as a result of unforeseen or changing circumstances. And as stated above the use of more efficient fossil fuel devices and/or the substitution of a less GHG fossil fuel for another fossil fuel may only buy time unless the use of renewable energy is expanded considerably. Also, it must be noted that the maximum period for any CDM project is 21 years at present.

[9] Bearing the above footnote in mind, the issue of long term may be of importance in the context of GHG sinks or sequestration, but many of these initiatives are combined sink/production projects where renewable goods are the primary concern. It also relates to the possibilities of delayed or displaced "leakage" in which the reductions achieved by CDM projects eventually are offset by emissions from other related activities. This could occur due to an increase in a GHG emitting activities elsewhere that is a direct result of the CDM project activity. However, the issue of long term has to be weighed against population increase and economic expansion, especially in developing countries. These two factors invariably lead to an increased demand for goods and services – more energy, food and construction materials etc. – as witnessed during the settlement and industrialization of the USA. Therefore, it is important to judge CDM projects in dynamic rather than static contexts, especially as the World's population is expected to increases to about 10 billion from the current 6 billion before it stabilizes. If this is done, the project may lead to reduced emissions in related activities, rather than a halting of emissions.

[10] This baseline usually includes the measurement of conditions before a CDM project commences. This is sometimes referred to as a baseline survey and is a measure by which the success of the project is judged.

[11] Article 12.6 of the Kyoto Protocol states that "the clean development mechanism shall assist in arranging funding of certified project activities as necessary". In this paper, the certification of CDM initiatives is referred to as the validation of CDM project activities. This is to avoid confusion from the distinct and different action of certifying ERs. Validation of a CDM activity does not guarantee that the project's ERs will be certified. Certification of a project's ER can be done only after project implementation and after the actual ERs have been generated. Therefore, for the purpose of this paper, validation is not the same as certification. Validation can be viewed as licensing a project to proceed to implementation.

[12] This is the fate of many currently circulating AIJ pilot project proposals. These project proposals cannot be closed because of a number of financial reasons. Under current rules, carbon reduction units from AIJ pilot projects may not be used for compliance purposes under the Kyoto Protocol. In addition, there is a lack of financial instruments to capitalize the little value these carbon reduction units have and project proposals often do not fit into the portfolio of investors and/or are not well prepared.

[13] For example the International Performance Measurement and Verification Protocol, Vine et al. (2001).

[14] An examination of growing *Jatropha curcas* for plant oil indicated that the cost of the oil was between 2.6 and 4 times the cost of kerosene and 6–10 times the cost of diesel in Zimbabwe and India respectively. However, if the oil was sold for soap manufacture, it could compete with other plant oil products or tallow, Openshaw (2000a,b).

[15] A more complete coverage of Monitoring, Evaluation, Reporting and Verification is given in Vine et al. (2001).

References

AED: 1997, *Greenhouse Gas Assessment Handbook,* Prepared for the Global Environment Division, Environment Dept. World Bank. IRG (EEM), Washington, DC, USA.

Bode, J-W., de Beer, J., Blok, K. and Ellis, J.: 2000, *An Initial View on Methodologies for Emission Baselines: Iron and Steel Case Study,* Paris, OECD/IEA.

Bosi, M.: 2000, *An Initial View on Methodologies for Emission Baselines: Electrical Generating Case Study',* Paris, OECD/IEA.

Bouwman, A.F. (ed.): 1990, *Soils and the Greenhouse Effect,* New York, Wiley.

Ellerman, A.D., Jacboy, H.D. and Decaux, A.: 1998, 'The effects on developing countries of the Kyoto Protocol and carbon dioxide emissions trading', *Policy Research Working Paper* 2019, the World Bank, Washington.

Ellis, J.: 1999, 'Emission baselines for clean development mechanism projects: lessons from the AIJ pilot phase', *Proceedings of Workshop on Baseline for CDM,* Tokyo, Japan.

Friedman, S.: 1999, 'The use of benchmarks to determine emissions additionality in the clean development mechanism, *Proceedings of Workshop on Baseline for CDM,* Tokyo, Japan.

Haites, E.: 1998, 'International emissions trading and compliance with GHG limitation commitments', Working Paper W70, International Academy of the Environment, Geneva, Switzerland.

Hassing, P. and Mendis, M S.: 1998, 'Sustainable development and GHG reduction', *Issues and Options; The Clean Development Mechanism,* UN Development Programme, New York.

Hassing, P. and Mendis, M.S.: 1999, 'An International Market Framework for CDM Transactions', Internal paper prepared for DGIS (The Netherlands Ministry of Development Cooperation), IRG (EEM), Washington, DC, USA.

Kaplan, M. and Mein, J.: 2000, *Matching Carbon Emissions Reduction to Financing: Building prototypes,* Aspen Colorado, USA.

Matsuo, N. et al.: 1998, 'Issues and Options in the Design of the Clean Development Mechanism', The Institute for Global Environmental Strategies (IGES), United Nations Development Programme, New York, USA.

Michaelowa, A.: 1999, 'Baseline settings in the AIJ pilot phase', *Reports on AIJ Projects and Contributions to the Discussions of the Kyoto Mechanisms,* Bundesumweltministerium, Berlin.

Openshaw, K.: 2000a, 'A review of *Jatropha curcas*: an oil plant of unfulfilled promise', *Biomass and Bioenergy* **19**, 1–15.

Openshaw, K.: 2000b, 'Benin: a baseline survey of organic carbon in woody biomass and soils on different land-use types in the PGFTR project area', Alternative Energy Development/International Resources Group for the Government of Benin-World Bank/GEF PGFTR Project, IRG (EEM), Washington, DC, USA.

Puhl, I.: 1998, 'Status of research on project baselines under the UNFCCC and the Kyoto Protocol', OECD and IEA Information Paper, Paris, France.

PWC (Price Waterhouse Cooper): 2000, 'A business view on key issues relating to the Kyoto Mechanism', London, PWC.

Sandor, R.: 2000, 'CDM – simplicity is the key', *Environmental Finance, Supplement: The Carbon Challenge – Industry, Climate Change and Kyoto,* pp. 19–21.

Vine, E.L., Sathaye, J.A. and Makundi, W.R.: 2001, 'An overview of guidelines and issues for the monitoring, evaluation, reporting, verification, and certification of forestry projects for climate change mitigation', *Global Environmental Change* **11**, 203–216.

Werksman, J.: 1999, 'Responding to non-compliance under the climate change regime', OECD Information Paper ENV/EPOC(99)21/FINAL, OECD, Paris, France.

Western, D.J.: 1981, 'A survey of natural wood supplies in Kenya', Kenya Rangeland Ecological Monitoring Unit, Box 47146, Nairobi, Kenya.

WBCSD (World Business Council for Sustainable Development): 2001, *The Greenhouse Gas Protocol – A Corporate Accounting and Reporting Standard,* Geneva, Switzerland, 64 pp.

WBSSP (World Bank Strategy Study Program): 1997, *Host Country Institutional Framework,* http://www.esd.worldbank.org/cc.

ENERGY USE AND CO_2 PRODUCTION IN TROPICAL AGRICULTURE AND MEANS AND STRATEGIES FOR REDUCTION OR MITIGATION

PAUL L.G. VLEK*, GABRIELA RODRÍGUEZ-KUHL and
ROLF SOMMER

*Zentrum für Entwicklungsforschung (ZEF), Walter-Flex-Straße 3, 53113 Bonn, Germany
(*author for correspondence, e-mail: P.Vlek@uni-bonn.de; fax: +49 (0) 228 73-18 89
tel.: +49 (0) 228 73-18 65)*

(Accepted in Revised form 15 January 2003)

Abstract. Carbon dioxide emissions due to fossil fuel consumption are well recognized as a major contributor to climate change. In the debate on dealing with this threat, expectations are high that agriculture based economies of the developing world can help alleviate this problem. But, the contribution of agricultural operations to these emissions is fairly small. It is the clearing of native ecosystems for agricultural use in the tropics that is the largest non-fossil fuel source of CO_2 input to the atmosphere.

Our calculation show that the use of fossil energy and the concomitant emission of CO_2 in the agricultural operational sector – i.e. the use of farm machinery, irrigation, fertilization and chemical pesticides – amounts to merely 3.9% of the commercial energy use in that part of the world. Of this, 70% is associated with the production and use of chemical fertilizers. In the absence of fertilizer use, the developing world would have converted even more land for cultivation, most of which is completely unsuitable for cultivation. Current expectations are that reforestation in these countries can sequester large quantities of carbon in order to mitigate excessive emissions elsewhere. But, any program that aims to set aside land for the purpose of sequestering carbon must do so without threatening food security in the region. The sole option to liberate the necessary land for carbon sequestration would be the intensification of agricultural production on some of the better lands by increased fertilizer inputs.

As our calculations show, the sequestration of carbon far outweighs the emissions that are associated with the production of the extra fertilizer needed. Increasing the fertilizer use in the developing world (without China) by 20%, we calculated an overall net benefit in the carbon budget of between 80 and 206 Mt yr^{-1} dependent on the carbon sequestration rate assumed for the regrowing forest. In those regions, where current fertilizer use is low, the relative benefits are the highest as responding yield increases are highest and thus more land can be set aside without harming food security. In Sub-Saharan Africa a 20% fertilizer increase, which amounts to 0.14 Mt of extra fertilizer, can tie up somewhere between 8 and 19 Mt of CO_2 per year (average: 96 t CO_2 per 1 t fertilizer). In the Near East and North Africa with a 20%-increased fertilizer use of 0.4 Mt yr^{-1} between 10 and 24 Mt of CO_2 could be sequestered on the land set aside (40 t CO_2 per 1 t fertilizer). In South Asia this is 22–61 Mt CO_2 yr^{-1} with an annual additional input of 2.15 Mt fertilizer (19 t CO_2 per 1 t fertilizer).

In fact, carbon credits may be the only way for some of the farmers in these regions to afford the costly inputs. Additionally, in regions with already relatively high fertilizer inputs such as in South Asia, an efficient use of the extra fertilizer must be warranted.

Nevertheless, the net CO_2 benefit through implementation of this measure in the developing world is insignificant compared to the worldwide CO_2 output by human activity. Thus, reforestation is only one mitigating measure and not the solution to unconstrained fossil fuel CO_2 emissions. Carbon emissions should, therefore, first of all be reduced by the avoidance of deforestation in the developing world and moreover by higher energy efficiency and the use of alternative energy sources.

Key words: carbon dioxide, carbon sequestration, global warming potential, greenhouse policy, liming, methane, nitrous oxide, soil carbon, trace gas flux.

Environment, Development and Sustainability **6:** 213–233, 2004.

1. Introduction

The recent conclusion by the intergovernmental panel on climate change (IPCC) "that there has been a discernible human influence on global climate" (IPCC, 2001a) is one more call for action on the reduction of green house gas (GHG) emissions.

Global industrial energy consumption as well as energy consumption in the agricultural sector showed a steady increase over the last decades. Worldwide about one fifth of the annual anthropogenic (GHG) emission comes from the agricultural sector (excluding forest conversion), producing about 50% and 70% of anthropogenic methane (CH_4) and nitrous oxide (N_2O) emissions and about 5% of anthropogenic emissions of carbon dioxide (Cole et al., 1996).

According to IPCC's Special Report on Emission Scenarios (SRES; IPCC, 2000a), global CO_2 emissions due to industrial energy consumption are projected to increase from $6\,\mathrm{Gt\,C\,yr^{-1}}$ in 1990 to $9.0\text{--}12.1\,\mathrm{Gt\,C\,yr^{-1}}$ by 2020, $11.2\text{--}23.1\,\mathrm{Gt\,C\,yr^{-1}}$ by 2050 and $4.3\text{--}30.3\,\mathrm{Gt\,C\,yr^{-1}}$ by 2100 depending on the underlying scenarios. The CO_2 emission due to agricultural land use activities (including forest conversion) added another $1.1\,\mathrm{Gt\,C\,yr^{-1}}$ in 1990, which is projected to change to 0 to $1.5\,\mathrm{Gt\,C\,yr^{-1}}$ by 2020, -0.4 to $+0.8$ by 2050 and -2.1 to $+0.4\,\mathrm{Gt\,C\,yr^{-1}}$ by 2100. Though, possibly loosing importance during the 21st century, the conversion of native ecosystems to agricultural use in the tropics currently is the largest non-fossil fuel source of CO_2 input to the atmosphere. Thus, the best mitigation strategy may be to sequester some of this lost carbon through the restoration of the natural vegetation. This, of course, presumes that this effort not be undone by continued deforestation. Therefore, to avoid land conversion in order to halt the release of carbon from pristine ecosystems appears to be a widespread consensus, although consensus and practise appear still far apart (FAO, 2000a).

Production increases have historically been achieved through land expansion, yield increases and land intensification. Although FAO (2000a) estimates that only 35% of the land suitable for crop production is currently under production, there is a widespread perception that the long global expansion of arable land is slowing down and that only little more could be brought under cultivation. The estimated overall arable land expansion in developing countries will be only about 12.5% within the 35 years from 1995 to 2030, compared to an expansion of 41% during the same period in the past (1961/63–1995/97; FAO, 2000a).

Between 1700 and 1990, trends in forest and farmlands moved in opposite direction: from 6.2 billion to 4 billion hectares of land under trees, and from 265 million to 1.4 billion hectares under crops (Engelman, 1995). Despite the fact that in developing countries an annual increase of arable land of 0.34% is expected by FAO (2000a) between 1995/97 and 2030, the cropped area per person will nonetheless decrease with one hectare feeding six persons in the developing countries by 2030 (0.17 ha/person). This is only double the 0.07–0.08-hectare benchmark, which Smil (1993) argues is the minimum unit of arable land per person required to feed a country's population, on an essentially vegetarian diet and without intensive use of fertilizers. This benchmark would need to be much higher if a meat-based diet were the global norm (Engelman, 1995). Today, 0.42 ha are required to feed a person on a high-meat diet using mechanized and input-intensive agriculture (Von Weizsäcker et al., 1997).

With 90% of the population growth in the coming decades taking place in the developing world, these regions are under enormous pressure to produce food. During the next thirty years crop yield increases are still expected to be the main source for growth in crop production accounting for 69% in developing countries. The annual rate of growth in yields, however, is expected to be only half (1.0% for the period 1995/97–2030) of what it was during the period 1961/63–95/97 (2.1%; FAO, 2000a). Increases in crop production will go hand in hand with a rapid increase of inputs, especially the use of fertilizer. It is estimated that today more than half (55–58%) of the world's fertilizer is used on cereals, 12% on oil crops, 11% on grass and the remaining on a range of crops (FAO, 2000b; Harris, 1998). The total world consumption of 134 Mt in 1995/97 is expected by FAO (2000b) to increase to 182 Mt in 2030 at a rate of 1.34 Mt (0.9%) per annum. Approximately 60% (112 Mt) will be consumed in developing countries, including China, the average growth during the next thirty years being double as much as in developed countries. Gilland (1998) expects even higher fertilizer consumption. According to his estimates between 1998 and 2050 the annual consumption of nitrogen fertilizer alone would increase by 1.5 to 2.0 Mt. Soh and Isherwood (1999) even estimated an annual fertilizer increase of 3.3 Mt during the period 1998–2003.

A variety of measures have been developed and proposed to mitigate GHG emissions in the agricultural sector – from reducing agricultural consumption of fossil fuels, production of bio-fuels, manure management practices, improved rice production practices to increasing stocks of organic C in soils and biomass. The decline of natural resources per head of the population (e.g. land per person), the limited yield-growth potential, low efficiencies in developing countries and the feasibility of local farmers in these countries to implement mitigation are widely recognized constraints in achieving mitigation. Assuming a full implementation of these mitigation measures, Rosenberg et al. (1998) showed that the effectiveness of many mitigation practices and technologies is nonetheless uncertain and that, with the rapid increase in fertilizer use and crop production, they are unlikely to balance increases in N$_2$O and CH$_4$ emissions. Thus, a need for closer analysis arises, as even comprehensive studies regarding an environmental performance of agricultural production do not take into account all production processes of sectors contributing to agriculture.

In the first part of this paper we assess the contribution of various practises in tropical agriculture, other than land conversion, in terms of energy use and to the emission of CO$_2$. In the second part, we evaluate the trade-offs in reversing the trend in land conversion through intensification on the remaining land. We estimate the CO$_2$ balance that can be expected when part of the (marginal) land in the developing world currently cultivated is restored to forest and natural grassland, and fertilizer use on the remaining land is augmented to compensate for lost production.

We distinguished the following five tropical regions: Sub-Saharan Africa, Near East and North Africa, South Asia, East Asia, as well as Latin America and the Caribbean. Only those countries were included into the analysis (see annex), which are the most significant fertilizer consuming countries according to the last IFA/IFDC/FAO (1999) study, as only for these countries fertilizer consumption could be distinguished according to its use in rice, wheat and maize production.

2. Energy use in tropical agriculture

Energy consumption in developing countries is growing parallel to the growth in modern sectors, such as industry, motorized transport, and urban areas. The significance of the energy consumption by the agricultural sector as a source of GHG emissions is closely related to the level and intensity of agricultural production. Basically, energy consumption increases with a transition from traditional to more energy-oriented agricultural production methods. In developing countries, commercial energy in agricultural production is mainly used for the manufacture and operation of farm machinery, equipment for pump irrigation, and for the production and application of mineral fertilizers and chemical pesticides. The energy use to date in these sectors and in the different regions varies greatly. The details are given in the following.

2.1. Farm machinery

Till now, human power has been the main source of energy for agricultural operations in developing countries (73%), followed by draught animal (20%) and machinery power (7%; Stout, 1990). Nevertheless, the power – mainly for seeding, weeding and harvesting – that can be provided by human beings is quite limited. An average adult produces only about 2700 kJ of energy on a continuous, 10 h-working-day basis, i.e. 75 W (Faidley, 1992). Draft animals are used extensively in many developing countries for land preparation, water lifting, crop harvesting, threshing and transport.

The number of tractors in use in developing countries increased to 5.5 millions in 1998. The manufacture of tractors and other farm machinery is fairly energy intensive, requiring about 83.7 MJ to produce 1 kg of machinery (Leach, 1976). The size of tractors and the total weight of other farm machinery associated with these tractors vary from region to region. In developing countries, tractors are mainly used for tillage and transportation. Therefore, the average tractor size is small and the associated equipment adds little. The total weight is estimated to be only 6 t (Stout, 1990).

Energy consumed in field operation is affected by many factors including weather, soil type, depth of tillage operation, field size, speed, degree of mechanization, and management ability. The annual fuel consumption of a tractor varies with its size and with the agricultural operation it performs. In developing countries consumption is estimated at 3 t diesel fuel per tractor and year (Stout, 1990).

In 1998, as much as 918 PJ of energy is used for farm machinery. About 76% is consumed for operation and 24% for manufacture, assuming a replacement rate of 8% per year (Table I). In terms of energy consumption, agricultural operations done by farm machinery are most important in South Asia with 33%, followed by in Latin America and Near East/North Africa, each with 27% of the total energy use.

2.2. PUMP IRRIGATION, EQUIPMENT, MANUFACTURE AND OPERATION

Irrigation energy consists of two parts: energy for pumping the water, either from groundwater or surface source, and energy for distribution. The irrigation equipment consists of pumps, engines, pipes and other material, such as equipment for sprinkler and drip irrigation.

The average energy required to produce this equipment is assumed to be 83.7 MJ per kg of equipment (Stout, 1990). The required energy per hectare of irrigated land varies with the depth of the water being pumped, the type of irrigation system, and the water requirements of the crops. When pressurized systems such as sprinkler or trickler are used, additional pumping energy is expended. In addition to pumping energy, fuel is required to provide water supply and construct the field irrigation system. Fuel requirements for such operations are estimated to vary from 200 kg of diesel fuel per hectare and year in Africa and Asia to 180 kg ha^{-1} yr^{-1} in Latin America and the Caribbean (Stout, 1990). This is equivalent to about 8.38 and 7.45 GJ per hectare, respectively[1].

The developing countries account for three-quarter of the world's irrigated area (FAO, 2000a). In 1998, pump irrigation in developing countries consumed 105 PJ of energy, about 90% was utilized for operation and 10% for manufacture, assuming a replacement rate of 10% per year (Table I). The Near East and North Africa required 73 PJ for irrigation, i.e. 70% of the total energy use for irrigation in developing countries. It is expected that the area equipped for irrigation in developing countries will expand by about 23% over the next five years (FAO, 2000a). The increasing investment and operating costs of pump irrigation systems and the increasing global scarcity of freshwater make a most effective use compelling.

2.3. FERTILIZERS

The consumption of mineral fertilizers in developing countries in the second half of the 20th century has been increasing at very high rates, though starting from a low fertilizer consumption base (FAO, 2000a). In 1998, the share of global fertilizer consumption in all developing countries had increased to 61.1%, while the world consumption showed an overall decline for the total of the three major nutrients (FAO, 2000b).

TABLE I. Annual commercial energy used for agricultural operation, namely farm machinery, pump irrigation, mineral fertilizer and pesticide production.

	Unit	Sub-Saharan Africa	Near East/North Africa	East Asia	South Asia	Latin America & Caribbean	All	% of item	% of total
Farm machinery									
Tractors in operation	Million	0.2	1.5	0.6	1.8	1.5	5.5		
Energy for manufacture	PJ	8	59	22	74	60	223	24	6
Energy for operation	PJ	24	184	70	231	187	696	76	20
Total	PJ	31	243	92	305	247	918		26
% of region		3	27	10	33	27			
Pump irrigation									
Total irrigated area	1000 km²	50	272	191	826	183	1522		
Area irrigated by pumps	1000 km²	0.9	80	5	0.3	29	116	10	0.3
Energy for manufacture	PJ	0.1	7	0.5	0.02	3	11		
Energy for operation	PJ	0.7	67	5	0.2	22	94	90	2.7
Total	PJ	1	73	5	0	26	105		3
% of region		1	70	5	0	24			
Mineral fertilizer									
Nitrogen consumption	Mt	0.9	3.7	5.4	14.5	4.7	29		
Phosphate consumption	Mt	0.5	1.5	1.6	4.8	3.4	12		

Potash consumption	Mt	0.3	0.3	1.7	1.5	3.1	7		
Energy for N-fertilizer prod.	PJ	72	286	422	1126	364	2270	91	64
Energy for P-fertilizer prod.	PJ	7	20	22	66	46	163	7	5
Energy for K-fertilizer prod.	PJ	2	3	15	14	27	60	2	2
Total	PJ	82	309	459	1205	437	2492		70
% of region		3	12	18	48	18	70		
Chemical pesticides									
Insecticide consumption	kt	9	12	27	47	48	143		
Fungicide consumption	kt	10	10	14	11	44	90		
Herbicide consumption	kt	11	8	28	10	92	149		
Energy for Insecticide prod.	PJ	0.9	1.2	2.6	4.5	4.6	14	37	0.4
Energy for fungizide prod.	PJ	1.0	1.0	1.4	1.0	4.2	9	23	0.2
Energy for herbicide prod.	PJ	1.1	0.8	2.7	1.0	8.9	14	39	0.4
Total	PJ	3	3	7	7	18	37		1
% of region		8	8	18	18	48			
Sum	PJ	117	629	563	1517	727	3552		
% of region		3	18	16	43	20			

Nitrogenous fertilizer is by far the most important mineral fertilizer, both in the amount of plant nutrient used in agriculture and in energy requirements. The principal products are ammonia, urea, ammonium nitrate, urea/ammonium-nitrate solution, di-ammonium phosphate and ammonium sulphate. Nitrogen fertilizers are energy-intensive. One kilogram of nutrient-N requires about 77.5 MJ for its manufacture, packaging, transportation, distribution, and application (Stout, 1990). Manufacture requires pure gaseous nitrogen and hydrogen. Pure gaseous nitrogen is simple and inexpensive to produce compared to hydrogen. The main sources of hydrogen for fertilizer production are natural gas and coal (Helsel, 1992). Transportation routes in some developing countries can be very energy demanding.

World demand for nitrogen fertilizer is expected to increase at a rate of 1.9% in all regions with the exception of West Europe. Most of the growth will be in Asia (82%), with limited reserves of good agricultural land, as opposed to Latin America (8%) and in Africa (5%) (FAO, 2000b). Data for China, where fertilizer use has a marked impact on the world consumption, were not available in a crop-disaggregated form.

The consumption of nitrogen fertilizer in the developing world is highest in South Asia with 14.5 Mt in 1998 requiring 1126 PJ of energy (Table I). This is 16-fold higher than the consumption in Sub-Saharan Africa (0.9 Mt, 72 PJ). On the whole, in 1998 nitrogen fertilizer production required 91% of the energy used for mineral fertilizer production.

Phosphate is much less energy-intensive than nitrogen fertilizer. Most rock phosphates, which are obtained through mining operations, must be refined to develop the more common phosphate fertilizers. The energy requirement to mine, concentrate, process, package, transport, distribute, and apply 1 kg of nutrient is estimated at about 13.8 MJ (Stout, 1990). Energy sources for their production are steam and electricity, which can be generated from any convenient fuel. About 25% of phosphate fertilizers are single superphosphate, which also contain calcium and sulphur, 70% are acidulated with the help of sulphuric acid as a first step (Helsel, 1992). World demand of phosphate is expected to grow at an average of about 1.8%, especially in Asia with an increase of 6.3% (FAO, 2000b).

Phosphate constituted about 7% of the energy used for mineral fertilizers in 1998, which is equal to 163 PJ. Again, South Asia and Sub-Saharan Africa were the major and minor consumers in the developing world, respectively.

Potash fertilizers are developed primarily from underground mining of potassium salt ores. The mining and refining is very capital-intensive and is responsible for all or most of the energy required in the production of potash fertilizers (Helsel, 1992). The total energy required to mine, concentrate, package, transport, distribute, and apply 1 kg of nutrient is estimated to be about 8.8 MJ (Stout, 1990). The annual growth rate in world demand of Potash is forecast to be 2.3% per year, while the annual average increase in East and South Asia is expected to growth between 3% and 5% in the next years (FAO, 2000b).

Potash constituted about 2% of the commercial energy used for mineral fertilizers in 1998 (60 PJ of energy).

Overall, 2492 PJ of energy were used for mineral fertilizer production in the developing world in 1998, of which about on half (48%) was expended in South Asia alone, but only 3% in Sub-Saharan Africa.

2.4. PESTICIDES

The worldwide use of agricultural pesticides increased from an equivalent value of US$ 20.5 billion in 1983 by on average 3% per year to US$ 27.5 billions in 1993. The growth rate ranged regionally from −1.2% for Eastern Europe to 6.3% for North America. Pesticide use reached around US$ 34.2 in 1998 corresponding to a slightly increased average growth rate of 4.4% per year regionally, ranging between 4.0% (North America) and 5.4% (Latin America; Yudelman et al., 1998). Pesticides consumption is likely to grow faster in the developing countries than in the developed world (Morrod, 1995), though the introduction and spread of new pesticides may occur more rapidly in the latter. The environmental implications of such growth are difficult to assess because, first of all, reliable data on the actual quantitative pesticide consumption worldwide are not available and, furthermore, the cost per unit weight has risen considerably as pesticides have become more biologically effective. It is therefore difficult to determine whether increased use has been associated or not with lower application rates of active ingredients per hectare.

Table I lists the average amounts of pesticides consumed in the developing countries in 1996, excluding Cambodia, Korea, Philippines, Mexico, Lesotho, Malawi and Nigeria, where data (FAOSTAT) were not available.

Given the minute contribution of the biocides to the overall energy requirements in tropical agriculture, there is little point in assessing the consumption trends for the near future. In fact, the developments in demand for organically grown food, biotechnology in the field of 'smart' pesticides and IPM that lead to reductions in biocide use, may be fully balanced out by the increased adoption of minimum tillage in the tropics (FAO, 2001) which demands the increased use of herbicides.

In all considered developing countries 37 PJ of energy were used in 1996 for pesticides assuming that the production of 1 kg of active pesticide-ingredient needs 97.6 MJ. The Latin American and Caribbean countries alone were responsible for 48% of this consumption, especially through the use of herbicides.

2.5. RELATIVE IMPORTANCE OF THE DIFFERENT AGRICULTURAL OPERATIONS AND RESULTING GHG EMISSIONS

Comparing the different main, energy-consuming operations in agriculture of the developing countries as listed above, it turns out that the production of mineral fertilizers consumed around 70% of the total energy of 3553 PJ used in 1998. Nitrogen fertilizers alone claimed 64% of the commercial energy used in agriculture. Farm machinery was the second most important energy consumer (26%), while pump irrigation (3%) and pesticides (1%) were of minor importance. The amounts, type and percentages of commercial energy used in various operations varied from region to region. South Asia is the major consumer in this balance outranking Sub-Saharan Africa, Near East/North Africa and East Asia together.

The primary concern is the emission of CO_2 as a result of the intensification measures requiring energy. These data have been summarized in Table II, whereby the conversion from energy to CO_2 took into account which source of energy is normally used in generating the necessary energy. This has been oil for all items except for the production of nitrogen fertilizer, where gas is the prevailing energy source. To provide 1 GJ of energy, the combustion of oil and natural gas produces 75 and 50 kg of CO_2, respectively.

The total emissions amount to 210 Mt of CO_2 per year, with most of it (56%) emitted from Asia. Due to the more efficient energy/CO_2 ratio of natural gas, mineral fertilizer is slightly loosing importance and according to our calculation contributes 62% to the emissions. In this respect, the improvement of production techniques basically opens possibilities for more energy-efficient agricultural operations. This is becoming practice for instance in the production of nitrogen fertilizer in Germany, where latest techniques bring down the energy required for 1 kg of fertilizer from currently 77.5 MJ to below 59 MJ, i.e. by at least 24% (Patyk and Reinhard, 1997; Scholz, 1997). Applying this number to the whole world, the CO_2- emission due to fertilizer production would therewith be reduced by 27 Mt, the percentage share of total by about 6% from 62% to 56%. Doing so, however, is of little justification, as for instance in the USA energy inputs for nitrogen fertilizer production remained at the above-given level or even slightly increased (IPCC, 2001b). Thus, in the historical context that developed countries are the first able to invest in energy-efficient techniques, it is rather unlikely that developing countries have already taken up latest techniques for N-fertilizer production.

Nevertheless, relating commercial energy use by agricultural operation to the overall industrial energy consumption, it becomes clear that it has a share quite small. Only between 0.9% and 6.5%, and on the average merely 3.9% of the total energy consumption of the developing world enters the operational sector of the agriculture (Table III). Thus, it can also be concluded that the annual CO_2 emissions due to agricultural operations is small in relation to the emission due to the commercial energy use. However, in some countries, especially in Latin America, most electrical energy – as part of commercial energy – is produced by hydropower, with on-stream zero-CO_2 output, so that percentage share of agriculture-related CO_2 emission is actually somewhat higher then calculated in Table III.

It thus appears hardly the point of attack for major reductions in the emission of GHG in the developing world to suppress the use of any of the technologies that have contributed to food security in many parts of the world.

TABLE II. Use of commercial energy inputs to agricultural operations and resulting total CO_2 emission in different regions of the developing world in 1998 (pesticides: 1996).

	Unit	Sub-Saharan Africa	Near East/ North Africa	East Asia	South Asia	Latin America and Caribbean	All regions	% share of input
Farm machinery	PJ	31	243	92	305	247	918	33
	Mt CO_2	2.33	18.25	6.90	22.85	18.52	69	
Pump irrigation	PJ	0.8	73	5	0.3	26	105	4
	Mt CO_2	0.06	5.51	0.37	0.02	1.92	8	
Mineral fertilizer	PJ	82	309	459	1205	437	2492	62
	Mt CO_2	4.4	16.0	23.9	62.3	23.7	130	
Chemical pesticides	PJ	3	3	7	7	18	37	1
	Mt CO_2	0.22	0.22	0.50	0.49	1.33	3	
All inputs	PJ	117	629	563	1517	727	3552	
	Mt CO_2	7	40	32	86	45	210	
% share of each region		3	19	15	41	22		

TABLE III. Commercial energy use in developing countries in 1997 and the share of agricultural operations.

	Sub-Saharan Africa	Near East/ North Africa	East Asia	South Asia	Latin America and Caribbean	All
Total population, 1997 (Million)[a]	373.8	283.9	514.5	1255.3	464.9	2892.5
Commercial energy use, 1997 (PJ)[b]	12,658	15,828	16,499	23,299	22,909	91,194
Energy in agriculture, Table II (PJ)	117	629	563	1517	727	3552
% share	0.9	4.0	3.4	6.5	3.2	3.9

[a] According to World Bank, World Development Indicator (2000).
[b] Calculated as commercial energy use per capita (regional average taken for countries without data) times the population.

3. The GHG trade-offs in intensifying tropical agriculture

In its effort to meet the human demand for food, the world relies on both, the expansion and intensification of agricultural production. Expansion takes place through the conversion of forest and natural grassland to agricultural land. Intensification of agriculture is based on concentrated livestock production and increasing use of fertilizer and other inputs. Both add to the agricultural sector's GHG emissions. The question we ask here is whether the cost of further intensification in terms of fertilizer-related CO_2 emissions can be justified by the C sequestration on agricultural land that is recovered for a conversion to forest and natural grassland.

3.1. FERTILIZER PRODUCTIVITY AND RECOVERABLE LAND

To calculate how much land is recoverable, information is needed about the productivity (p) for fertilizers, which can be written as:

$$p = \frac{y_1 - y_0}{f_1}, \tag{1}$$

where, y_0 (kg ha^{-1}) is yield obtained without fertilizer use, y_1 (kg ha^{-1}) is the actual yield and f_1 is the fertilizer use (kg ha^{-1}) for each region.

Wheat, rice and maize are chosen for the calculation of these production functions in all developing countries, because of their dominance as the primary source of basic nutrition and their importance regarding fertilization and land use (IFA/IFDC/FAO, 1999). The calculation is based on the use of mineral fertilizer by each of these major crops and regions in the years 1990–1998. For simplification, we divide the amount of the crop produced per country by their area planted to obtain the current average yield (y_1). By the division of the sum of total fertilizer used on that crop by the area planted we obtain the current fertilizer use in the region (f_1). The estimation of non-fertilizer yield (y_0) for each region is obtained through backward extrapolation of a yield function generated from values of the last 30 years. Yields for the major crops for each country were derived from the FAO-Database (1998) and plotted to give exponential curves. By extrapolation to before the green revolution (1965), yield levels (y_0) were estimated for a time that fertilizer use was presumably insignificant. We are well aware of the shortcomings of these estimates, but the errors are of minor importance for the purpose at hand.

The average fertilizer productivity shows considerable variation across the regions. Productivity for rice varied from 15 to 34, for wheat from 11 to 18 and for maize from 14 to 23 kg grain per kg fertilizer (Table IV).

We were interested in calculating the added production per region for a given increase in fertilizer use. In order to remain within realistic growth rates for the coming 5–10 years we chose an increase of 20% in nutrient consumption. Although most inputs obey the law of diminishing returns, we assumed that within a range of 20% the fertilizer productivity could be assumed constant.

TABLE IV. Production, fertilizer use, productivity, expected yield (-increase) and recoverable land for the major crops rice, wheat and maize in developing countries.

	Sub-Saharan Africa	Near East/ North Africa	East Asia	South Asia	Latin America & Caribbean
Rice					
Recent yield (y_1) (kg ha^{-1})	1772	6327	3300	3333	3158
Former yield (y_0) (kg ha^{-1})	1258	3289	1620	1461	1276
Fertilizer use (f_1) (kg ha^{-1})	15	163	69	126	106
Productivity (p_1)	33.9	18.6	24.6	14.8	17.8
Expected yield (y_2) (kg ha^{-1})	1868	6931	3659	3703	3538
Yield increase (%)	5.4	9.5	10.9	11.1	12.0
Recoverable land (%)	5.1	8.7	9.8	10.0	10.7
Wheat					
Recent yield (y_1) (kg ha^{-1})	1612	1843	n.a.	2507	2600
Former yield (y_0) (kg ha^{-1})	706	827	—	930	989
Fertilizer use (f_1) (kg ha^{-1})	53	67	n.a.	140	88
Productivity (p_1)	17.0	15.2	n.a.	10.5	18.3
Expected yield (y_2) (kg ha^{-1})	1789	2047	—	2692	2918
Yield increase (%)	11.0	11.1	—	7.4	12.2
Recoverable land (%)	9.9	10.0	—	6.9	10.9
Maize					
Recent yield (y_1) (kg ha^{-1})	1488	4651	1684	1571	2624
Former yield (y_0) (kg ha^{-1})	809	1387	883	963	859
Fertilizer use (f_1) (kg ha^{-1})	30	184	51	45	80
Productivity (p_1)	23.0	17.2	18.6	13.7	22.0
Expected yield (y_2) (kg ha^{-1})	1635	5176	2021	1701	2971
Yield increase (%)	9.9	11.3	20.0	8.3	13.2
Recoverable land (%)	9.0	10.1	16.7	7.6	11.7

Higher productivity can be achieved at a given level of fertilizer use with improved crop and/or fertilizer management. However, there are economic, educational, and social constraints to improve fertilizer productivity. Many tropical farmers have access to a limited selection of fertilizers, applicators and knowledge.

The first approximation of the agricultural land that can be set aside for natural regeneration is based on the fertilizer productivities in Table IV. We calculated the expected yield (y_2) by increasing fertilizer use by 20% in the different regions as:

$$y_2 = p(f_1 + 0.2 \, f_1) + y_0 \qquad (2)$$

E.g. in the case of rice production in Sub-Saharan Africa:

$$y_2 = 33.9 \, (15 \, \text{kg ha}^{-1} + 3 \, \text{kg ha}^{-1}) + 1258 \, \text{kg ha}^{-1}$$

The actual yield (y_1) in Sub-Saharan Africa for rice of 1772 kg ha^{-1} would thus increase to the expected yield (y_2) of 1868 kg ha^{-1} which equals a yield increase of 5.4%. Hence, an increasing fertilizer use of 20% for rice in Sub-Saharan Africa would allow the re-conversion of 5.1% of the rice area to natural vegetation without loss in overall rice production (Table IV). If the least productive land is taken out of production, the area set aside might be even larger.

The same calculation applied to rice, wheat and maize in the tropics, demonstrates that with an increase of 20% of fertilizer use on all these crops, 22.9 million hectare of arable land could be taken out of cultivation without loss in cereal production (Table V).

3.2. CO_2 BALANCE

The estimates of carbon sequestration due to regeneration of the natural vegetation vary notably in the literature. Afforestation and reforestation in tropical regions can potentially achieve carbon sequestration rates in aboveground and belowground biomass of 4–8 t ha^{-1} yr^{-1} (Dixon et al., 1994; Nabuurs and Mohren, 1995; Nilsson and Schopfhauser, 1995; Brown et al., 1996). ISRIC (1999) considers an average rate of carbon sequestration in soil of 0.3 t ha^{-1} yr^{-1} with a maximum of around 1 t ha^{-1} yr^{-1}. Lal et al. (2000) suppose that a restoration of soils in the tropics has a potential to sequester C at the rate of 1.5 t ha^{-1} yr^{-1}. The wide variation in vegetation carbon density in the low latitudes introduces considerable uncertainty in these estimations. Watson et al. (1996) estimated a potential carbon sequestration of 1.1–2.6 t ha^{-1} yr^{-1} in low latitudes by natural and assisted forest regeneration. This includes above-and belowground vegetation, soil carbon, and litter.

Assuming that 20% more fertilizer use in developing countries leads to 22.9 Million hectare recoverable land that can be used for re-vegetation, we obtain a potential carbon sequestration between 25.1 and 59.4 Mt yr^{-1} for all the regions combined, depending on the assumed sequestration rate, with an average of 42.3 Mt yr^{-1} (Table VI).

TABLE V. Recoverable area after a 20%-increase in fertilizer use in rice, wheat and maize production in developing countries (planted area taken from IFA/IFDC/FAO, 1999).

	Sub-Saharan Africa	Near East/ North Africa	East Asia	South Asia	Latin America & Caribbean	Total per crop
Rice						
Planted area (Million ha)	4.8	1.2	42.3	45.9	5.7	
Recoverable area (%)	5.1	8.7	9.8	10.0	10.7	
Recoverable area (Million ha)	0.2	0.1	4.1	4.6	0.6	9.7
Wheat						
Planted area (Million ha)	2.4	23.6	—	27.0	9.5	
Recoverable area (%l)	9.9	10.0	—	6.9	10.9	
Recoverable area (Million ha)	0.2	2.4	—	1.9	1.0	5.5
Maize						
Planted area (Million ha)	16.9	1.9	11.9	7.7	28.9	
Recoverable area (%)	9.0	10.1	16.7	7.6	11.7	
Recoverable area (Million ha)	1.5	0.2	2.0	0.6	3.4	7.7
Total recoverable area (Million ha)	2.0	2.7	6.1	7.0	5.0	22.9

TABLE VI. Potential carbon sequestration by forest regeneration on land taken out of agricultural production.

	Sub-Saharan Africa	Near East/ North Africa	East Asia	South Asia	Latin America & Caribbean	Total
Recovered area (Million ha)	2.0	2.7	6.1	7.0	5.0	22.9
Potential carbon sequestration by regenerating forest (Mt C yr^{-1})						
Low rate (1.1 t ha^{-1} yr^{-1})	2.2	2.9	6.7	7.7	5.5	25.1
High rate (2.6 t ha^{-1} yr^{-1})	5.2	6.9	16.0	18.3	13.1	59.4
Average	3.7	4.9	11.3	13.0	9.3	42.3

The total fertilizer use in developing countries reached 48 Mt in 1998. For the limited number of the most significant fertilizer consuming countries according to the IFA/IFDC/FAO (1999) study, which were included in our study, rice, wheat and maize accounted for 20.4 Mt. A 20% increase in N, P_2O_5 and K_2O would add an additional amount of 4.1 Mt fertilizer (Table VII).

Our calculations based on these numbers provide a conservative picture of the actual area that may be reclaimed by nature through intensification on the remaining land, as some countries were omitted from the study. The same holds true for the CO_2 emission, but the relative gains and losses in the CO_2 balance would remain

TABLE VII. Fertilizer use for rice, wheat and maize in developing countries and related CO_2 output as well as additional output by a 20%-increase of fertilizer use.

	Sub-Saharan Africa	Near East/ North Africa	East Asia [Mt]	South Asia	Latin America & Caribbean	Total
N-Fertilizer consumption	0.41	1.35	2.24	7.70	1.97	13.7
CO_2 output by N-fertilizer production	1.59	5.23	8.68	29.82	7.63	52.9
P_2O_5-Fertilizer consumption	0.23	0.74	0.67	2.23	1.16	5.0
CO_2 output by P-fertilizer production	0.24	0.77	0.69	2.31	1.20	5.2
K_2O-Fertilizer consumption	0.05	0.02	0.20	0.84	0.63	1.7
CO_2 output by K-fertilizer production	0.03	0.01	0.13	0.55	0.42	1.1
Total fertilizer consumption 1998	0.69	2.11	3.11	10.77	3.76	20.4
Total CO_2 output 1998	1.86	6.01	9.50	32.69	9.25	59.3
20% consumption	0.14	0.42	0.62	2.15	0.75	4.1
20% CO_2 output	0.37	1.20	1.90	6.54	1.85	11.9
Consumption + 20%	0.83	2.53	3.73	12.92	4.51	24.5
CO_2 output + 20%	2.23	7.21	11.40	39.22	11.10	71.2

approximately the same. The regional differences are clear from Table VII. A 20% increase in fertilizer use amounts to a mere 0.14 Mt in Sub-Saharan Africa but 2.15 Mt in South Asia.

The CO_2 release due to fertilizer production and use can be calculated as follows: for all developing countries, nitrogen accounts for nearly 70% by weight of total fertilizer use on rice, wheat and maize. Gas is the fossil fuel used for nitrogen production and oil for phosphate and potash. Thus, emission for the production of 20.4 Mt of fertilizer in 1998 amounted to 59.3 Mt CO_2. An increase by 20% would raise this to 71.2 Mt CO_2 (Table VII). Hence, an increase of 20% fertilizer use for maize, wheat and rice in the tropics produces 11.9 Mt more CO_2. The regional differences remain proportional to the current level of fertilizer use. Thus, South Asia would require the emission of 18 times the amount of CO_2 needed for Sub-Saharan Africa to produce the extra fertilizer.

To evaluate the environmental impacts of additional fertilizer use in the tropics, we have to compare the CO_2 emission due to 20% more fertilizer use for rice, wheat and maize, with the CO_2 sequestration in the area reclaimed by natural vegetation across the regions. To that end, we multiply the high and low carbon sequestration rates in the low latitudes (Table VI) by 3.67 to obtain the carbon equivalent of CO_2 sequestration. The complete balance is presented in Table VIII.

TABLE VIII. CO$_2$ emission for 20% additional fertilizer use in the production of rice, maize and wheat and CO$_2$ sequestration on reforested land taken out of cultivation without loss of overall production.

	Sub-Saharan Africa	Near East/ North Africa	East Asia	South Asia	Latin America & Caribbean	Total
CO$_2$-emission by 20% fertilizer production (Mt)	0.37	1.20	1.90	6.54	1.85	11.86
Recovered area (Mill. ha)	2.0	2.7	6.1	7.0	5.0	22.9
CO$_2$ sequestration by forest regeneration on recovered areas (Mt)						
Low rate	8.1	10.7	24.8	28.4	20.3	92.3
High rate	19.2	25.3	58.5	67.2	47.9	218.1
CO$_2$ balance (Mt)						
Low	7.7	9.5	22.9	21.9	18.4	80.4
High	18.8	24.1	56.6	60.6	46.1	206.3
Average	13.3	16.8	39.8	41.2	32.3	143.4

The balance of CO$_2$ emission and sequestration under assumption of increasing fertilizer use is highly positive for all of the regions and results in usable sequestration rates somewhere between 80.4 and 206.3 (average 143.4) Mt CO$_2$ yr^{-1}.

The area recovered per unit of CO$_2$ emission for the additional fertilizer is far from uniform. Whereas in Sub-Saharan Africa 1 Mt of CO$_2$ could set aside 5.4 million ha of land, this is only 1.1 million ha in South Asia. The remaining regions fall somewhere in between. The carbon sequestered on this land has been calculated for high and low production potentials of the land but will vary widely even within the regions. Despite these gross simplifications, it is clear that intensification on the more resilient and productive land will yield great benefits in terms of carbon sequestration, at least in the first couple of decades. These benefits will eventually subside as the vegetation on this land reaches a climax state. Once again however, the benefits to the worldwide carbon balance seem miniscule given the current annual CO$_2$ emission due to human activity of over 6 Gt C and growing (IPCC, 2000a). In this light, the more positive conclusion would be that the intensification strategy would allow developing countries to claim substantial benefits under the Kyoto protocol-CDM.

4. Conclusions

Carbon dioxide emissions due to fossil fuel consumption are well recognized as a major contributor to climate change. In the debate on dealing with this threat, expectations are high that agriculture based economies of the developing world can help alleviate this problem even though their contribution to these emissions are modest. The conversion of native ecosystems to agricultural use in the tropics is believed to be the largest non-fossil fuel source of CO$_2$ input to the atmosphere. The use of fossil energy and the concomitant emission of CO$_2$ in the agricultural operational sector amounts to less than 4% of the commercial energy use in that part of the world. Of this, around 70% is associated with the production and use of chemical fertilizers. In the absence of its use, the developing world would have converted even more land for cultivation, most of which is actually unsuitable for cultivation. Current expectations are that reforestation in these countries can sequester large quantities of carbon in order to mitigate excessive emissions elsewhere.

Reforestation is only a mitigating measure and not a solution to fossil fuel CO_2 emissions. Carbon emissions should be reduced by higher energy efficiency and the use of alternative energy sources. Any programme that aims to set aside land for the purpose of sequestering carbon must do so without threatening food security in the region. With 90% of the population growth in the coming decades taking place in the developing world, these regions are under enormous pressure to produce food. Under these conditions, the sole option to liberate the necessary land for carbon sequestration would be the intensification of agricultural production on some of the better lands. As our calculations show, the sequestration of carbon far outweighs the emissions that are associated with the production of the extra fertilizer needed. In fact, carbon credits may be the only way for some of these farmers to afford these costly inputs. The production of more fertilizer increases the CO_2 emission, but its use reduces the need of further expansion into forested areas and may allow land to be set aside for revegetation or reforestation. Our calculations show that this would provide a net benefit in the carbon budget.

In those regions, where current fertilizer use is low, the relative benefits are the highest. In Sub-Saharan Africa, a 20% increase amounts to 0.14 million tons of fertilizer which can tie up somewhere between 7.7 and 18.8 million tons of CO_2. In the Near East and North Africa one would need an increased use of 0.4 million tons to meet the target between 10 and 24 million tons of CO_2 and in South Asia 2.2 million tons to meet the target between 22 and 61 million tons of CO_2. These benefits will eventually subside as the vegetation on this land reaches a climax state. Once again however, the benefits to the overall and regional carbon balance seem miniscule given the current annual CO_2 emission due to human activity from the developing world (excluding China) of more than 6 billion tons. This is true particularly for Asia and Latin America where the bulk of these emissions occur. The benefits in the intensification–revegetation strategy may lie in the potential to augment the overseas development assistance budgets devoted to agriculture in some of these regions, which have been steadily declining over the past decade. With a carbon credit of around 10 US dollars per ton of C (3.67 tons of CO_2) under the Kyoto protocol-CDM, Africa could earn up to 51 million dollar per year by applying 0.14 million tons of fertilizer. At 200 US dollars a ton, the fertilizer would cost 28 million dollars. At 300 dollars per ton, a more realistic figure given the high transportation costs in Africa, the carbon credit would be more or less used up. However, the implementation of such a strategy might finance a fertilizer subsidy or credit scheme. The situation looks less favorable in those regions where fertilizer use is currently higher.

Inexpert use of fertilizer may lead to low fertilizer use efficiency and to associated water and air pollution. These symptoms are already common in China and India, where important fertilizer losses are taking place by leaching and volatilization (FAO, 2000b). In such regions, the extra fertilizer needed may in fact be obtained by more efficient use of fertilizer. Proper timing of fertilizer to match the nutrient demand of the crop, controlled-release fertilizers, nitrification inhibitors, optimization of tillage, irrigation and drainage, and ultimately the use of precision farming can lead to substantial savings (Vlek and Fillery, 1984; IPCC, 1996). This is particularly true in those regions where N application rates rise above $150\,\mathrm{kg\,ha^{-1}}$ and exceed crop nutrient demand. Few of such areas can be found in Sub-Saharan Africa and Latin America.

Notes

[1] According to the general conversion factor for energy, where the combustion of 1 t of oil provides 41.87 GJ.

References

Brown, S., Sathaye, J., Cannell, M. and Kauppi, P.: 1996, Management of forest for mitigation of greenhouse gas emissions, in R.T. Watson, Zinyowera, M.C. and R.H. Moss (eds.), *Climate Change 1995. Impacts, Adaptations and Mitigation of Climate Change: Scientific-Technical Analyses*', Contribution of Working Group II to the Second Assessment Report of the Intergovernmental Panel on Climate Change, Cambridge, United Kingdom and New York, NY, USA, Cambridge University Press, pp 773–797.

Cole, V., Cerri, C., Minami, K., Mosier, A., Rosenberg, N. and D. Sauerbeck.: 1996. Agricultural options for mitigation of greenhouse gas emissions, in R. Watson, M. Zinyowera and R. Moss (eds.), *'Climate Change 1995: Impacts, Adaptations and Mitigation of Climate Change: Scientific-Technical Analyses'*, Contribution of Working Group II to the Second Assessment of the IPCC, Cambridge, Cambridge University Press.

Dixon, R.K., Brown, S., Houghton, R.A., Solomon, A.M., Trexler, M.C. and Wisniewski, J.: 1994, 'Carbon pools and flux of global forest ecosystems', *Science* **263**, 185–190.

Engelman, R.: 1995, 'Feeding Tomorrow's People from Today's Land', *Environmental Conservation* 22(2).

Faidley, L.W.: 1992, 'Energy and Agriculture', in R.C. Fluck (ed.), *Energy-in-farm Production*, Energy in World Agriculture 6, Amsterdam, Elsevier.

FAO: 1998, *FAOSTAT 1998*, FAO Statistical Database on CD-ROM, Rome.

FAO: 2000a, *Agriculture: Towards 2015/30*, Technical Interim Report, Rome.

FAO: 2000b, *Current World Fertilizer Trends and Outlook to 2004/2005*, Rome.

FAO: 2001, *Conservation Agriculture; Case Studies in Latin America and Africa*, Rome, FAO, 69 pp.

Gilland, B.: 1998, *World Population and Food Supply, 1996 2050*. Comment at the NAS Colloquium on Plants and Population: Is there Time? Dec. 1998, UC Irvine. http://www.lsc.psu.edu/nas/The%20Program.html.

Harris, G.: 1998, '*An Analysis of Global Fertilizer Application Rates for Major Crops*' Paris, France. International Fertilizer Industry Association. http://www.fertilizer.org/CROPS/CROPS/harris.htm.

Helsel, Z.R.: 1992, 'Energy and Alternatives for Fertilizer and Pesticide use', in R.C. Fluck (ed.), *Energy in World Agriculture 6*. Amsterdam, Elsevier, pp. 177–201.

IFA/IFDC/FAO: 1999, *'Fertilizer Use by Crop'* Fourth Edition, Rome, Italy.

IPCC: 1996, *'Technologies, Policies and Measures for Mitigation Climate Change'*, Cambridge, Cambridge University Press.

IPCC: 2000a, *'Emissions Scenarios. A special Report of Working Group III. Summary for Policymakers'*, UNEP, WMO, 20 pp.

IPCC: 2001a, *'Climate Change 2001: Impacts, Adaptations, and Vulnerability. Contribution of Working Group II to the Third Assessment Report of the Intergovernmental Panel on Climate Change (IPCC)'*, Cambridge, Cambridge University Press, 1000 pp.

IPCC: 2001b, 'Climate Change 2001: Mitigation – Contribution of Working Group III to the Third Assessment Report of the Intergovernmental Panel on Climate Change (IPCC)', Cambridge University Press. 700 pp.

ISRIC: 1999, 'Global Change: Management Options for Reducing CO_2-Concentration in the Atmosphere by Increasing Carbon Sequestration in the Soil', Report: 410 200 031, Wageningen.

Lal, R., Kimble, J.M. and Stewart, B.A.: 2000, Advances in Soil Science. Global Climate Change and Tropical Ecosystems, Washington, CRC Press.

Leach, G.: 1976, 'Energy and Food Production. Intern. Institute for Environment and Development', Guildford, Surrey.

Morrod, R.: 1995, 'The role of pest management techniques in meeting future food needs: Improved conventional inputs', Paper presented to the IFPRI workshop on "Pest Management, Food Security, and the Environment: The Future to 2020", Washington, D.C., May 1995.

Nabuurs, G.J. and Mohren, G.M.J.: 1995, 'Modelling analyses of potential carbon sequestration in selected forest types', Canadian Journal of Forest Research 25, 157–172.

Nilsson, S. and Schopfhauser, W.: 1995, 'The carbon-sequestration potential of a global afforestation program', Climatic Change 30, 267–293.

Patyk, A. and Reinhard, G.: 1997, 'Energy and material flow analysis of fertiliser production and supply', 4th Symposium for Case Studies in Life Cycle Analysis, Brussels, December 1996, pp. 73–86.

Rosenberg, N.J., Cole, C.V. and Paustian, K.: 1998, 'Mitigation of greenhouse gas emissions by the agricultural sector: An introductory editorial', Climatic Change 40, 1–5.

Scholz, V.: 1997, 'Methods for calculating the energy demand of vegetable products presented on solid biofuels', Agrartechnische Forschung 3, 11–18.

Smil, V.: 1993, 'Global Ecology: Environmental Change and Social Flexibility', London, Routledge, 240 pp.

Soh, K.G. and Isherwood, K.F.: 1999, A review of the current agricultural and fertilizer situation', Paper presented at the 67th IFA Annual Conference, 17–20 May 1999, Manila. the Philippines, Paris, IFA.

Stout, B.A.: 1990, 'Handbook of Energy for World Agriculture', London & New York, Elsevier Applied Science.

Vlek, P.L.G. and Fillery, I.R.P.: 1984, Improving nitrogen use efficiency in wetland rice soils', Paper read before The Fertiliser Society of London on the 13th December 1984. The Fertiliser Society Proceedings No. 230.

Von Weizsäcker, E., Lovins, A.B. and Hunter-Lovins, L.: 1997, Factor Four: Doubling Wealth – Halving Resource Use; the New Report to the Club of Rome', London. Earthscan, 322 pp.

Watson, R., Zinyowera, M.C. and Moss, R. (eds.): 1996, 'Climate Change 1995. Impacts, Adaptations and Mitigation of Climate Change: Scientific Analyses. Contribution of Working Group II to the Second Assessment Report of the Intergovernmental Panel on Climate Change', Cambridge, Cambridge University Press, 861 pp.

Worldbank: 2000, 'World Development Indicator, 2000', Database on CD-ROM.

Yudelman, M., Ratta, A. and Nygaard, D.: 1998, 'Pest management and food production: looking to the future' Food, Agriculture and the Environment Discussion Paper 25, Washington, D.C. IFPRI.

Annex: List of countries considered in this study

Sub-Saharan Africa

Angola	Mauritius
Côte d' Ivore	Nigeria
Ethiopia	Senegal
Guinea	South Africa
Kenya	Tanzania
Lesotho	Togo
Madagascar	Zambia
Malawi	Zimbabwe
Mauritania	

Near East/North Africa

Algeria	Morocco
Egypt	Saudi Arabia
Iran	Syria
Jordan	Turkey
Lebanon	

South Asia

Bangladesh	Korea
India	Laos
Nepal	Malaysia
Pakistan	Myanmar
Sri Lanka	Philippines
East Asia	Thailand
Cambodia	Vietnam
Indonesia	

Latin America and Caribbean

Argentina	Guatemala
Bolivia	Honduras
Brazil	Mexico
Chile	Nicaragua
Colombia	Panama
Costa Rica	Paraguay
Dominican Republic	Peru
Ecuador	Uruguay
El Salvador	Venezuela

GHG MITIGATION POTENTIAL AND COST IN TROPICAL FORESTRY – RELATIVE ROLE FOR AGROFORESTRY

WILLY R. MAKUNDI and JAYANT A. SATHAYE

Lawrence Berkeley National Laboratory, 1 Cyclotron Rd, Berkeley, CA 94720
(*author for correspondence, e-mail: WRMakundi@lbl.gov; fax: 510-486-6996; tel: 510-486-6852)

(Accepted in Revised form 15 January 2003)

Abstract. This paper summarizes studies of carbon mitigation potential (MP) and costs of forestry options in seven developing countries with a focus on the role of agroforestry. A common methodological approach known as comprehensive mitigation assessment process (COMAP) was used in each study to estimate the potential and costs between 2000 and 2030. The approach requires the projection of baseline and mitigation land-use scenarios derived from the demand for forest products and forestland for other uses such as agriculture and pasture. By using data on estimated carbon sequestration, emission avoidance, costs and benefits, the model enables one to estimate cost effectiveness indicators based on monetary benefit per t C, as well as estimates of total mitigation costs and potential when the activities are implemented at equilibrium level. The results show that about half the MP of 6.9 Gt C (an average of 223 Mt C per year) between 2000 and 2030 in the seven countries could be achieved at a negative cost, and the other half at costs not exceeding $100 per t C. Negative cost indicates that non-carbon revenue is sufficient to offset direct costs of about half of the options. The agroforestry options analyzed bear a significant proportion of the potential at medium to low cost per t C when compared to other options. The role of agroforestry in these countries varied between 6% and 21% of the MP, though the options are much more cost effective than most due to the low wage or opportunity cost of rural labor. Agroforestry options are attractive due to the large number of people and potential area currently engaged in agriculture, but they pose unique challenges for carbon and cost accounting due to the dispersed nature of agricultural activities in the tropics, as well as specific difficulties arising from requirements for monitoring, verification, leakage assessment and the establishment of credible baselines.

Key words: carbon offsets, COMAP, cost effectiveness, greenhouse gases, LULUCF potential.

1. Introduction

Terrestrial ecosystems play an essential role in the global carbon cycle.[1] Tree growth serves as an important means to capture and store atmospheric carbon dioxide in vegetation, soils and biomass products. This form of carbon storage may not be permanent due to the likelihood of release by anthropogenic and natural disturbances or processes.[2] However, the use of biomass products from sustainably managed forests to substitute for unsustainably harvested forest products or fossil-based products, or for fossil fuels, offers an opportunity for the permanent removal of GHG emissions from the atmosphere.

A recent assessment of the land use, land-use change and forestry (LULUCF) options suggests that the total global technical potential for biologically feasible afforestation and reforestation activities between 1995 and 2050 will average between 1.1 and 1.6 Gt C yr^{-1}, of which 70% will be in the tropics (IPCC, 2000a).

Environment, Development and Sustainability **6**: 235–260, 2004.
© 2004 *Kluwer Academic Publishers.*

An assessment of potential sequestration from additional activities in improved land-use management and other land-use changes suggests that by 2010, it may exceed $1.3\,\mathrm{Gt\,C\,yr^{-1}}$, rising to about $2.5\,\mathrm{Gt\,C\,yr^{-1}}$ by 2040 (IPCC, 2000b). The LULUCF technical potential for carbon sequestration and emission reduction estimated by the IPCC Report represents about a sixth of the estimated $6.3 \pm 0.6\,\mathrm{Gt\,C}$ average annual carbon emissions from fossil fuel combustion and cement production (IPCC, 2000c). However, given the economic, social, and institutional barriers facing these options, the achievable potential from the LULUCF options may be considerably lower than the technical potential.

The technical potential for carbon sequestration reported in the IPCC Second Assessment Report (SAR) amounts to between 8.7 and 12.1 years worth of afore-mentioned average annual carbon dioxide emissions from fossil fuel combustion and cement production between 1995 and 2050 (Brown et al., 1996). Of this potential, 40–61 Gt C is estimated to be in tropical countries plus China, representing between 6.4 and 9.7 years of carbon emissions from fossil fuel use and industrial emissions.

In general, the forestry MP varies across countries depending on the suitability of their land for forestation, the levels of current and future carbon dioxide-emitting activities, potential for substitution in carbon-intensive services and products, and of other options for reducing deforestation. Based on a recent survey of emissions and very preliminary sequestration estimates in the energy and forestry sectors of select developing countries (Sathaye and Ravindranath, 1998), it seems that the estimated MP in LULUCF far exceeds the emissions from the respective energy sectors (see Table I).

TABLE I. Carbon emissions from forestry and energy sectors.

Country	National emissions in 1990 (Mt C)	E&I[a] sector in 1990 (Mt C)	Forestry sector 1990 (Mt C)	Forestry sector 2020 (Mt C)	Total mitigation potential (Mt C)	Ratio[b] MP/E&I
China	507	556	−61	−105	9740	17
India	146	141	1	21.0	8753	60
Indonesia	38	38	−94	−106	1745	41
S. Korea	62	66	1	—	119	2
Mongolia	5	3	1	−0.3	317	83
Myanmar	−2	1	−2	−1.4	582	647
Pakistan	20	17	2	19.0	161	9
Philippines	35	10	22	0.6	2380	205
Thailand	45	21	21	6.0	1259	54
Mexico	127	74	53	—	4115	55

Source: Sathaye and Ravindranath (1998).
[a] Energy sector and industrial processes (E&I) in 1990.
[b] Ratio of MP in forestry to the annual emissions from the energy and industrial sector. The column represents the number of years the forestry sector could offset the country's emissions at 1990 levels.
Note: *Sum of Col. (2) and (3) need not equal col. (1) because the other sector emissions (agriculture, waste management, etc.) may not be included in the energy and forestry figures.

The amount of time it takes to tap this potential depends on the mix of forestry mitigation options that is suited to each country. Reducing deforestation potentially could be achieved over a short time span if appropriate socio-economic incentives were established and maintained to halt activities that cause deforestation and the misuse of forest resources. Forestation would take longer simply because tree growth takes many years to reach maturity, depending on species and site conditions. One of the forestation options in the tropics, which is likely to be very attractive, is agroforestry since it combines wood production with agricultural or pastoral activities.

Agroforestry is attractive on account that it intervenes the carbon emission cycle at many points. First it sequesters carbon in vegetation and possibly in soils depending on the pre-conversion soil C. Secondly, the more intensive use of the land for agricultural production reduces the need for slash and burn or shifting cultivation, which contribute significantly, to deforestation. Thirdly, the wood products produced under agroforestry serve as substitute for similar products unsustainably harvested from the natural forest. Also, to the extent that agroforestry increases the income of farmers, it reduces the incentive for further extraction from the natural forest for the purpose of income augmentation.

A number of scientific and policy questions are being asked in international and national debates by—national governments and climate change negotiators, potential investors in GHG mitigation activities, local communities and other stakeholders. How much additional carbon stock might be created, and how much emissions reduction might be achieved through these mitigation activities? What is the cost effectiveness and total cost of implementing these mitigation activities? Which forestry mitigation options are the most important for developing countries, and local communities?

This paper addresses some of these issues through a summary evaluation of the results from studies in seven countries – Brazil (Fearnside, 2001); China (Xu et al., 2001); India (Ravindranath et al., 2001); Indonesia (Boer, 2001); Mexico (Masera et al., 2001); Philippines (Lasco and Pulhin, 2001) and Tanzania (Makundi, 2001). In addition, we highlight the relative potential for agroforestry in GHG mitigation. In examining the agroforestry option, results from two other separate studies— Vietnam (UNEP, 1998a,b) and Tanzania (Makundi and Okiting'ati, 1995) are also discussed. The paper illustrates the potential and costs of various mitigation options across countries, and provides some observations on how the analysis of MP and costs of forestry mitigation options could be improved to provide more realistic estimates of both.

The studies focus on quantifying the benefits of forestry practices, and generally do not identify policy changes or incentives necessary for their implementation. The potential barriers to implementation, and monitoring of carbon stock, raise complex issues with institutional, socio-economic, public policy, gender role, and economic ramifications that would need to be addressed in order for these technically feasible options to be realized. The specific coverage of agroforestry sheds some light on a group of mitigation activities, which may find extensive applicability in developing countries.

2. Analytical approach

Estimating MP across countries requires the use of a consistent and comprehensive analytical framework. The COMAP model (Sathaye et al., 1995b), which was used in the studies summarized in this paper, has been extensively used in mitigation assessments by developing countries (see for example the US Country Studies Program). The approach requires the projection of land-use scenarios for both a baseline and for a mitigation case. In parallel, it requires data on a per hectare basis on carbon sequestration in vegetation, detritus, forest products, soils and also on GHG emission avoidance activities. In order to estimate the net monetary benefit per ha or per t C, the model requires data on costs and benefits associated with all mitigation activities under consideration. These estimates are then combined with the land use scenarios in order to estimate cumulative or annual carbon flows and monetary costs and benefit over a specified future period, thus giving an estimate of potential and cost of mitigation activities in LULUCF for each country.

2.1. Mitigation options and their characterization

The first step in mitigation assessment involves the characterization of mitigation activities which typically includes information on the carbon stored in various pools, the biomass growth and decay rates, fate of the biomass, and the option's costs and benefits. In LULUCF there are three main types of mitigation activities—emission reduction, sequestration, and substitution. Each GHG sub-sector in LULUCF has some or all of these mitigation opportunities.

In agriculture, the emission reduction and substitution opportunities exist in rice cultivation, animal husbandry, biogas use for energy, fertilizer application and cultivation methods while offering carbon sequestration in agricultural tree crops, soil carbon storage and agroforestry. In rangelands and grasslands, emission reduction opportunities arise from improved range and fire management and improved animal husbandry while biomass replenishment and enhanced carbon storage in soils increase carbon sequestration. Emission reduction in waste management from land use mostly involves the use of animal and farm waste for biogas production for community energy needs.

The forestry sub-sector has more extensive mitigation activities, including forest conservation and protection, efficiency improvements and substitution of fossil fuels and other carbon intensive products. Forestry also sequesters carbon through increased vegetation cover (forestation), increased carbon storage in soils, and conversion of biomass to long-term products. Agroforestry combines both sequestration and emission reduction depending on the use of the wood products from the activity

TABLE II. Summary of mitigation options by country.

Study country	Options included in the study
Brazil	Afforestation (short- and long-rotation)
China	Afforestation (short- and long-rotation)
	Agroforestry
	Regeneration
	Bioenergy
	Forest protection
India	Afforestation (short- and long-rotation)
	Regeneration
	Forest protection
Indonesia	Forest plantation and timber estate
	Afforestation
	Reforestation
	Enhanced natural regeneration
	Forest protection
	Bioelectricity
	Reduced impact logging
Philippines	Afforestation (short- and long-rotation)
	Natural regeneration
	Forest protection
	Bioenergy
Mexico	Long- and short-rotation plantations
	Forest restoration
	Agroforestry
	Sustainable forest management
	Bioenergy
Tanzania	Community short-rotation woodlots
	Long-rotation softwood plantations
	Long-rotation hardwood plantations

and the complementary effect on forest protection through avoided deforestation and unsustainable use of forest resources. Table II shows the list of specific mitigation options analyzed in this study. These options were not exhaustive, but rather those which the researchers believed had a high likelihood of being implemented and could be analyzed with the data and resources at hand.

2.1.1. Comparison of parameters for the mitigation options

The mean annual increment MAI refers to the average rate of biomass carbon growth over the life of a forestation option and they vary depending on species, site productivity and management regime. The MAI for the regeneration options varied from as low as $0.8 \, t \, C \, ha^{-1} \, yr^{-1}$ in China to about $3 \, t \, C \, ha^{-1} \, yr^{-1}$ in the Philippines, and for long-rotation plantations from $1.6 \, t \, C \, ha^{-1} \, yr^{-1}$ in China to as high as $11.1 \, t \, C \, ha^{-1} \, yr^{-1}$ in Tanzania. The short-rotation plantations have higher rates ranging from $3.8 \, t \, C \, ha^{-1} \, yr^{-1}$ in China to $19.2 \, t \, C \, ha^{-1} \, yr^{-1}$ in Tanzania. To the extent that data permitted, each study accounted for the increase in soil carbon, which was estimated to range from $0.5 \, t \, C \, ha^{-1} \, yr^{-1}$ in China to $3 \, t \, C \, ha^{-1} \, yr^{-1}$ in India.

Among the various forestation options studied, the rotation period varies from as short as 7–8 years for short-rotation in Mexico and India to as long as 50 years in the case of restoration plantations in Mexico. Generally the long-rotation plantations have periods ranging between 25 and 40 years. Regeneration options in each country have much longer periods to maturity, lasting as high as 80 years in northeastern China.

The cost of planting is relatively similar and stable over time and reflects the overall income levels in the country. Costs tend to be higher in Mexico (about $400–500 ha^{-1}), and lower in India, the Philippines, China and Tanzania (between $150 and $300 ha^{-1}). Costs are higher for long-rotation plantations in each country. The life-cycle costs of these options, excluding harvesting, are only somewhat higher since the annual recurring cost of plantations tend to be small relative to the initial cost. The recurring costs include the cost of monitoring of carbon stocks.

In Indonesia, due to three- to four-fold drop in the value of the Indonesian currency (the Rupiah) since 1997, current costs in US dollars are significantly lower. Initial establishment costs range between $18 ha^{-1} for enhanced natural regeneration and about $50 ha^{-1} for a short-rotation plantation. However, once the devaluation effects run through the monetary, factor and product markets, the long-term cost structure may well return.[3]

The costs of forest protection/conservation (excluding opportunity costs) and management options tend to be lower than those for forestation. Forest protection costs range from as low as $5 ha^{-1} in the Philippines based on government budgets to as high as $41 ha^{-1} in southeast China. Experience in the countries shows that the lower values are clearly inadequate to accomplish conservation goals, and after factoring in the opportunity cost of land and labor, costs in every study country exceed the monetary benefits of forest protection/conservation.

Agroforestry options have investment costs varying from negligible in Tanzania where the opportunity cost of rural labor is miniscule, to medium cost in China at $80–140 per ha and as high as $273 per ha in Mexico. The cost of establishing and carrying out agroforestry activities may be elevated mostly due to the cost of the agricultural inputs.

3. Land-use context: Historical trends and future scenarios

Mitigation activities in the LULUCF compete for land in a zero-sum game. Land which is protected is essentially withdrawn from other potential uses, some of which may have different MP and cost. Forestation options face the staunchest competition for land due to the inherent need to reserve the most fertile land for agricultural production. Many factors drive the land use distribution in a country, including demographic variables (population growth rate, rural/urban population ratios); economic factors (incomes, export of primary products, growth rate); biophysical factors (climate, soil fertility); and land use intensity (shifting vs. permanent agriculture, selective cutting vs. clear cutting). Agroforestry relieves the pressure on natural forests somewhat given its ability to combine both wood and agricultural production on the same parcel of land.

3.1. HISTORICAL LAND-USE PATTERNS

The study countries constitute a very large land area of the world. Individually, the land area ranges from over 963 million ha for China, closely followed by Brazil's 845 million ha, to 30 million ha for the Philippines (Table III). The forested area varies considerably, with Indonesia having as high as 57% of the land area in forests, followed by Brazil and Tanzania with 46% each. China has the lowest proportion (11%) of the land area under forest cover.

The India, Indonesia, Mexico, and the Philippines studies focused on the entire forested area in each country, while the other countries covered only a portion of the forested area on which the activities were most likely going to be implemented. The Brazil study covered forestation in the Amazon region, while China study focused on the three most forested regions out of five in the country, the northeast, southeast and southwest. The Tanzania study focuses on the miombo woodlands, which constitute about 95% of the forested area in the country and accounted for about 90% of the annual deforestation (Makundi and Okiting'ati, 2002).

The rate of deforestation is a highly complex and contested figure in any country, and thus difficult to compare across countries. The magnitude of deforestation is substantial even in countries where forest resources are not abundant. The rate of deforestation for Indonesia has been reported to range from 0.75 to 1.5 million ha per year in the 1995–1997 period (Table III) (MOF, 1996; Walton and Holmes, 2000). The rate for Brazil has fluctuated from 1.1 to 2.9 M ha from the late 1980s to early 1990s (Fearnside, 1997; INPE, 1998; 1999) The estimate of deforestation for Tanzania and Mexico is about 0.750 and 0.720 million ha per year, respectively, though the official figures claim a lower rate in both countries (Makundi and Okiting'ati, 2002; Masera et al., 1997). Slowing deforestation would clearly reduce emissions but implementing options and enforcing policies to achieve this is often thwarted by, among other factors; the high opportunity cost of land and the lack of comparable alternative opportunities to earn a living in rural areas.

Is there enough land available for GHG mitigation activities in the developing countries? At first glance, the prohibitively high population densities and low agricultural productivity in some of the study countries might seem too restrictive to allow land to be used for forestation. As Table III indicates, however, estimates of degraded lands or wasteland available for forestation, (without considering economic, social, cultural, and other barriers), amount to several tens of millions of hectares. For comparison, Table III also shows the potential estimated by Trexler and Haugen, (1995) for regeneration, farm forestry (agroforestry) and plantation options for the period 1990–2040. This land either originally contained forests or has been left fallow and agriculture is no longer practiced for various social and economic reasons. Much of this land is suitable or could be made suitable for forestation

TABLE III. Historical land-use patterns, deforestation, and forestation potential.

Country	Total land area (kha)[*]	Forested area (kha)	Deforestation rate in study area (kha yr^{-1})	Land suitable for forestation	
				This study (kha)	Trexler and Haugen (kha)[p]
Brazil	845700	390000[n]	1113–2906[m]	85000[l]	85000
China[a]	963296	158941[k]	60	31953[d]	Not estimated
India	328760	63300[b]	274[c]	53200	35000
Indonesia[f]	192401	104500	750–1500	31000[e]	13600
Mexico	196700	115652	720[i]	21000[j]	35500
Philippines	30000	5200	99[h]	4400[g]	8000
Tanzania	89161	41857	750	7500[o]	11100
Total	2556857	837593	Not applicabl.	234053	188200

[*] *Source* = FAO Forest Resource Assessment 2000 (FAO, 2001).
[a] Includes forests with at least 20% crown cover. Data are for 1998.
[b] Data for 1995.
[c] Data for 1995–1997.
[d] Degraded lands in three study regions in 2000.
[e] Unproductive land, grasslands and critical lands.
[f] Annual average 1990–1997 (includes transmigration, agricultural development, forest fire and shifting cultivation; excludes illegal logging).
[g] Grassland areas, sub-marginal forests and brushlands.
[h] Annual average for 1995–1998 period.
[i] Early 1990s. Forest area includes semi-arid vegetation, which accounts for 66 Mha.
[j] Degraded forest landforestland.
[k] Of the total forest land, forests in study area = 115.6 million ha (three regions only).
[l] Estimated potential for natural regeneration, farm forestry and plantations from Trexler and Haugen (1995).
[m] From (Fearnside, 1997; INPE, 1998; 1999).
[n] Forests and 'cerrados' located in the Amazon region only.
[o] 3.5 mi. ha for short rotation community woodlots, and 2.5 mi. ha (50% of the fallow area) for reforestation and 1.5 mi. ha for all other forestation including agroforestry, long rotation plantations, non-forest tree crops (wattle, rubber, oil palm, etc.).
[p] Figures from Trexler and Haugen (1995). Estimated potential in regeneration, farm forestry and plantations between 1990 and 2040.

programs in the study countries. This may require a change of management from individual farmers to that by private companies and commensurate harvesting, or include incentives to individual farmers to re-orient their land use practices. China and India both import wood products with a value of several hundreds of millions of dollars (Kadekodi and Ravindranath, 1995; Zhang et al., 2000), and forestation programs on such lands could offset at least part of this drain on their foreign exchange reserves, while simultaneously providing rural socioeconomic benefits if the programs were sustainably managed.

3.2. FUTURE LAND-USE SCENARIOS

A baseline scenario, and one or two alternative mitigation scenarios were constructed for each study country for the period 2000–2030. The baseline scenario represents a set of assumptions about likely changes in land-use and land-cover patterns in the country based on historical data and emerging demographic and economic trends. In the mitigation scenarios, activities such as afforestation or forest protection are explicitly identified, and simulated using the COMAP model in order to estimate the change in the number of hectares and associated carbon stock for each type of land use throughout the period under consideration, as well as costs and selected benefits. In general, this simulation is based on projected demand for forest products, with the mitigation scenario examining an alternative to produce the products at lower GHG emissions implication or at an increased carbon sequestration level.

Several of the countries have ambitious government plans which were intended to meet the country's needs for forest products and services (Table IV). Invariably, these plans have been only partially implemented because of lack of resources, economic and policy incentives, and social reasons. In these cases, the mitigation scenario studied here used the forestry sector targets set forth in the government plans as the basis for setting the mitigation level. The analysts based their estimate of the level on the past history of implementation success of similar previous plans. Table V below shows the carbon and cost in baseline and mitigation scenarios

TABLE IV. Land area scenarios for mitigation activities (Mha).

Country	2000–2012	2000–2030	Mitigation scenario description
Brazil			
Forestation	6.8	19.8	Based on Trexler and Haugen (1995)
China			
Forestation	7.6	19.7	Technical plan scenario—60% of the government plan in the northeast, southeast and southwest regions
Forest protection	5.1	13.5	
India			
Forestation	12.2	29.5	Sustainable forestry scenario that is designed to meet 2010 biomass demand through domestic LULUCF activities
Forest protection	3.6	8.5	
Indonesia			
Forestation	11.6	29.2	Same as the scenario for India
Forest protection	0.5	1.1	
Philippines			
Forestation	0.6	1.7	Scenario assumes 50% of the rate of land development under the government plan
Forest protection	0.07	0.13	
Mexico			
Forestation	3.0	9.1	More effective and wider implementation of baseline scenario activities to meet domestic biomass demand
Tanzania			
Forestation	0.4	1.7	Meets 50% of demand for woodfuel, sawlogs and chiplogs, by 2024
Total			
Forestation	42.2	110.8	
Forest protection	9.2	23.2	

presented in this summary paper. The scenarios for each country are described below:

Brazil: One mitigation scenario was analyzed in a preliminary analysis using the COMAP model. This limited scenario is based on land use projections from

TABLE V. Carbon stock in mitigation and baseline scenarios (Mt C) and cumulative costs.

Country	1990	2000	2012	2030	Cumulative costs (millions of 1998 US $)	
					2000–2012	2000–2030
Brazil					590	1206
Baseline scenario[a]	0	0	0	0		
Mitigation scenario	0	0	87	448		
Increment	0	0	87	448		
China					589	1390
Baseline scenario	9714	11115	11197	11321		
Mitigation Scenario	9714	11115	11236	11532		
Increment	0	0	39	211		
India[b]					615	1194
Baseline scenario	5610	5731	5727	5720		
Mitigation scenario	5610	5731	6053	6680		
Increment	0	0	326	960		
Indonesia					4950	8601
Baseline scenario	18680	17450	16500	16140		
Mitigation scenario	18680	17450	17228	18650		
Increment	0	0	728	2510		
Philippines					82	151
Baseline scenario	1300	1130	965	805		
Mitigation scenario	1300	1135	990	881		
Increment	0	5	25	76		
Mexico[c]					—	—
Baseline scenario	24029	23397	22927	22586		
Mitigation scenario	24029	23434	23520	24376		
Increment	0	37	593	1790		
Tanzania					49	165
Baseline scenario	128	128	128	128		
Mitigation scenario	128	130	181	332		
Increment	0	2	53	204		
Total					6875	12707
Baseline scenario	59461	58951	57444	56700		
Mitigation scenario	59461	58995	59208	62451		
Increment	0	44	1764	5751		
Increment+Brazil			1851	6199		

Notes: Increment (T) = Mitigation(T-2000) − Baseline (T-2000) NA = Not applicable.
[a]Baseline carbon stock for Brazil assumed to be at equilibrium for the areas where afforestation will take place.
[b]For Mexico, only life cycle cost was given without the year of occurrence, as such discounting could not be done.
[c]For India, cost of forest protection is not included. The India study envisions a national program to halt deforestation that otherwise would have converted virtually all forests to other uses by 2030. The estimated cost for such a program is large, and amounts to over $10 billion by 2030, an amount which is not included in the table.

Trexler and Haugen, 1995, for regeneration and plantation activities only; other options like avoided deforestation or forest management are not included since corresponding cost data were not readily available to the authors. The baseline scenario assumes that future land would have remained in its current state. We focus on two selected activities among many others that could be implemented. The total land area under mitigation amounts to 19.8 Mha by 2030.

China: Two alternative scenarios were analyzed. One scenario reflects government plans which call for forest area to be increased by 27.3 Mha from 1999 to 2010 and by another 46 Mha from 2011 to 2030, and 18.9 and 35 Mha of new nature reserves would be established during the respective periods. In addition, 13 Mha are planned to be established between 1999 and 2010 under agroforestry. Table IV shows a second more conservative scenario that would achieve 60% of the goal of the government plan. The land available for regeneration is an order of magnitude higher than that for short- and long-rotation plantations by 2030, and a small amount of land is slated to be added to areas already under protection.

India: Two alternatives, a sustainable forestry scenario, which is shown in Table IV, and a commercial forestry scenario were evaluated. The first one is designed to meet the incremental national biomass demand between 2000 and 2015, and includes increased forest protection and regeneration options. The second one focuses on meeting the increased biomass demand primarily through commercial forestry. The wasteland available for forestation is quite large, almost 30 Mha, and the amount of land which could benefit from additional protection is 8.5 Mha.

Indonesia: Two alternative scenarios, a government-plan scenario and a mitigation scenario, were analyzed. The first scenario projects forestation rates similar to those in the government plan as laid out in Repelita VI (1998–2003), although historically these have been rarely achieved. The mitigation scenario assumes that the rate of timber plantation establishment is increased such as to meet all wood demand by 2010 (Table IV). Short-rotation plantations, enhanced natural regeneration, long-rotation reforestation, and reduced impact logging options dominate the 29.2 Mha of land available for forestation activities.

Philippines: As in the case of Indonesia, the government's forestry master plan scenario and a mitigation scenario are analyzed. The master plan assumes aggressive tree planting to meet local demand for wood products. The second scenario assumes a forestation rate, which is 50% of the government plan scenario (Table IV). The total land area available for mitigation in this scenario is relatively small, about 1.7 Mha by 2030; much of this is concentrated in short- and long-rotation plantations. Another 0.1 Mha is identified for protection.

Mexico: One alternative mitigation scenario is analyzed, which assumes improved penetration of all mitigation activities. In this scenario, 2030 deforestation rates will have been reduced to 25% of current ones, native forests are managed more efficiently with improved survival rates, plantations make Mexico self-sufficient in paper and cellulose products, and bioenergy plantations play a prominent role (Table IV). Restoration plantations, i.e., plantations established to restore degraded

land, and management of temperate forests constitute the bulk of the land requirements for mitigation activities. By 2030, a total area of 9.1 Mha would be under
some form of mitigation activity in this scenario.

Tanzania: The main mitigation scenario analyzed involves implementation of
the Tropical Forest Action Plan (TFAP) for establishing community short rotation
woodlots to meet 50% of the demand for wood fuel, sawlogs and chiplogs. Two
versions of this scenario are analyzed. The first scenario which is reported here
involves the conversion of 1.7 Mha of woodlands to short-rotation plantations ter-
minating in 2024, assuming that the demand for these products will have peaked, and
the plantations are managed in perpetual rotations (Table IV). Other less extensive
afforestation scenarios for long-rotation industrial softwood and hardwood were
also analyzed.

4. Summary of results

4.1. CARBON STOCK SCENARIOS[4]

The vegetation carbon stock in the study countries varies with the largest stock in
Brazil, followed by Mexico, Indonesia, China, India, Tanzania and the Philippines.
The land-use and land-cover change scenarios lead to significant opportunities for
improving the biomass and carbon pools in the future, that increase with the time
period of study.

Table V shows the changes in the live vegetation carbon stock under the two
scenarios. Except in China and Tanzania, the total carbon stock declines between
2000 and 2030 in the baseline scenario as deforestation is anticipated to continue into
the future. Slowing deforestation thus constitutes an important opportunity to reduce
or avoid emissions. By 2012, the difference in carbon stock varies between 53 Mt C
in Tanzania to 728 Mt C in India. The India figure is deceptively high because a
large part of the forest is assumed to need protection. The cumulative potential by
2012 compared to 2000 amounts to 1807 Mt C which increases to 6155 Mt C by
2030. On an annual average basis, the potential for the seven countries amounts to
about 125 Mt C yr^{-1} between 2000 and 2012, and 218 Mt C yr^{-1} between 2013 and
2030.

This study estimates the cumulative potential in Brazil for short- and long-rotation
plantations to be 87 Mt C by 2012 increasing to 448 Mt C by 2030, but it did not
evaluate the potential for avoidance of emissions from deforestation. A recent report
(Da Motta, et al., 1999) suggests that this potential is of the order of 2718 Mt C and
that for natural forest management amounts to another 735 Mt C. Combined with
the estimate in this study, the total cumulative potential in Brazil for the four options
would add up to about 3900 Mt C or almost 70% of the cumulative amount estimated
for all other study countries combined.

Table V shows the cumulative carbon potential and the associated cost of the mitigation scenario assessed from the year 2000 to the end of the first Kyoto Protocol

commitment period (2012), and also for 2000–2030. The cost estimate indicates the value of resources needed to implement a mitigation scenario without regard to its monetary benefits or cost of baseline forestry activities. Thus these estimates overstate the likely actual net cost of these mitigation options, and do not address who would pay these costs and receive monetary benefits form timber harvest or other revenues—a combination of public and private entities. Note that the cumulative cost in 2000–2012 is more than half of that for 2000–2030, mostly due to the effect of discounting on the skewed profile of costs and benefits of forest mitigation.

Except for India, the cost figures include the opportunity cost of land for the forest protection option. Other costs for the baseline scenario are not deducted. These projected costs may be compared with historical data for the amount of money allocated from the government budget to the forestry sector in each country.

4.2. ECONOMIC IMPLICATIONS OF THE POTENTIAL

The activities noted in Table II form the basis for the mitigation carbon scenarios shown in Table V. In this section, we focus on two topics, (1) cost-effectiveness of mitigation options and the potential for carbon sequestration and emissions avoidance, and (2) present value of the cumulative costs of mitigation scenarios. The latter information is useful for potential investors and government policy makers in assessing the investment needed for a regional or national scenario that contains a mix of mitigation options.

Much of the economic analysis of climate change mitigation options in the forestry and other sectors has focused on the estimation of the cost effectiveness (costs or net benefits per t C) of options (Brown et al., 1996; IPCC, 2001c). This estimation permits a ranking of options by their costs or net benefits, which provides policy makers with information about the comparative importance of each option.

4.2.1. Cost effectiveness of mitigation options

The cost effectiveness indicator of mitigation activities i.e., an option's *cost per t C,* depends on the extent to which all factors contributing to net costs and changes in carbon stock have been included, and the time period over which these are measured. The reporting of costs of LULUCF mitigation options has largely been limited to the estimation of investment or establishment cost per ha or per t C (Brown et al., 1996; IPCC, 2001c *op. cit.*). The IPCC Report (IPCC, 2001a *op. cit.)* provides additional information on the net present value (NPV) per t C for selected forestry options in developing countries. An estimate of the establishment cost and NPV per t C for Brazil, India, China, Malaysia, Mexico, Tanzania and Thailand was published in 1995 (Sathaye and Makundi, 1995). The data and information reported in this set of studies expands that approach to the estimation of costs by reporting the annualized cost per t C for a specified period, and the mitigation carbon potential relative to a baseline scenario.

For estimating cost effectiveness of options the approach used here involves accounting for all the cost elements and non-carbon benefits of an option, annualize these for a specified period (2000–2030), and then express the net costs in terms of the average annual carbon emissions avoided or carbon sequestered, i.e., the annualized net cost (benefit) per t C (henceforth referred to as cost per t C). We report this parameter for the mitigation option after deducting the cost per t C estimated for the baseline scenario. The latter represents the foregone opportunity cost of the baseline option. This approach to estimating the cost is comparable to that described by UNEP for energy projects (UNEP, 1998a,b). The estimated value may be compared with a potential international price of carbon, or the cost per t C for mitigation activities in other sectors such as energy, industry and waste management.

An important caveat is worth noting in using this approach. Carbon flows of forestry projects unlike those from energy projects vary over time. An energy mitigation project is assumed to provide constant annual emissions reductions, but the amount of carbon sequestered in a forestry project varies annually and reaches equilibrium after a species reaches maturity or is harvested within a sustained yield management regime (Sathaye et al., 1995a). Also, the cost and benefits for non-carbon inputs and outputs come at different times, with most of the costs being incurred at the beginning of the project while benefits e.g. timber are realized at the end of rotation period. Averaging annual carbon flows over a defined time period is thus an artifact that permits the cost per t C for forestry projects to be compared with that for energy projects.

The cost per t C was estimated for each country in the study for the options listed in Table II. The cost per t C was matched with the cumulative vegetation carbon sequestered (above and below ground) or emissions avoided between 2000 and 2030.

A discount rate of 10% real (after accounting for inflation) is used for China, Indonesia, Mexico, Tanzania and Brazil, and 12% real for India and the Philippines. These rates reflect the rates used by multilateral banks to evaluate energy and forestry projects in the study countries. Private discount rates are likely to be much higher, e.g., approaching 18% real in Brazil (Meyers et al., 2001). On the other hand, for environmental projects a rate of 6% has been suggested by the Indian Planning Commission (Kadekodi and Ravindranath, 1995). There is a significant school of thought in the literature which generally advocate the use of lower discount rates in evaluating long-term, environmental and social programs or projects than those used for short-term, commercial or private projects (Sathaye and Makundi, 1998).

A negative cost indicates that the direct revenue generated by the mitigation option from the sale of timber and other products exceeds its costs, including the price or opportunity cost of land. The carbon potential at a negative cost per t C varies across countries. This potential depends on the options selected for study in each country, the magnitude and time profile of the baseline and mitigation carbon, its costs, the prices and yields of timber and non-timber products. The time profile of the above monetary and carbon factors has a significant impact on the estimated costs because of the aforementioned high discount rates.

In China, because of the high price that timber and non-timber products are assumed to fetch relative to costs, all nine options (three different ones in each of the three study regions) are estimated to have a negative cost per t C, and for similar reasons, the costs are negative for Brazil. On the other hand for India, cost per t C is negative only for the regeneration option largely because its cost of planting is very small. Short-rotation plantations and regeneration offer negative cost opportunities in the Philippines. Short-rotation plantations also have negative costs in Mexico, Indonesia and Tanzania. In Mexico, long-rotation plantations, forest management and bioenergy are estimated to be negative cost options too. All other options are estimated to have positive costs. Forest protection is the highest cost option in three countries that evaluated this option (India, the Philippines and Indonesia), mostly due to the high opportunity cost.

4.2.2. The mitigation supply curve

Figure 1 shows a step curve representing the potential supply schedule for carbon in forestry mitigation activities in the seven studied countries. Each segment represents an amount of carbon which can be sequestered or protected at a specific net discounted cost between 2000 and 2030.

The larger countries dominate the combined potential for carbon sequestration. The combined cost curve for all options across the study countries shows that about half the cumulative carbon potential may be realized at a negative cost, which is about 150 Mt C yr^{-1} which can be sequestered or protected at a net non-carbon benefit over 30 year period. Coincidentally, this finding is similar to that reported for the energy sector in the TAR (IPCC, 2001c). The IPCC Third Assessment Report (Metz and Davidson, 2001) shows that about half the technology potential worldwide could be tapped at a negative cost and the other half at a cost ranging up to $100 per t C. The positive cost potential may be seen as the minimum carbon price that would be needed to implement these options, without counting the benefits associated with the reduction of atmospheric carbon. In this study, under a carbon price of say $20 per t C, the cumulative potential between 2000 and 2030 amounts to about 5 Gt C. Though the mitigation curve depicted below is dominated by conventional forestation and forest protection activities, two countries, China and Mexico included some agroforestry options in their assessment. However, this option has a potential for much broader role in carbon sequestration and emission reduction given the extensive likely application in the rural and farm economies of developing countries.

250

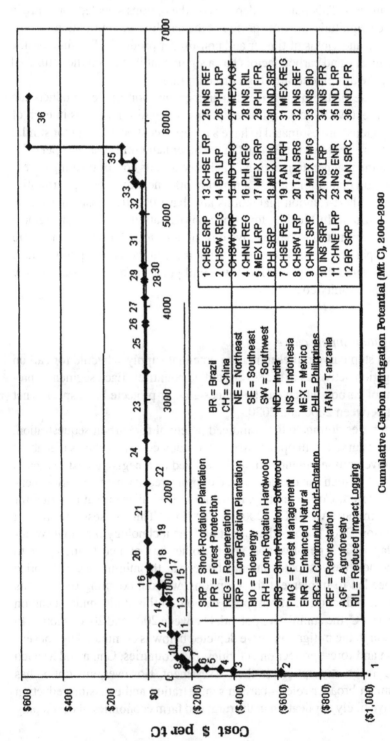

Figure 1. Forestry MP (Brazil, China, India, Indonesia, Mexico, Philippines and Tanzania).

5. Role of agroforestry

In general, agroforestry involves growing or managing tree crops with agricultural crops. The term is also used to include silvi-pastoral activities that involve growing or management of tree crops in the same land area with a significant practice of animal husbandry. Wood from agroforestry projects involving harvesting can also be utilized in such a way that the emission reduction is enhanced compared to the reference scenario. Such uses may include:

(i) more efficient charcoal production (kilns), packaging (briquettes), utilization (cook stoves), improved use of charcoal for industry (e.g. steel, tobacco and tea curing) etc.,

(ii) use of the sustainably grown biomass to replace fuel wood from depletable natural forests,

(iii) use of sustainably grown biomass for fossil fuel substitution, including, wood-fuel, ethanol and bio-electricity as well as substituting high emission content products e.g. cement and steel with wood from agroforestry.

There are a variety of agroforestry practices, which have been in existence in different countries. Such practices which may be expanded for the benefit of GHG mitigation include but not limited to:

– inter-cropping for the purpose of producing both agricultural and forest products,
– boundary and contour planting for demarcation and protection against wind and soil erosion as well as agricultural and wood products,
– shifting cultivation which is followed by succession vegetation, with or without fallow,
– taungya system (inter-cropping in the forest before canopy closure) applied as an integral part of forest management, in both natural and plantation forestry,
– pastro-silviculture for producing both forest and animal husbandry products from the same management unit,
– non-timber tree farms such as those established for rubber, tannins, bamboos, rattan, are in many cases considered under agro-forestry since they are seldom included in plantation forestry,
– Orchards and woody fruit trees may also be classified under agroforestry.

5.1. METHODOGICAL STEPS FOR ASSESSING AGROFORESTY
    FOR GHG MITIGATION

(i) Identification and description of types of applicable agroforestry activities with potential for carbon sequestration or emission reduction compared to a reference case. For each type of activity to be analyzed, a baseline has to be described. In some cases the baseline is shifting cultivation, in other cases it is farming of permanent annual crops, though in some cases farming of perennials (or sparse tree crops) can also be improved by increasing the woody biomass density.

(ii) Assessment of the current and future land area available for agroforestry. Large agricultural areas may not be amenable to significant inter-cropping due to crop husbandry requirements e.g. shade intolerant crops, mechanized farming, etc.

(iii) Assessment of the current and future product demand from the agroforestry activities help in the determination of the land area and wood production scenarios under each agroforestry activity. The wood products from these activities will be supplemental to the supply from the country's forest estate. In many cases the trees will be harvested at maturity, and the demand for the agricultural crop and wood product will drive the extent of carbon sequestration via such activities. However, in some cases e.g. shade trees and shelterbelts, the trees may not be harvested at maturity. In this case the estimate for the area which will be put under agroforestry is determined strictly by the demand for the required service, e.g. shade trees, wind breaks, demarcation, etc.

(iv) Estimate the GHG impact per hectare for each agroforestry activity. The carbon accounting for agroforestry is essentially based on individual tree measurements, which can then be used to generate allometric equations for wider area application. If the trees are not for periodic harvesting, the carbon sequestration will be estimated on the basis of maximum growth to maturity, possibly including any emission reduction from avoided erosion. This is equivalent to plant and store approach (Makundi, 1995). If the tree crop will be harvested at maturity, usually the planting is staggered so as to ensure a stable tree cover through harvesting only a small proportion of the tree crop—ideally a rotation-th of the crop annually or periodically. This is equivalent to carbon accounting for perpetual rotations (Makundi, 1995 *op. cit.*).

(v) Estimate unit costs and benefits for each mitigation activity. The direct costs and benefits should only involve the tree crops, though we may consider indirect impacts if they are compelling. For example, the intercropping of nitrogen-fixing leguminous tree species e.g. *leucaena leucocephala* or *acacia spp.* increases the productivity of the agricultural crop. This increase may need to be estimated and counted in the incremental benefit column. The opportunity cost of the land may not be an issue if the reference case involves conventional agricultural activities. The non-market and intangible costs and benefits should be imputed if possible, otherwise they should be itemized and described for the purpose of supplemental criteria for ranking different mitigation options.

(vi) Using the estimates of GHG impact and cost and benefits, compute the cost-effectiveness indicators per unit area and per t C e.g. present value of cost, net present value (NPV), initial cost, etc.

5.2. CONTRIBUTION OF AGROFORESTRY

As indicated above, only China and Mexico evaluated agroforestry among the study countries. However, Vietnam had undertaken a mitigation assessment under a separate study (UNEP, 1998a,b) in which agroforestry was evaluated. Similarly, in a different study for Tanzania, three types of agroforestry activities were evaluated

(Makundi and Okiting'ati, 1995) and their cost effectiveness estimated. Both these studies used the same methodological approach (COMAP) which was also used in the seven studies reported in this paper. In discussing the role of agroforestry for mitigation assessment, results from these earlier studies will be discussed together with those of China and Mexico.

In the three regions in China which were studied and reported above, 42% of the 78 million ha of land available for mitigation activities is classified as suitable for agroforestry (Table VI). In the mitigation scenario analyzed, 19.5 million ha of this available land was put under agroforestry (Table VII). However, due to the low carbon density of agroforestry systems (46.5 t C per ha^{-1}), of the 2093 Mt C which will accumulate in the mitigation scenario, only about 292 Mt C will be in agroforestry—which is 14% of the potential in Chinese LULUCF sector.

In the China study, since the benefits from the agricultural crop was not considered, an estimate of NPV would be misleading when compared to the other options, and as such we made no effort to the individual cost and benefit items for the agroforestry option. The total investment cost presented for this option was comparatively low, ranging between $80 and $140 per ha with a weighted average of $108 per ha. (Table VII), compared to the other options which had an investment ranging from $280 to $11420 per ha. The range is wide due to the high investment cost in bioenergy programs. The low investment per unit in agroforestry can be attributed to the dispersed rural nature of agroforestry and utilization of cheap farm labor on land which was of low opportunity cost.

Mexico's analysis of change in the carbon stock showed large increases in emissions from deforestation in the unmanaged temperate coniferous and degraded forestlands, but had a substantial gain in the tropical and temperate hardwood forests (Table VIII). Agroforestry shows a cumulative sequestration of 271 Mt C by 2030, about 21% of all MP in the LULUCF in Mexico. This is comparable to the annual agroforestry potential for China which was estimated at 10 Mt C (See Table VII).

TABLE VI. China—Land area for mitigation activities in 2000 (10^3 ha).

Activities	Land areas in regions for activities				
	Northeast	Southeast	Southwest	Total	Percent (%)
Short rotation[a]	681	3789	2446	6916	9
Long rotation[a]	1135	3789	3057	7981	10
Regeneration[a]	2270	5305	4891	12466	16
Forest conservation[b]	4057	5682	3735	13474	17
Agroforestry[c]	9743	16141	6599	32483	42
Bioenergy[a]	483	2273	1834	4590	6
Totals	18369	36979	22562	77910	100

[a] Calculated based on land available and relative proportion in 1990, forest area (1989–1993) (CMOF,1994), forest area (1994–1998) issued by CMOF in 1999 and long-term forestry plan (CMOF, 2000).
[b] Forest conservation area refers to that of existing dense forests. Figures in the table are calculated based on forest nature reserve planning in long-term forestry plan (CMOF, 2000) mentioned above and actual regional proportion of nature reserves in 1993 and 1997.
[c] Land for agroforestry development is from the agricultural land, and was estimated to be 60% of the total arable area.

TABLE VII. Land requirements, potential carbon benefits and investment costs in Agroforestry *versus* other mitigation options in China (2000–2030).

Region	Mitigation option	Cumulative area increment (kha)	Carbon pool increment (Mt C)	Percent of C-pool in agroforestry (%)	Investment costs ($ ha^{-1})[a]
Northeast	All other options	3151	601.0		380–11400
	Agroforestry	7192	107.8	15	80
	Sub-total		708.8		
Southeast	All other options	18007	601.6		290–11420
	Agroforestry	7488	112.3	16	140
	Sub-total		713.9		
Southwest	All other options	9195	598.1		280–11380
	Agroforestry	4820	72.3	11	100
	Sub-total		670.4		
Total	All other options	30353	1800.7		
	Agroforestry	19500	292.4	14	108
Total China	All mitigation options	49853	2093.1		

Agroforestry C-density (tC ha^{-1}) is 14.99 compared to 59.33 tC ha^{-1} for all other options.
[a]The high values are from the bioenergy programs of which cost includes that of generation equipment and annual maintenance, allocated on a per hectare basis.

Table IX shows the costs for selected mitigation options, with agroforestry having a relatively low investment cost estimated at $173 per ha or $273 per ha^{-1} life cycle cost. The only lower cost option was temperate forest management at $94 per ha life cycle cost. The cost per unit for agroforestry is also low ($4–$10 per t C) despite a low sequestration carbon density (27–66 t C per ha). The reasons for the low cost are similar to those for other countries, specifically low cost of inputs like labor in rural areas. The exclusion of the value of the agricultural produce also makes the NPV non-comparable to the other options and was not reported.

Two other studies for which this framework has been applied (Vietnam and Tanzania) produced similar results to those reported here in terms of relative cost and potential. In a 1998 UNEP study on mitigation analysis for Vietnam, one of the options studied would involve planting scattered trees in and around farms in the countryside (ALGAS, 1998). The results showed an estimated potential to sequester about 48 t C per ha at a discounted cost of $59.2 per ha, giving a discounted unit cost of $1.35 per t C (Table X). The total amount of carbon sequestered under agroforestry was estimated at 78.7 Mt C representing about 6% of the four mitigation options considered. The cost of implementing this option is much higher than that of protection and is comparable to enhanced natural regeneration, though it is about a third of the cost for a ton of Carbon sequestered under reforestation. The protection option is relatively cheaper since it does not include opportunity cost of the protected forest area.

A different study which was done for Tanzania (Makundi and Okiting'ati, 1995) compared the cost effectiveness per t C of three different agroforestry schemes also compared to a wood fuel plantation under two different management regimes. The

TABLE VIII. Mexico—net carbon sequestration (Mitigation-Reference Scenario) 1990–2030 (kt C).

Land use/cover class	2000	2008	2012	2030
Unmanaged forests				
Temperate conifer	7944	−156119	−210086	−496351
Temperate broadleaf	6510	40650	61541	146479
Tropical evergreen	17337	107929	157237	297955
Tropical deciduous	19569	111431	161469	339915
Semi-arid forests	1989	12935	12622	−464
Degraded forest lands	−2777	−101612	−194991	−632390
Plantations				
Long rotation	0	3520	8263	6783
Short rotation	0	5630	19525	78949
Restoration plantations	0	89627	165499	565377
Bioenergy plantations	0	25668	61946	263005
Managed forests				
Temperate conifer	−293	187461	260293	620081
Tropical evergreen	0	0	0	0
Protected forests				
Temperate	0	31568	55494	180051
Tropical evergreen	0	26233	45019	141493
Tropical deciduous	0	44864	77424	209097
Wetlands	0	0	0	0
Semi-arid forests	0	5938	19494	99433
Other uses				
Agriculture	−13296	−71971	−108129	−290958
Pasture	−15576	−126702	−195392	−468556
Agroforestry	15939	63873	92622	270969
Total	37348	300921	489099	1304549

study showed that inter-cropping between short rotation tree species and corn was the most cost effective with NPV ($4.67 per t C). The longer rotation regimes having roughly the same low NPV ($0.29 per t C and $0.27 per t C, respectively) regardless of whether they were using inter-cropping or boundary scheme (Table XI). However, the longer rotation agroforestry schemes had significantly more carbon sequestered compared to the short rotation. The two plantation forestry options had three times the carbon sequestration potential than agroforestry option using same rotation and species. The government run option was the least cost effective, with a negative NPV ($ −0.43 per t C) compared to the community/government partnership which sequestered the same amount of carbon but with an NPV of $3.4 per t C, second only to the short rotation agroforestry option.

Though there was no attempt to extrapolate these results for the whole country, the fact that 80% of the population is dependent on agriculture and most live in similar ecosystems (miombo woodlands) to the area where these options were implemented, would suggest a very large potential at a positive net present value. It is noteworthy that for the country, a mixture of these schemes would be necessary since the use of the wood products is species dependent, and different agricultural practices may require different agroforestry schemes.

256

TABLE IX. Mexico—costs of selected mitigation options.

Forest mitigation option	Investment ($ ha⁻¹)	Maintenance ($ ha⁻¹)	Monitoring ($ ha⁻¹)	Life cycle cost ($ ha⁻¹)	NPV ($ ha⁻¹)	Project cycle/ rotation (yr)	Carbon density (t C ha⁻¹)	Increment in carbon stock (t C ha⁻¹)	Carbon cost ($ t C⁻¹)
Short-rotation plantation	415	1708	8	2131	497	7	154	61	35.1
Long-rotation plantation	394	998	N/A	1392	5780	20	191	98	14.2
Restoration plantation	438	391	8	837	N/A	50	180	87	9.6
Agroforestry systems	173	101	0.00	274	N/A	16	120–159	27–66	4.1–10.0
Temperate forest management	5	57	32	94	78	50	234	141	0.7
Bioenergy	1224	1707	8	2940	345	7	281	188	15.6

TABLE X. Vietnam—COMAP output for 4 forestry mitigation options.

Category	Incremental carbon stock (Mt C)	Mitigation potential (t C ha^{-1})	Present value of benefit		Present value of cost	
			($ t C^{-1})	($ ha^{-1})	($ t C^{-1})	($ ha^{-1})
Enhanced natural regeneration	87.3	47.2	1.27	55.55	1.04	45.42
Reforestation	209.6	107.3	5.51	577.10	3.35	351.19
Natural forest protection	862.5	132.7	0.69	91.04	0.33	43.45
Scattered trees[a]	78.7	47.7	9.25	404.25	1.35	59.17
Total potential	1238.10					
Weighted average		116.97	2.09	190.73	0.96	96.69

Source: UNEP 1998—Vietnam Final Report.
[a]Most of the scattered trees option constitute of farmers and communities planting in agricultural lands.

TABLE XI. Tanzania—woodfuel plantations and agroforestry for carbon sequestration (costs and benefits in 1986 USD).

	Govt. fuel plantation	Govt./public partnership	Eucalyptus and maize	Boundary gravellia and maize	Inter-cropping gravellia and maize
Project life (Yr)	6	6	6	20	20
Initial investment	287	200			
Other cost Yr 1	53				
Yr 2	20	13	33	27	27
Revenues from fuelwood only	600	600	187	213	260
Sequestered C (t C ha^{-1})	47	47	15	23	73
NPV ($ t C^{-1})	−0.43	3.40	4.67	0.29	0.27

Source: Makundi and Okiting'ati (1995).

6. Conclusions

In this paper, we report a summary of results of carbon mitigation and associated costs and benefits in forestry in some developing countries, with a specific emphasis on the relative role of agroforestry as a mitigation activity. The studies applied a common analytical framework to estimate the potential and produce cost effectiveness indicators for the purpose of comparing and ranking the options. The paper also reports on the estimated cumulative amount of carbon which can be sequestered (or emissions avoided) by the first commitment period of the Kyoto Protocol (2000–2012), and for a longer period (2000–2030), as well as the costs and benefits of undertaking the activities. The paper then examines the relative role of agroforestry in the potential and costs in those countries where agroforestry activity was analyzed. To complement this assessment, the results of two other studies separately conducted in two developing countries using the same analytical framework are presented, with a focus on the agroforestry options.

The results show that about half the MP of 6.9 Gt C (an average of 223 Mt C per year) between 2000 and 2030 in the seven countries could be achieved at a negative cost, indicating that non-carbon revenue is sufficient to offset direct costs of about half of the options. The other half can be offset by a cost of less than $100 per t C. The agroforestry options analyzed bear a significant proportion of the potential at medium to low cost per t C when compared to other options. The role of agroforestry in these countries varied between 6% and 21% of the MP, though the options are much more cost effective due to the low wage or opportunity cost of rural labor. If the value of additional agricultural output caused by the application of agroforestry was included in the analysis, then the option would be more cost effective. Also, agroforestry options are attractive due to the large number of people engaged in agriculture and can be implemented at a very small practical scale. The potential and cost presented above face many barriers.

The main barriers are technical (skills and know how of managing tree crops and their products), financial (capital and credits) and institutional (different interests between farmers and forest departments and industry). These barriers would require targeted policies and incentives in order to smoothly implement the mitigation options. The forestry policies mostly require formulation and implementation of effective forest protection and conservation policies which will serve as incentives to agroforestry activities. Policies which impede slash-and-burn or shifting cultivation may encourage the adoption of sedentary agroforestry. Also, there has to be complementarity between the forest sector production and the agroforestry policies since individual farmers engaged in agroforestry may be unable to compete with the commercial production forestry, especially for wood products harvested from natural forests.

Another set of non-forest sector policies will also be essential in order to breach the gap between the technical and achievable potential. These include land tenure policies e.g. public *versus* private lands, agricultural and rural development policies which encourage agroforestry and promote the products, tax incentives and access to credit, trade policies to protect the output from agroforestry, including aggressive marketing for export, tariffs, etc. However, agroforestry options pose unique challenges for carbon and cost accounting due to their dispersed nature and their dependence on the specific farm economy. Other important concerns arise from the pertinent issues regarding monitoring, verification, leakage and the establishment of credible baselines.

Notes

[1] The IPCC reports an estimated 1146 Gt C stored within the 4.17 billion ha of tropical, temperate and boreal forest areas, about a third of which is stored in forest vegetation (IPCC, 2000c). Another 634 Gt C is stored in tropical savannas and temperate grasslands.

[2] Carbon emissions from land-use change worldwide during 1989–1998, for instance, are estimated to be 1.7 ± 0.8 Gt C yr^{-1} (IPCC, 2000c). This is offset by terrestrial uptake of carbon dioxide and results in a net terrestrial uptake of 0.2 ± 1.0 Gt C yr^{-1}.

[3] For example, examination of data from Tanzania where the currency was systematically devalued thirty-fold between 1986 and 2000, (from 27 to 800 Shillings per US dollar), shows the establishment cost for a forest plantation in the same locality (Sao Hill) changed from US \$217 to US \$200 ha^{-1} (Makundi, 2001). The price of forest products shows similar stability over the period. This would tend to support the use of a pre-devaluation cost structure, since the current costs and prices are transitional and may be more reflective of the short-term shock associated with massive currency devaluation, than the underlying cost structure of a plantation program which is a long-term activity.

[4] The COMAP model version 3 computes the equilibrium carbon stock in live and decomposing vegetation, soils and products. It also computes the annual live vegetation carbon stock from 1990 to 2030. We report on the changes in the annual stock in this section.

References

Asia Least-cost Greenhouse Gas Abatement Strategy (ALGAS): 1998, '*Final ALGAS Summary Report*', Asian Development Bank, Manila.

Boer, R.: 2001, 'Conomic assessment of mitigation options for enhancing and maintaining carbon sink capacity in Indonesia', *Mitigation and Adaptation Strategies for Global Change.* **6**, 313–334.

Brown, S., Sathaye, J., Cannell, M. and Kauppi, P.: 1996, 'Chapter 24: Management of forests for mitigation of greenhouse gas emissions', in R. Watson, M. Zinyowera and R. Moss (eds.), *Climate Change 1995: Contribution of Working Group II to the Second Assessment Report of the Intergovernmental Panel on Climate Change*, Cambridge, UK, Cambridge University Press.

Chinese Ministry of Forestry (CMOF): 1994, *Chinese Forest Resource Statistics 1989–1993*, Beijing (in Chinese).

Chinese Ministry of Forestry (CMOF): 2000, 'Long-term Forestry Planning', in J. Liu (ed.), *National Ecological and Environmental Planning*, Chinese Commercial United Press, Beijing, pp. 178–215.

Da Motta S., Young, C. and Ferraz, C.: 1999, 'Clean development mechanism and climate change: Cost effectiveness and welfare maximization in Brazil', *Report to the World Resources Institute*, Institute of Economics, Federal University of Rio de Janeiro, Brazil, pp 42.

FAO (Food and Agricultural Organization of the United Nations): 2001, *Global Forest Resources Assessment 2000: Main Report* (FAO Forestry Paper 140). Rome, Italy.

Fearnside, P.M.: 1997, 'Monitoring needs to transform Amazonian forest maintenance into a global warming mitigation option', *Mitigation and Adaptation Strategies for Global Change* **2**, 285–302.

Fearnside, P.M.: 2001, 'The potential of Brazil's forest sector for mitigating global warming under the Kyoto Protocol', *Mitigation and Adaptation Strategies for Global Change* **6**, 355–372.

INPE (Instituto Nacional de Pesquisas Espaciais): 1998, '*Amazonia: Deforestation 1995–1997*', INPE, São José dos Campos, SP, Brazil. Document released via internet (http://www.inpe.br).

INPE (Instituto Nacional de Pesquisas Espaciais): 1999, '*Monitoramento da Floresta Amazônica Brasileira por Satélite/Monitoring of the Brazilian Amazon Forest by Satellite: 1997–1998*', INPE, São José dos Campos, São Paulo. Document released via internet (http://www.inpe.br).

Intergovernmental Panel on Climate Change (IPCC): 2000a, 'Chapter 3: Afforestation, reforestation and deforestation (ARD) activities', in R. Watson, I.R. Noble, B. Bolin, N.H. Ravindranath, D.J. Verardo and D.J. Dokken (eds.), *Land Use, Land-use Change, and Forestry. A Special Report of the IPCC*, Cambridge, UK, Cambridge University Press, pp 377.

Intergovernmental Panel on Climate Change (IPCC): 2000b, 'Chapter 3: Additional Human- Induced Activities – Article 3.4 ', in R.T. Watson, I.R. Noble, B. Bolin, N.H. Ravindranath, D.J. Verardo and D.J. Dokken (eds.), *Land Use, Land-use Change, and Forestry. A Special Report of the IPCC*, Cambridge, UK, Cambridge University Press, pp. 377.

Intergovernmental Panel on Climate Change (IPCC): 2000c, 'Summary for Policy Makers', in R.T. Watson, I.R. Noble, B. Bolin, N.H. Ravindranath, D.J. Verardo and D.J. Dokken (eds.), *Land Use, Land-use Change, and Forestry. A Special Report of the IPCC*, Cambridge, UK, Cambridge University Press.

Intergovernmental Panel on Climate Change (IPCC): 2001a, 'Chapter 4: Technological and economic potential of options to enhance, maintain and manage biological carbon reservoirs and geo-engineering', in B. Metz, O. Davidson, R. Swart and J. Pan (eds.), Contribution of *Working Group III to the Third Assessment Report of the IPCC*, Cambridge, UK, Cambridge University Press.

Intergovernmental Panel on Climate Change (IPCC): 2001b, 'Chapter 5: Barriers, opportunities, and market potential of technologies and practices', in B. Metz, O. Davidson, R. Swart and J. Pan (eds.),

Contribution of *Working Group III to the Third Assessment Report of the IPCC*. Cambridge, UK, Cambridge University Press.

Intergovernmental Panel on Climate Change (IPCC): 2001c, 'Chapter 3: Technological and economic potential of greenhouse gas emissions reduction', in B. Metz, O. Davidson, R. Swart and J. Pan (eds.), Contribution of *Working Group III to the Third Assessment Report of the IPCC*. Cambridge, UK, Cambridge University Press.

Kadekodi, G.K. and Ravindranath, N.H.: 1995, 'Macro-economic analysis of forestry options on carbon sequestration in India', *Working Paper E/173/95*, Institute of Economic Growth, Delhi University Enclave, Delhi.

Lasco, R.D. and Pulhin, F.B.: 2001, 'Climate change mitigation activities in the Philippine forestry sector: Application of the COMAP model', *Mitigation and Adaptation Strategies for Global Change* 6, 313–334.

Makundi, W.: 1995, 'Chapter 11: Forestry', in: J. Sathaye and S. Meyers (eds.), *Greenhouse Gas Mitigation Assessment: A Guidebook*, Dordrecht, Netherlands, Kluwer Academic Publishers.

Makundi, W.R.: 2001, 'Greenhouse gas mitigation potential in the Tanzanian forest sector', *Mitigation and Adaptation Strategies for Global Change*, 6, 335–353.

Makundi W.R. and Okiting'ati, A.: 1995, 'Carbon flows and economic evaluation of mitigation options in Tanzania's forest sector', *Biomass and Bioenergy* 8, 381–393.

Makundi, W.R. and Okiting'ati, A.: 2003, Greenhouse gas emissions and carbon sequestration in the forest sector of Tanzania', *Climatic Change* (Submitted).

Masera, O.R., Ceron, A.D. and Ordonez, A.: 2001, 'Forestry mitigation options for Mexico: Finding synergies between national sustainable development priorities and global concerns. *Mitigation and Adaptation Strategies for Global Change* 6, 291–312.

Masera, O.R., Ordonez, M.J. and Dirzo, R.: 1997, 'Carbon Emissions from Mexican Forests: Current Situation and long-term scenarios, *Climatic Change* 35, 265–295.

Metz, B. and Davidson, O. (eds.): 2001, 'Mitigation—Working Group III Contribution to the Intergovernmental Panel on Climate Change (IPCC)', *Third Assessment Report*, Cambridge University Press.

Meyers, S., Sathaye, J., Lehman, B., Schumacher, K., van Vliet, O. and Moreira, J.: 2001, 'Preliminary assessment of potential CDM early start projects in Brazil' *LBNL Report* – 46120, Berkeley, USA.

Ministry of Forestry (MOF): 1996, 'Forestry statistics of Indonesia 1994/95', Agency for Forest Inventory and Land Use Planning, Ministry of Forestry, Jakarta, Indonesia.

Ravindranath, N.H., Sundha, P. and Sandhya, R.: 2001, 'Forestry for sustainable biomass production and carbon sequestration in India', *Mitigation and Adaptation Strategies for Global Change* 6, 233–256.

Sathaye, J.A. and Makundi, W.R. (eds.): 1995, 'Forestry and climate change', *Special Issue: Biomass and Bioenergy* 8, 279–393.

Sathaye, J.A. and Makundi, W.R.: 1998. 'Chapter 7—Long-term and cross-sectoral cost issues', in *Mitigation and Adaptation Cost Assessment—Concepts, Methods and Appropriate Use*, UNEP-UCCEE, Roskilde, Denmark.

Sathaye, J.A., Makundi, W.R. and Andrasko, K.E.: 1995b. 'A comprehensive mitigation assessment process (COMAP) for the evaluation of forestry mitigation options', *Biomass and Bioenergy* 8, 345–356.

Sathaye, J., Meyers, S., Allen-Diaz, B., Cirillo, R., Gibbs, M., Hillsman, E., Makundi, W. and Ohi, J.: 1995a, *Greenhouse Gas Mitigation Assessment: A Guidebook*, Dordrecht, Netherlands, Kluwer Academic Publishers.

Sathaye, J. and Ravindranath, N.H.: 1998, 'Climate change mitigation in the energy and forestry sectors of developing countries', *Annual Review of Energy and Environment* 23, 387–437.

Trexler, M. and Haugen, C.: 1995, *Keeping it Green: Evaluating Tropical Forestry Strategies to Mitigate Global Warming*. Washington, DC, World Resources Institute.

UNEP, 1998a: *Mitigation and Adaptation Cost Assessment—Concepts, Methods and Appropriate Use*. Roskilde, Denmark, UNEP-UCCEE.

UNEP, 1998b: 'Asia least cost greenhouse gas abatement strategy—ALGAS Vietnam final report', *ADB/GEF-UNDP project*, Asia Development Bank, Manila.

Walton, T. and Holmes, D.: 2000, 'Indonesia's forests are vanishing faster than ever', *International Herald Tribune*, 25 January 2000.

Xu, D., Zhang, X.Q. and Shi, Z.: 2001, 'Mitigation potential for carbon sequestration and emission reduction through forestry activities in southern and eastern China', *Mitigation and Adaptation Strategies for Global Change* 6, 213–232.

Zhang, P., Shao, G., Zhao, G., le Master, D., Parker, G., Dunning J. Jr. and Li Q.: 2000, 'China's forest policy for the 21st century. *Science* 288, 2135–2136.

CLIMATE VARIABILITY AND DEFORESTATION–REFORESTATION DYNAMICS IN THE PHILIPPINES

TOLENTINO B. MOYA* and BEN S. MALAYANG III

School of Environmental Science and Management, University of the Philippines Los Baños, Philippines
*(*author for correspondence, e-mail: totimoya@laguna.net, mmmmm@laguna.net; fax: (6349)536 2251; tel.: (6349)536 2251, 536 3080)*

(Accepted in Revised form 15 January 2003)

Abstract. Changes in the conditions in the 'warm pool' in the Pacific region are reflected in the changes in the local climate system of the Philippines. Both El Niño and La Niña episodes in the Pacific Oceans introduce high variability into the local climate pattern, especially rainfall, in the Philippines. Whereas El Niño appears when annual rainfall is ≥10% lower than normal annual rainfall, La Niña occurs when annual rainfall is at least equal to the normal. About 15.7 million ha of forest cover had been lost between 1903 and 1998, but only 1.64 million had been reforested in the same period, indicating the presence of unbelievably low ecological stability. Apart from this, the denuded forests freed about 8.24×10^9 Mg C into the atmospheric greenhouse pools. Neither deforestation nor reforestation was undertaken with deliberate regard to the occurrences of El Niño or La Niña. Very high rates of deforestation were observed to coincide with or precede strong El Niño or Niña episodes, thus confounding further the ecological instability of denuded forest systems, especially those with slope ≥18%. Similarly, the reforestation cycle indicates that saplings are at most 5 years old every time an El Niño or a La Niña occurs; in most reforestation schemes, saplings are only 1–2 years old when these events occur. These reforested areas are vulnerable to drought in El Niño years and to high runoff erosion during La Niña years. Because they are young, saplings in reforested areas dry easily and pose hazards to forest fires, which were observed to destroy larger tracts of forest cover during El Niño more so when annual rain ≥10% below the normal. In retrospect, the study indicated that had forests been exploited with conscious regard to the recurrence of El Niño or La Niña episodes, ecological impacts could at least be toned down. In the same vein, reforestation should have been more successful it were implemented with due considerations to extreme climate variability. Once trees were planted, the weather elements become more crucial than politicians' meddling and other socio-economic factors to the growth and development of reforested sites.

Key words: deforestation, ecological imbalance, El Niño, forestland, La Niña, reforestation.

1. Introduction

The Philippines consists of about 7100 islands and extends for about 1800 km from about 21°N to 4°N latitude and from around 117°E to 127°E longitude. Most of these islands were believed to have been partially, if not wholly covered with forest vegetation at the start of the 20th century. These forests have played vital ecological functions especially for agriculture: such as stabilizing soils, conserving nutrients, and moderating water supplies. Unfortunately, deforestation and poor land use practices in many parts of the country have weakened these support services at a rapid rate. The total forest cover was estimated to be only about 5.1 million ha in 2001 – or 17% of the total land area – far below the 60% required for ideal ecological balance (Bengawan, 2001). Should deforestation remains unabated, it is predicted that all forests will be gone by 2025. Of the 15.9 million ha of forest cover cleared between 1903 and 2001, only about 1 ha has been reforested per 10 ha deforested. This wide-scale reduction in national forest cover explains to a large extent the ecological disruption that the country experiences in terms of drought, flooding, and climate change. The magnitude and intensity of the ecological impacts are influenced further by topography, soil characteristics, geologic conditions, land use after forest clearing, and rainfall patterns.

Inter-annual climate variability, manifested in the more frequent occurrence of El Niño or La Niña events, has altered rainfall patterns in many locations in the world especially those in the tropics. Hence, changes in climatically driven vegetation, like forests, could be anticipated. The timing of El Niño and La Niña occurrences could be predicted with high accuracy now, this information can be utilized therefore in the sustainable exploitation and management of forests (deforestation and reforestation cycles) to begin to stabilize ecology in denuded forest ecosystems. Forest regeneration through reforestation has been an important component in rehabilitating degraded forestlands, where predictive information on the occurrence of climate variability – El Niño or La Niña can be used.

The Philippine government has been undertaking reforestation projects on degraded forestlands,[1] especially on critical watersheds, for about three decades now. Despite these efforts, it was calculated that only one-tenth of deforested areas had been replanted even if reforestation were totally successful. Nonetheless, a review published in 1994 found that only about 42% of the total area planted had become successful, productive plantations (cf ACIAR, 2000).[2] To start with, the reforestation programs were ill-conceived, then these were managed based on insufficient data and information (Korten, F. cf Bengawan, 2001). Furthermore, politicians meddled in the program, government foresters became contractors, tree species planted were of the commercial type, and the reforestation targets were not reached in many parts of the country (Bengawan, 2001). Obviously, enormous efforts in reforestation will be still required to hardly start ecological stabilization of large denuded forestland. The interacting influence of powerful weather disruption, particularly El Niño, on forest clearing and regeneration has been barely reported to date. This study aims to analyze the influence of inter-annual climate variability, manifested by El Niño and La Niña episodes, on deforestation–reforestation dynamics in the Philippines.

2. ENSO-driven inter-annual climate variability

El Niño Southern Oscillation (ENSO) is a naturally occurring climate cycle. Extreme phases of ENSO introduce inter-annual variability into climate patterns of many locations around the globe but most especially those in the tropics.

2.1. GLOBAL PATTERNS

El Niño and La Niña result from interaction between the surface of the ocean and the atmosphere in the tropical Pacific. They are extreme phases of ENSO; which both refer to large-scale changes in sea-surface temperature (SST) across the eastern tropical Pacific (IRI, 2002a). Usually, SST readings off South America's west coast range from 15°C to 25°C, while they exceed 30°C in the 'warm pool' located in the central and western Pacific. During El Niño, the warm pool expands to cover the tropics, however during La Niña the easterly trade winds strengthen and intensify the cold upwelling along the equator and the West coast of South America causing SSTs along the equator to fall as much as 3.9°C below normal (IRI, 2002b). These ENSO events bring in inter-annual variability to and disrupt the climate patterns in many places around the globe, particularly in the tropics. For instance, parts of the Philippines and Indonesia are prone to drought during El Niño, but are vulnerable to flooding during La Niña. According to the Climate Prediction Center, La Niña events by and large last approximately 9–12 months and some episodes may persist for as long as 2 years. Similarly, typical El Niño events may stay longer than a year.

El Niño and La Niña events are classified by different criteria. Some classification systems use the strength and sign of the Southern Oscillation Index (SOI), while others use SST anomalies for many Pacific regions. Normalized anomalies (deviations from the average) in SSTs in the 'warm pool' are used to detect the occurrence of El Niño or La Niña events. Threshold values of normalized SST deviation are used to affirm the existence of El Niño or La Niña event in a given year. The criterion for an El Niño (warm episode) is a deviation of at least 1°C, while that for a La Niña (cold episode) the threshold was relaxed to 0.8°C (Livezey, 1998 cf Schaefer et al., 1999).

Another scale that is widely used to differentiate El Niño or La Niña strength is the SOI. The index is based on the surface (atmospheric) pressure difference between Darwin, Australia (131°E, 12°S) and Papeete, Tahiti (149°W, 17°S). The SOI is given in normalized units of standard deviation and it can be used as an intensity scale. For example, SOI values for the 1982–1983 El Niño were about 3.5 standard deviations, which were roughly twice as strong as the 1991–1992 El Niño, with only about 1.75 in SOI units. Null (2002) compared and integrated four classification lists to categorize specific years as El Niño and La Niña events; the results are shown in Figure 1.

2.2. ENSO IMPACTS IN THE PHILIPPINES

An El Niño event is manifested in the Philippine climate by drier-than-normal weather conditions that can last for one or more seasons, bringing drought to many parts of the country. On the other hand, a La Niña event is characterized by delayed onset of rains, monsoon breaks, and an early to near-normal onset of rainy season

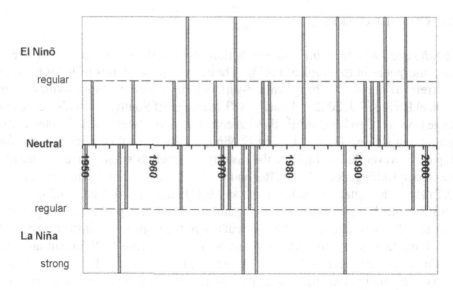

Figure 1. El Niño, La Niña and neutral years; 1950–2000.

and a longer rainy season (Jose, 2002). El Niños' impacts on the Philippine climate system generate enormous strain on water resources; because water inflows into major watersheds and reservoirs are inextricably linked with rainfall amounts, stored water in reservoirs and other impoundments are certainly reduced. As a consequence, water appropriated for households and irrigated agriculture is severely curtailed; in rare instances, water for agriculture has been totally cut in favor of domestic and industrial water supply during severe El Niño-driven drought.

During the 1997–1998 El Niño episode, drought lowered crop yield in about 961 thousand ha of rice and corn areas. Production losses from the combined rice and corn areas have been calculated to be equal to 1.81 million tons valued at about US$ 247 680 (PCARRD, 2001). The same study reported that about 153 thousand ha of rice areas sustained yield loss due to drought amounting to 628 thousand tons in the 1982–1983 El Niño. For forest resources, risks of forest fires intensify during El Niño events because of alteration of fire regimes. Forest fires throughout the Philippines destroyed about 51.3 thousand ha of forest cover; 2.5 thousand ha of which were reforested areas, fruit plantations and primary virgin forests during the 1991–1992 El Niño (The Xinhua News Agency, 1992). In a span of two decades (1980–1999), about 345 thousand ha out of the cumulative total 480 thousand ha of forest cover lost, were attributed to forest fires, making fire a leading cause of deforestation (NSO, 2000).

Although other areas benefit from substantial amounts of rain during La Niña, thousands of hectares of prime agricultural lands are destroyed by floodwaters. Rapid runoff rates erode top soils from bare or grass-covered forestland and deposit silt and debris in rivers, reservoirs, and other major waterways during extreme rainfall events.

3. Forest resources of the Philippines

3.1. DEFORESTATION

In 1903, the Philippines was covered with about 21 million ha of forests sprawled on about 70% of the national total land area (Figure 2). But in 2001, only about 5.1 million ha of forest cover – or 17% of the national total land area – remain intact. Forest cover had been removed at 40 thousand ha year^{-1} between 1903 and 1918, then the clearing rate more than doubled to 100 thousand ha year^{-1} in the next 20 years until 1939. Since second World War in 1940 until 1966, deforestation rate returned to about 41 thousand ha year^{-1}, when timber is harvested for revenue, and deforested areas were subsequently cleared for agriculture. Extremely high rates of forest cover removal occurred at 1.78 million ha in 1973, 1.01 million ha in 1984, 1.38 million ha in 1986, and 0.97 million ha in 1988. In sum, 15.9 million ha of forests have disappeared in a span 98 years; it follows that on the average, 162 thousand ha of forests were cleared per year. Legislation to phase out raw log exports was first introduced in 1973 as an attempt to curtail overcutting, but was never implemented. Rather it was replaced with a modified scheme, which served to concentrate ownership of timber licenses in the hands of a few Marcos supporters (Bengawan, 2001). For much of the 20th century, the Philippines was Asia's greatest exporter of rainforest timber. The deforestation process was unsustainable and environmentally damaging, and now logging is banned and reforestation has become a government priority.

Over time the conversion of forests into croplands has been the leading cause of deforestation in the tropics. Population growth, inequitable land distribution, and the expansion of export agriculture have reduced cropland available for subsistence farming, forcing many farmers to clear virgin forest to grow food. Also, fuel wood

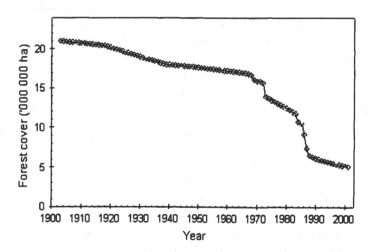

Figure 2. Deforestation trend of Philippine forest cover between 1903 and 2001.

gathering, commercial timber harvesting, and cattle ranching contributed to the great expansion of denuded forestland. Much of the clearing has occurred in the uplands (forestlands) as a result of demands from a rapidly growing populations, which need land both for farming and wood.

In the Philippines, unchecked illegal logging remains the main culprit for continued deforestation. Government negligence has prompted the devastation of the forests. Much of the forests that remain have been invaded by commercial loggers, too because the government has passed laws favorable to logging concessions and feebly implemented forest protection policies over the years (Legarda cf Bengawan, 1999).

3.2. PROBLEMS ASSOCIATED WITH DEFORESTATION

With acute ecological imbalance attributable to large denuded forest land, goes the disintegration of the ecological integrity of many areas. Simultaneously, El Niño and La Niña occur more frequently, thus aggravating any ecological impacts deforestation brings about to society. The combined effects of these (seemingly) twin phenomena, can impair the health of forests and compromise their function to maintain the stability of soil and water resources. In the Philippines, deforestation is the major reason behind flooding, acute water shortages, rapid soil erosion, siltation, and mudslides that have proved to be costly not only to the environment and properties but also in human lives (Legarda cf Bengawan, 1999). Severe soil erosion reduces the productive capacity of soil in sloping denuded forestland, which generates conditions favorable to growth and development of *Imperata cynlindrica* grass, aka cogon in the Philippines. Besides, soil erosion transfers sediment to river channels, which further aggravates local flooding or speeds up the silting of reservoirs downstream.

On the plus side, deforestation lowers the amount of water lost through evapotranspiration (ET), thereby increasing available water supply in certain areas, only if land use practices following deforestation will not compact soils. Poor crop production practices or overgrazing on denuded forestlands weaken soil's ability to absorb rainwater resulting in greater amount of water running off than soaking into the subsurface, where it can be stored and released more gradually. Depending on the intensity and duration of rainfall, the loss of soil infiltration capacity can also increase flooding. Occurring more frequently now, La Niña could step up the risks due to flooding because of potentially high runoffs from the bare forestlands.

Other impacts of large-scale deforestation include national shortage of wood, which has made substantial timber imports necessary. Or, there exists a growing reliance on coconut trees for construction timber, which impinges on the country's coconut industry. The loss of forest cover meant the loss or fragmentation of natural habitats, which endangers biodiversity. Moreover, deforestation frees up substantial amount of carbon that eventually end up in the atmosphere and contribute to global warming. Assuming that the total 15.9 million ha forests that were lost between 1903 and 2001 were all natural forests, and that natural forests harbor above-ground carbon density of $518\,\mathrm{Mg\,C\,ha^{-1}}$ (Lasco, 2001), then deforestation in the Philippines have contributed about $8.24 \times 10^9\,\mathrm{Mg\,C}$ to the atmospheric greenhouse pools.

3.3. REFORESTATION EFFORTS

Forestry statistics show that reforestation by the State started in 1960 when about 55 thousand ha were replanted with trees out of the total 3.8 million ha forest cover that had been cleared since 1903. Thereafter, reforestation proceeded at about 9 thousand ha year^{-1} until 1975. From 1976 onwards, timber licensees and other private sectors participated in reforestation efforts and jack the rate up to more than 60 thousand ha year^{-1}. Highest rates of reforestation equaling about 131 thousand ha in 1989 and about 192 thousand ha in 1990 were observed.

The cumulative total forest cleared from 1903 to 1998 amounted to about 15.66 million ha while the total area reforested equaled to about 1.64 million ha. If reforestation were fully (100%) successful, this is tantamount to 1 ha reforested for every 10 ha deforested. This is much lower than the ratio of 1 ha reforested : 1.67 ha deforested to maintain 60% ecological balance. Still the ratio is much lower than that required for 40% ecological sustainability – 1 ha reforested : 2.5 ha forest cleared. Considering that major reforestation efforts by government organizations, private companies, and farmers have been far from generally successful, the ecological balance in Philippine forest ecosystems could be even lower. A review published in 1994 found that only about 42% of the total area planted had become successful, productive plantations (ACIAR, 2000). Discounting the reforestation success rate to 42%, the reforestation : deforestation ratio further slides to 1 ha reforested : 23 deforested; to a large extent, this accounts for the ecological disasters that country frequently experiences lately.

Although it is a formidable, if not an impossible task to do, ecological stabilization of large denuded forest systems should start now. Unfortunately, the stabilization work, with reforestation as a major component, will be surely confounded by rainfall aberrations spawned by El Niño and La Niña episodes. This calls for a more creative and collective endeavor to weaken the challenges the State presently contends with in protecting the remaining forests and in implementing a large-scale and successful reforestation.

4. Methodology

Two major data sets were used in this paper: (1) database on inter-annual climate variability, and (2) deforestation and reforestation dynamics in the Philippines. Most of the statistical analyses were performed through the use of STATISTICA v5.5 (StatSoft, 1998).

4.1. MEASURES FOR EL NIÑO AND LA NIÑA

El Niño and La Niña events were classified by a number of different criteria. Some classification systems used the strength and sign of the SOI, while others used the SST anomalies for a variety of Pacific regions. The Japan Meteorological Agency devised a new SST-based index for determining El Niño and La Niña (Japan Meteorological Agency, 1991) by: (1) calculating monthly SST anomalies averaged for the area 4°N to 4°S and 150°W to 90°W, and (2) applying a 5-month running mean of the data. The 5-month running mean of SST anomalies is implemented to smooth out possible intra-seasonal variations. A data file (jmasst1949-today.filter-5) containing 5-month running mean of the anomalies for the years 1949–2002 was obtained and used in this study.

Another widely used scale to differentiate El Niño strength is known as the SOI, which is based on the surface (atmospheric) pressure difference between Tahiti and Darwin, Australia. The SOI is given in normalized units of standard deviation and it can be used as an intensity scale. For example, SOI values for the 1982–1983 El Niño were about 3.5 standard deviations, so by this measure that event was roughly twice as strong as the 1991–1992 El Niño, which measured only about 1.75 in SOI units. There exists a database containing monthly means of normalized SOI for the region of the South Pacific between Darwin, Australia (131°E, 12°S) and Papeete, Tahiti (149°W, 17°S) during January 1866 and December 1994. This database was compiled from a variety of sources by the Climatic Research Unit at East Anglia University. Like the SST database, the SOI data set was obtained from the internet and used in subsequent analyses in this study.

Others use a combination of several criteria to gauge the type and strength of the event to characterize a certain year based on ENSO occurrence. In this study, we adopted the El Niño-La Niña-Neutral year classification done by Null (2002) and IRI (2002) to determine the type climate variability experienced in any given year in the Philippines as shown in Figure 1. Monthly rainfall data from 50 synoptic weather stations in the Philippines covering the period 1961–1999 were also used in this study.

4.2. DEFORESTATION AND REFORESTATION

Data and information on forest clearing, reforestation, and even forest fires in the Philippines from published reports in both local and international sources were used in this study. Philippine forestry statistics were taken mainly from published reports by Forest Management Bureau of the Department of Environment and Natural Resources (DENR, 1998). These data and information were complemented with data and information taken from published reports by the Food and Agriculture Organization (FAO, 1999) and by the National Statistics of the Philippines (NSO, 2000).

Yearly data on forest cover loss between 1903 and 1960 were reconstructed from forest cover statistics available in years 1903, 1918, 1939, 1948, and 1961. Similarly, the annual forest data between 1984 and 2001 were reconstructed from available forest data in years 1983, 1986, 1988, 1990, 1994, and 2001. Annual forest loss rate between any two-time periods was calculated and the loss rate applied to determine the hectarage of forest cover lost annually within the same period. All information on forest cover in other years was taken from FAOSTAT (FAO, 1999).

4.3. ANALYTICAL TECHNIQUES

As they have been characterized globally or for areas outside the country, internationally-classified El Niño, La Niña, or Neutral years were further analyzed to see whether or not the local climate system tightly reproduced the same global classification under what circumstances. Of the local climate parameters, El Niño, La Niña, and Neutral episodes would greatly influence the normal rainfall and temperature patterns. However, even if temperature could be an equally important climate statistic, only rainfall statistics were considered in this study. To correlate the local climate system to the behavior of 'warm pool' in the Pacific Ocean, three measures of local rainfall deviation, in tandem with SST anomalies and SOI, were used: (1) standardized total annual rainfall (SARN), (2) percentage deviation from 35-year average annual rainfall (PDARN), (3) standardized monthly deviations from 35-year average monthly rainfall (SDMRN), and (4) normalized total rainfall from August to December of a preceding year to January–May of a succeeding year (SAUMYRN).

The rainfall statistics that best reflect the global El Niño and La Niña into the local climate systems was used in the subsequent analyses of deforestation and reforestation dynamics. El Niño, La Niña, and Neutral episodes were used as a grouping variable in further statistical analyses.

5. Results and discussion

5.1. INTER-ANNUAL CLIMATE VARIABILITY

Between 1950 and 2000, 15 years were classified as El Niño years – six of which occurred as strong events; 12 La Niña events – four of which were strong, and the rest of the years neutral years (Figure 1). Between 1950 and late 1980s, an El Niño or a La Niña event has generally occurred after every 3 or 4 consecutive neutral years. Nevertheless, the recurrence interval for El Niño or La Niña is unclear and can be difficult to predict. In 50 years, an El Niño episode has returned after 1–6 years of occurrence. Noticeably, El Niño has occurred more frequently in the 1970s and 1990s compared to other periods. Similarly, the return cycle for La Niña seems unpredictable, the incident recurring after 1, 2, 5, 8, and 13 years. It should be emphasized that La Niña occurred more frequently in the 1970s relative to any other periods.

Together, 5 La Niña and 4 El Niño events occurred 1 year after another in 1970s and 5 El Niño and 2 La Niña incidents occurred in the 1990s. These occurrences have significant impacts on the manner climatically dependent natural resources, like forests, should be exploited more so if they are located in sensitive biophysical setting. In the Philippines, forestlands are located in areas with slope $\geq 18\%$, which have profound effects on the ecological instability of soil and water resources obtaining from denuded forest ecosystems. In the same vein, these events could influence the performance of assisted forest regeneration activities.

5.2. ENSO VARIABILITY AND PHILIPPINE RAINFALL PATTERNS

Figure 3 shows that all El Niño years occur when SSTs >0 and percentage deviation of annual rainfall from its mean $\leq 10\%$. On the other hand, all La Niña events occurred at SSTs <0 and PDARN ≥ 0, indicating a tight and accurate separation among El Niño, La Niña, and Neutral episodes from the combined use of SST anomalies and PDARN. The same results were obtained when SARN and SST were used in characterizing the occurrence of these events. With SST and SDMRN in tandem, all El Niño episodes happened at SST >0 and SDMRN ≤ 0.2 standard deviation; in comparison all La Niña events were observed at SST <0 and SDMRN ≥ 0. A strong La Niña occurred at SST about -0.7 and SDMRN >0.5 standard deviation. Neutral events fall on either side of this classification. Combining PDARN with SOI to classify the events did not result in as tight and accurate classification for El Niño events as with joint PDARN and SST; an El Niño event occurred at SOI >0 and PDARN $\leq 10\%$. All La Niña events however were observed at SOI <0 (cold episodes) and PDARN ≥ 0 (wetter conditions). Similar result patterns were obtained when the events were classified using SARN and SOI, and SOI and SDMRN.

Definitely, the global occurrence of El Niño, La Niña, and Neutral events is captured in the Philippine local climate systems by measures of rainfall anomalies. El Niño occurs when total annual rainfall $\leq 10\%$ of the normal and either SOI <0 or SST >0 prevail. In similar fashion, La Niñas are confirmed here in the Philippines when annual rains $\geq$ normal average and SOI >0 or SST <0 are simultaneously observed. In La Niña years, the annual rainfall (2616 ± 289 mm) is significantly higher at 1%-probability than those during El Niño years (2222 ± 265 mm), and neutral years (2347 ± 225 mm).

The mean annual rainfall during El Niño years is not significantly different than that during neutral years, meaning that weather conditions during neutral years tend to be dry, too. As expected, similar results were obtained with mean separation of monthly rainfall data: monthly rainfall (221 ± 26 mm) during La Niña years is significantly higher than those during El Niño (189 ± 23 mm), and neutral years (199 ± 19 mm). Monthly rainfall during El Niño does not significantly differ than that in neutral years, meaning that weather in neutral years tends to be dry as well.

In an El Niño event, monthly rainfall lower than normal generally occurs as early as August of the beginning year and lasts until May of the ending year (Figure 4). For strong El Niño events, such as that occurred in 1982–1983, 1986–1987, and 1997–1998 monthly rainfall began to subside lower than normal as early as April

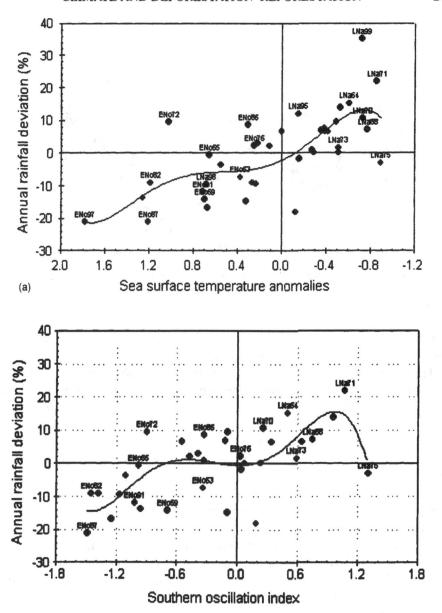

Figure 3. Qualitative analysis of the properties of El Niño and La Niña events in the Philippines based on percentage annual rainfall deviations SST anomalies, and SOI. Trend line was estimated by fifth order polynomial fitting. Point labels ENoyy and LNayy, respectively means El Niño and La Niña episodes that occurred in a given yy year.

of the beginning year and can run low until September of the following year. In comparison, La Niña events can start dry with monthly rainfall lower than normal between April and August of the beginning year, comes back up in September with rain equaling that of normal rainfall, and continues with monthly rainfall exceeding

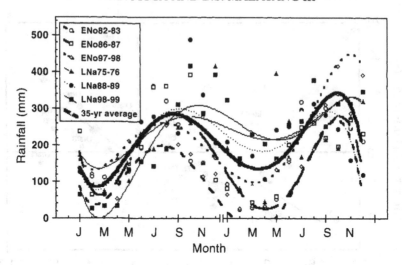

Figure 4. Comparison of monthly rainfall during El Niño and La Niña episodes with normal monthly rainfall in the Philippines. All trend lines were estimated by fifth order polynomial fitting. Point labels ENoyy-yy and LNayy-yy, respectively means El Niño and La Niña episodes that occurred in given a yy-yy year.

normal average until July of the ending year. Compared to normal years, the wet regimes are extended during La Niña years.

The disturbance in the normal monthly rainfall pattern due to El Niño or La Niña calls for a sweeping modification in the normal cropping calendar, which farmers used to base farming decisions. During El Niño, it is almost impossible to plant rice in rainfed areas; not only is rainfall inadequate but also is rice sensitive to variable moisture amounts. Where other biophysical factors are suitable, farmers should adopt less-water-demanding crops. With respect to reforestation in the uplands, the change in normal rainfall pattern should involve changes in management schemes. Usually, reforestation is done in the early parts of a year, so a strong and persistent El Niño that occurs after planting could kill seedlings, particularly those in sloping areas with thin topsoil. In comparison, during the La Niña events discussed here, the cropping calendar should be drastically shifted from May to November to September–July (following year) to coincide with adequate rainfall. Reforestation works could be done during La Niña events, but seedlings should be sufficiently covered with mulch to avoid mechanical injury during strong rains. In addition to mulching, management facilities and structures for high runoff should be provided. Similarly, exploitation of forest resources should be consistent with El Niño or La Niña episode to minimize potential ecological disasters.

5.3. DEFORESTATION–REFORESTATION DYNAMICS AND EL NIÑO/LA NIÑA EVENTS

The clearing of forest cover in the Philippines has been totally unrelated to the occurrence of important climate variability, such as El Niño or La Niña events.

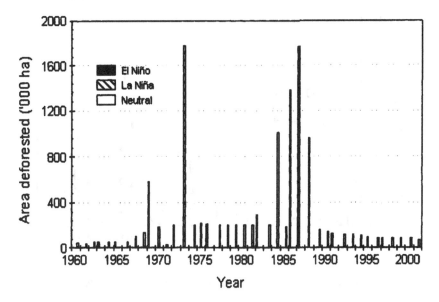

Figure 5. Annual rates of deforestation in relation to El Niño and La Niña occurrence; Philippines, 1961–2001.

Even during the 1970s when nine El Niño and La Niña events occurred 1 year after the other, very high rates of deforestation were perpetrated – about 1.8 million ha of forests were cleared in a strong La Niña year in 1973, which was followed by another strong and a moderate La Niña (Figure 5). The disaster floods that the Philippines endured in the 1970s are to a large extent accounted for by the large-scale defor- estation and the occurrence of La Niña episodes; in retrospect, these forests should control the floods generated from these wetter-than-normal conditions. Similarly, three very high rates of deforestation: 1.01 million ha in 1984, 1.38 million ha in 1986, and 1.76 million ha in 1987, preceded a strong La Niña incident in 1988, which caused significant flooding in the country. Great ecological disasters, such as flooding and droughts, could be minimized if deforestation had been carried out in regard to the occurrence of extreme climate phases such as El Niño or La Niña.

In comparison, assisted forest regeneration through reforestation, seemed to have been carried out more deliberately than deforestation in view that large reforestation schemes were done during neutral years compared to El Niño or La Niña years (Figure 6). However, reforestation schemes in neutral years 1976–1980 were fol- lowed by a strong El Niño episode in 1981. A total of 243 thousand ha reforested with trees ≤5 years old had been exposed to drought and fires that accompanied the strong 1981–1982 El Niño episode. A cycle of 2–3 year reforestation in neu- tral years have always been followed either by an El Niño or a La Niña episode; again exposing to droughts or floods assisted-regeneration forests with trees ≤3 years old. High reforestation rates of about 131 thousand ha in 1989, and 192 thou- sand ha in 1990 generated 323 thousand ha with trees ≤2 years old, which were subsequently exposed to a strong El Niño event in 1991. It would be difficult to

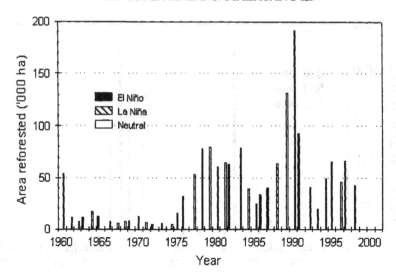

Figure 6. Annual reforestation rates in relation to El Niño and La Niña occurrence; Philippines, 1961–1998.

maintain high sapling survival rates under these conditions, that is why, reforestation projects were unsuccessful. These areas had been planted anyway despite political meddling and other socio-economic constraints, so the weather and other biophysical factors become important now for the reforested sites. Chances of success will improve if reforestation schemes are planned deliberately taking into account probable occurrences of El Niño or La Niña. Once they were planted, the climate and other biophysical factors directly determine the growth of saplings and influence the management of reforestation projects especially under extreme climate events.

5.4. EL NIÑO/LA NIÑA AND FOREST FIRES

It is likely that larger tracts of forest cover are destroyed by forest fires during El Niño years than any other years; but forests (small extent) gets burned, too during La Niña years even when rainfall is about 35% above normal average. Categorically, large tracts of forest cover, >50 thousand ha up to about 120 thousand ha, are destroyed by forest fires during El Niño years with PDARN ≤10% (Figure 7). Under drier-than-normal conditions, trees especially young ones could die from moisture deficit alone. Also, fire regimes are substantially altered to favor occurrence of fires, which destroy forest cover. During abnormally dry El Niño events, previously denuded forestlands, which have been covered with grass and shrubs, are essentially tinderboxes to accidental fires by upland people, or to intentional fires by commercial loggers who clear forests by fire. Considering that younger trees dry more easily than older ones, reforestation areas are more vulnerable to forest fires than natural forests are; this partly shrinks forest cover and contributes to the dismal failure of reforestation efforts.

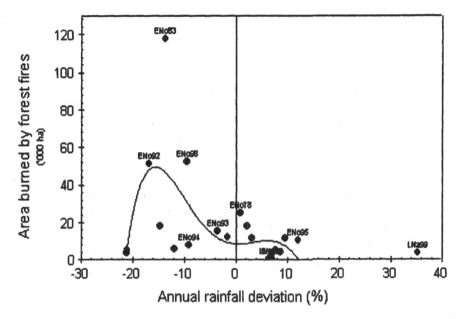

Figure 7. Extent of forest cover burned by forest fires at different intensity of droughts – proxied by percentage annual rainfall deviation. Trend line was estimated by fifth order polynomial fitting. Point labels ENoyy and LNayy, respectively means El Niño and La Niña episodes that occurred in a given yy year.

In summary, wide-scale deforestation should take into account the occurrence of El Niño or La Niña to mitigate if not to avert large ecological disturbance. Similarly, reforestation schemes must take into account the timing of occurrences of climate variability to increase chances of success in growing trees. Now, that the coming of an El Niño or La Niña can be more accurately predicted at least 1 year in advance, the information can be utilized for more successful reforestation. As there are virtually no more forests in the country to clear, the remnants should be protected and safeguarded from physical phenomena, such as El Niño and La Niña.

6. Conclusions

El Niño and La Niña episodes introduce variability into the Philippine climate system. During El Niño episodes, annual or monthly rainfall deviates from normal by $\leq 10\%$ and creates drier-than-normal weather conditions. La Niña incidents generally occur when annual or monthly rainfall is $\geq$ normal rainfall and produces wetter-than-normal weather conditions. Deforestation process proceeded without regard to biophysical factors, like climate, causing considerable ecological instability in most forest ecosystems. The total deforested area in the Philippines had contributed about 8.24×10^9 Mg C into the greenhouse pools in the atmosphere and is observed to spawn ecological impacts of disaster proportions during times of high climate variability. Obviously, the magnitude and intensity of the ecological disasters triggered by large-scale deforestation, such as flooding, droughts, and mudslides, should have been minimized if forest exploitation had been planned to include management leeway for the occurrence of El Niño or La Niña events.

Similarly, part of the explanation for the dismal failure to stabilize denuded forest ecology through reforestation, lies in the disruption of the local climate system by El Niño and La Niña episodes. Although reforestation schemes were apparently done with deliberate regard to climate variability – most of the reforestation schemes were implemented in neutral years – plausibly, the management scheme for the established young reforestation sites did not include tactics for big surprises such as El Niño or La Niña episodes. Due to more frequent occurrence of either El Niño or La Niña, the reforestation cycle always end up exposing trees at most 5 years old to the weather aberrations courtesy of these climate events. Young trees either died from droughts and forest fires during El Niño or are uprooted and carried away during La Niña episodes. Management schemes that include more creative solutions to address El Niño or La Niña impacts on reforestation can be formulated now that the coming of these climate events is being accurately predicted a year in advance. Adaptive mechanisms by local communities to environmental variability like El Niño or La Niña should be devised and integrated into future reforestation programs.

Notes

[1] By law, all lands in the country with slope $\geqslant 18\%$ are classified as (upland or) forestland whether or not they are covered with trees or any vegetation. A total of 15.65 million ha – or 53% of national total land area – of classified and unclassified forestland was reported in 1999.

[2] Based on 42% reforestation success, only 1 ha has been reforested per 23 ha deforested, which is way below than the 1 ha reforested for every 1.67 ha deforested.

References

ACIAR (Australian Center for International Agricultural Research): 2000, 'Bringing trees back in the Philippines', *Research Notes* **24(12/00)** page 3.

Bengawan, M.A.: 1999, 'Illegal logging wipes out Philippine forests', Environment News Service. http://ens-news.com/ens/oct1999/1999-10-11-01.asp

Bengawan, M.A.: 2001, 'Weak laws in the Philippines exacerbate deforestation', *The Earth Times*, 23 February 2001.

DENR (Department of Environment and Natural Resources): 1998, *Philippine Forestry Statistics*, Quezon City, Philippines, Forest Management Bureau.

FAO (Food and Agriculture Organization of the United Nations): 1999, *FAOSTAT*. Rome, Italy, FAO. http://www.fao.org

IRI (International Research Institute for Climate Prediction): 2002a, *El Niño and La Niña years for Atlantic hurricanes*. http://iri.columbia.edu/climate/ENSO/globalimpact/TC/Atlantic/ensoyear.htm

IRI.: 2002b, Cold and warm episodes by season. http://www.cpc.ncep.noaa.gov/products

Japan Meteorological Agency.: 1991, COAPS SST 1868–1997. Marine Department Report. fttp://www.coaps.fsu.edu/pub/JMA_SST_Index.

Jose, A.M.: 2002, 'ENSO Impacts in the Philippines', Resources for Impacts and Responses. http:// www.unu.edu/ env/ govern/ ElNIno/ CountryReports/ inside/ philippines/ TELECONNECTIONS/ TELECONNECTIONS_txt.html

Lasco, R.D.: 2001, 'Carbon stocks of forest ecosystems in Southeast Asia following deforestation and conversion', *APN (Asia-Pacific Network for Global Change Research) Newsletter* **VII(4)** 3–4. http://www.apn.gr.jp/pub027.htm#APN%20Newsletter% 20Vol.7,%20No.4%20October%202001

Livezey, R.E.: 1998, Personal communication (by Schaefer et al., 1999).

NSO (National Statistics Office): 2000, *Philippine Statistical Yearbook 2000*, Quezon City, Philippines, NSO.

Null, J.: 2002, *El Niño and La Niña Years: A Consensus List.* http://ggweather.com/enso/years.htm

PCARRD (Philippine Council for Agricultural Resources Research and Development): 2001, *El Niño Southern Oscillation: Mitigating Measures,* Los Banos, Philippines, PCARRD.

Schaefer, J.T. and Tatom, F.B.: 1999, *The Relationship Between El Niño, La Niña, and United States Tornado Activity.* http://www.spc.noaa.gov/publications/schaefer/el_Niño.htm

StatSoft.: 1998, Statistica v5.5. StatSoft Inc., OK, USA.

The Xinhua News Agency: 1992, 'Roundup: Severe drought ravages Philippines', 18 March 1992.